ART, SCIENCE, AND HISTORY IN THE RENAISSANCE

THE JOHNS HOPKINS HUMANITIES SEMINARS

The Seminars are sponsored by the Humanities Center of The Johns Hopkins University. Each year these interdisciplinary programs bring distinguished scholars from other institutions to take part in the study of some particular period or subject. To date the following books have come from the Seminars:

ASPECTS OF THE EIGHTEENTH CENTURY
Edited by Earl R. Wasserman (1965)

THE INDIVIDUAL AND SOCIETY IN THE MIDDLE AGES
by Walter Ullmann (1966)

ART, SCIENCE, AND HISTORY IN THE RENAISSANCE
Edited by Charles S. Singleton (1967)

ART, SCIENCE,
AND HISTORY IN THE
RENAISSANCE

EDITED BY CHARLES S. SINGLETON

THE JOHNS HOPKINS PRESS
BALTIMORE AND LONDON

THE JOHNS HOPKINS PRESS,
BALTIMORE, MARYLAND 21218
THE JOHNS HOPKINS PRESS LTD., LONDON

ISBN-0-8018-0602-X

ORIGINALLY PUBLISHED, 1968
SECOND PRINTING, 1970

Title page illustration.—Breydenbach, "Peregrinatio in Terram Sanctam," *detail (1486).*
Walters Art Gallery.

Jacket illustration.—I. Stradanus, The scientist and nautical science *(sixteenth century).*
National Maritime Museum.

PREFACE

his third Seminar of our series is concerned with a period which remains surely one of the most, if not indeed *the* most, controversial in the history of Western Europe. Do we accept it on its own terms and self-evaluation, through its own metaphor of *rináscita*? Do we attempt to take its measure from our own vantage point in time; or do we seek, through an exercise of the historical imagination, to come to this Renaissance, this Revolution as it were better termed, from the other side, from the Medieval conception of the world and man's place and activities in it?

Obviously, we do both. Nor is this brief prefatory note the place where even the basic questions of strategy in the study of such a period may be formulated. Here it shall simply be stated, for the record, that the Seminar met regularly throughout the year, to discuss these and other general questions as they seemed relevant and fruitful; and that the lectures contained in this volume were, of course, the bright spots on our calendar, then, as now, in retrospect. Each lecturer was left, in the utmost "Renaissance" freedom, to choose his own subject, though it was agreed in advance that the topics of Art, Science, and Humanism should form the main rubrics.

Two exceptions in the program should perhaps (again, for the record) be pointed out. The papers of Professors McMullin and Rubsamen were not delivered as lectures, but were heartily welcomed as contributions to our volume in the form in which they are here presented.

Charles S. Singleton

The Humanities Seminars
The Johns Hopkins University

CONTENTS

PART I: ART AND MUSIC

THE LEAVEN OF CRITICISM IN RENAISSANCE ART 🦢 E. H. GOMBRICH

f by the criticism of art we mean the detailed assessment of both the merits and the shortcomings of an individual work of art, the large bulk of writings on art bequeathed to us by the Renaissance will be found to be disappointingly poor in this respect.[1] The culture of the humanists was predominantly literary and rhetorical, and though they liked to congratulate themselves on the efflorescence of the arts their age had witnessed, they usually had to fall back on formulas culled from classical authors when they were confronted with a particular achievement. In their poetic tributes they were generally satisfied to ring the changes on the conventions evolved by the ancient writers of epigrams which praise the lifelikeness of a figure that "only lacks the voice." Even the more technical writings on art rarely go beyond general topics derived from the rhetorical tradition, such as the question of *decorum,* or the *paragone.* Later in the cinquecento the debate of the schools resulted in the discussion of the relative importance of *disegno* and *colore,* but though this antagonism afforded some opportunity for sniping at Michelangelo or Titian, even here the writers hardly came to grips with concrete problems of criticism.

I. THE NATURE OF THE EVIDENCE

But despite this largely negative evidence Renaissance artists behaved as if they were aware of the gadfly of criticism. There is a purposeful direction to be observed in the development of certain methods and problem solutions which clearly suggests the existence of rational standards by which works of painting and sculpture were judged. Only the existence of such standards and the awareness of failures as well as of successes can explain that spirit of rivalry and of experimentation that marks the "rebirth of the arts." Without a tendency to faultfinding there can be no such desire for the improvement of certain qualities.[2]

[1] Julius Schlosser Magnino, *La Letteratura Artistica* (Florence, 1956).
[2] For this approach, based on the methodology of K. R. Popper, see my book *Art and Illusion* (New York and London, 1960), and my papers "Visual Discovery through Art," *Arts Magazine,* Vol. XL, No. 1 (November, 1965), and "From the Revival of Letters to the Reform of Art; Niccolò Niccoli and Filippo Brunelleschi," in the forthcoming *Essays in the History of Art Presented to Rudolf Wittkower,* ed. Douglas Fraser, Howard Hibbard, and Milton J. Lewine (London: Phaidon Press, 1967).

It is the purpose of this paper to assemble a few texts to illustrate and test this interpretation. None of them is new to art historians, but perhaps they may still illuminate each other and confirm the role of that spirit of criticism which inspired the new conception of art. They may thus supplement and more closely define an argument I presented in an earlier paper on "The Renaissance conception of artistic progress and its consequences."[3]

In that paper I left the idea of progress rather undefined, taking it for granted that standards of lifelikeness such as the Renaissance inherited from antiquity were shared and understood by all. But this assumption, in a way, begs an important question. Those conventional terms of praise to which I alluded above are not confined to the Renaissance. Byzantine poets used them in profusion for works which to us look anything but lifelike,[4] and the same, after all, may be said of Boccaccio's famous eulogy of Giotto.[5] The point is precisely that without an articulate formulation of what constitutes lifelikeness there are no easy grounds for a negative judgment. The illusion given by paintings and works of sculpture, after all, is always a relative matter. In some respects Sluter or Jan van Eyck cannot ever be surpassed in this direction. We do not usually call their style Renaissance.[6] Is there any evidence for a different approach which marks that new movement?

To find a full awareness of what the Renaissance was about, we often do well to approach it from outside. The historian of the Italian Renaissance may be baffled when faced with the notorious problem of "periodization." But as a Polish historian reminded his colleagues when this perennial question once more came up for debate during the Rome Congress of the Historical Sciences, there is no such difficulty in countries which took over Renaissance standards and values from Italy in a developed form.[7] In Poland the adherents of the new movement talked differently, even dressed differently; in short, they emphasized the gulf that separated them from the old-fashioned traditionalists. Had they been interested in art, they would no doubt have formulated their objections to Gothic images as the German humanists, for instance, formulated their contempt of medieval Latin. In going to the periphery we gain, as it were, a ringside seat for watching the contest between the old and the new standards.

[3] A contribution to the XVIIth International Congress of Art History, Amsterdam, 1952, now reprinted in my *Norm and Form* (London, 1966).
[4] Cyril Mango, "Antique Statuary and the Byzantine Beholder," *Dumbarton Oaks Papers*, XVII, 1963.
[5] *Decamerone*, Giornata VI, Novella 5.
[6] J. Huizinga, "Renaissance en Realisme," *Verzamelde Werken*, IV (Haarlem, 1949), 276ff.
[7] C. Backvis, in *Atti del X Congresso Internazionale*, Rome, Sept. 4–11, 1955, pp. 536–38. His debating point came out more clearly in the discussion than in the printed version.

II. DÜRER'S TESTIMONY

Dürer, of course, saw himself both as a pupil of the Italian Renaissance and as its missionary. In the drafts and in the final formulations of his theoretical books he strove incessantly to explain to his readers how the new conception of art differed from the one they had learned. Careful not to offend their susceptibilities, he still insisted that there were rational grounds on which the traditional methods of German gothic painting could be criticized.

Up to now many able boys in our German lands were placed with a painter to learn the art, where, however, they were taught without any rational principle and solely according to current usage. And thus they grew up in ignorance, like a wild and unpruned tree. True it is, that some of them acquired a ready hand by steady practice, so that their work was produced powerfully, but without forethought and simply as it pleased them. Whenever knowledgeable painters and true masters saw such unconsidered work, they laughed at the blindness of these people, and not unfairly so, since nothing is less pleasing to a man of good sense, than mistakes in painting, however much diligence may have been employed in the work. But the fact that such painters were pleased by their own errors is only due to their never having learned the skill of measurement without which nobody can become a proper workman.[8]

Dürer's testimony here is precious for two reasons. It confirms that the painters of the older persuasion were "blind" to the "mistakes" they committed, and it insists nevertheless that those trained in the new standards are right in criticizing these mistakes as laughable. We could not wish for a better witness, for there is no doubt that Dürer here speaks from his own experience. Trained in the tradition of the Wolgemut workshop, he himself must have experienced the shock of discovering that there were so many mistakes in the paintings of this school. Once the power of perspective, of conjuring up a convincing interior, had been demonstrated to him, it was natural to ask how St. Luke could see the Madonna he was supposed to be painting on one of the panels of the Peringsdörffer Altar (Fig. 1). Once Mantegna's engravings had revealed to him the character of a correctly structured nude (Fig. 2), the conventions of German engravers (Fig. 3) must naturally have looked to him antiquated and simply wrong.

It was in fact in these two achievements that Dürer acknowledged the superiority of the Renaissance unreservedly. Asking a humanist friend to

[8] *Underweysung der Messung* (Nuremberg, 1525), fol. Aib. Quoted after K. Lange and F. Fuhse, *Dürers schriftlicher Nachlass* (Halle, a.S., 1893), p. 180. Unless otherwise stated the translations in this paper are my own. I have frequently sacrificed readability to accuracy.

Fig. 1.—Master of the Peringsdörffer Altar, "St. Luke Painting the Virgin"
(1487). Germanisches National Museum, Nuremberg.
(Photograph by the Museum.)

Fig. 2.—Andrea Mantegna, "Christ in Limbo," engraving. (B. XIII, 5.)

Fig. 3.—Master A. G. (Albrecht Glockenton?), "Christ in Limbo," engraving. (B. VI. 345, 12.)

Fig. 4.—Albrecht Dürer, "Instruction in Perspective." From the 1525 edition of his Unterweisung der Messung.

Fig. 5.—Albrecht Dürer, "Man Drawing a Nude." From the 1538 (posthumous) edition of his Unterweisung der Messung.

write him a Preface for the Treatise on Measurement, he begs him to re-member "that I highly praise the Italians for their nudes and most of all for perspective."[9] That "measurement," mathematics, could serve to elimi-nate mistakes in perspective Dürer had no doubt (Figs. 4 and 5). The prob-lem of the rendering of the nude was more elusive. For what was wanted was not just accuracy. The Italians seemed to possess another secret, the secret of beauty. Nothing is more moving in Dürer's writings than his struggle with this enigmatic problem. Was beauty also a matter of measure-ment? In his youth he had believed it, when Jacopo Barbari first showed him a figure of a man and of a woman based on measurements:

and at that time I would rather have seen what his opinion was than gained a new Kingdom . . . but I was still young at that time and had never heard of such things . . . and I well noticed that the aforementioned Jacopo did not want to explain his method to me clearly. But I took the matter upon myself and read Vitruvius who writes a little about the limbs of a man.[10]

Dürer had no doubt that the secret he was searching for had in fact been once known and laid down in books. Indeed the decline of art was to him a result of the loss of these books:

Many hundred years ago there were several great masters, mentioned in Pliny's writings, such as Apelles, Protogenes, Phidias, Praxiteles, Polycletus, Parrhasius and the others. Some of them wrote knowledgeable books on painting, but alas, alas, they are lost. For they are hidden from us, and we lack their great knowledge.

Nor do I hear that any of the present masters are at work writing or pub-lishing. I cannot think what causes this lack. And thus I want to put forward the little I have learned, as best I can, so that a better man than I am may find the truth and correct me by proving the errors of my present work. In this I shall rejoice, since even so I shall have been the cause of the truth coming to light.[11]

The criticism which Dürer here invites with such touching modesty concerns of course the rational foundations of art. It was the loss of these foundations which caused the art to be "extinct till it came to light again one century and a half ago."[12]

In discussing the causes of this loss, Dürer makes it perfectly clear that what he speaks about are the means rather than the ends of art. Those who destroyed the books had confused the two. Art is a tool, and any tool, for instance a sword, can be used in a good cause or in a bad one.[13] He thus would have pleaded with the Fathers of the Church:

---

[9] *Ibid.,* p. 254.   [10] *Ibid.,* pp. 342–43.   [11] *Ibid.,* p. 288.   [12] *Ibid.,* p. 339.   [13] *Ibid.,* p. 311.

Oh my beloved holy Lords and Fathers! Do not, for the sake of the evil they can wreak, lamentably kill the noble inventions of art which have been gained by so much labour and sweat. . . . Because the same proportions the heathens assigned to their idol Apollo, we shall use for Christ the Lord, the fairest of them all, and just as they took Venus as the most beautiful woman, we shall chastely use the same figure for the purest of the purest Virgins, the Mother of God. . . .[14]

Clearly then, beauty too, like perspective, is a means of art, and one for which standards exist, however hard they may be to define and however much Dürer was aware of his own ignorance of the secret. And here, too, the only way to improvement lay in listening to criticism. For painters are too easily enamored of their own products, like mothers are of their own children.[15] Let nobody, therefore, trust his own judgment in these matters. Even the unskilled will usually be able to spot a mistake, though they may not be able to tell how to correct it.[16] Faultfinders are again to be encouraged.

If this attitude could only prevail, art was sure to make progress far beyond the present. I know of no earlier account (and indeed of few later ones) in which this faith is more clearly expressed than in the following passage:

For I am sure that many a worthy man will still come forward, all of whom will be better in writing about this art, and in teaching it, than I am. For I myself think very little of my art. For I know my own shortcomings. Hence may anyone undertake to improve my shortcomings as best he can. Would God it were possible for me to see now the works and the skill of those future great masters who are not yet born, for I believe that would improve me. Oh, how often do I see great art and good things in my sleep, of a kind that never comes to me while awake. But when I wake up, I lose the recollection. . . .[17]

Socrates claimed that he knew more than others, for at least he knew how ignorant he was. Dürer had a right to repeat this proud Socratic claim. The Gothic masters of his youth had been blind to their mistakes. The Renaissance movement had given the artist sight. He now knew that there were rational standards of which his work fell short. Even where he felt he had improved the art, the road to perfection still lay before him and his successors. An acceptance of the Renaissance conception of art implied an acceptance of the notion of progress. Thus criticism implied a view of history, that view which Vasari, a generation after Dürer, enshrined in the first edition of the *Vite*.

[14] *Ibid.*, p. 316. [15] *Ibid.*, p. 304. [16] *Ibid.*, pp. 229, 289. [17] *Ibid.*, p. 305.

III. VASARI'S *VITE;*
THE POWER AND PERILS OF HINDSIGHT

To read Vasari with some of Dürer's formulation in mind is to see the task he faced in a new light. His ruling conception of art as a solution of certain problems certainly provided him with a principle of selection that alone enabled him to write history rather than a mere chronicle. It is well known that the history of the revival of art *da Cimabue in puoi* is for him a history of those contributions which advanced the arts toward his ideal. In a remarkable article Mrs. S. Leontief Alpers has shown that Vasari made a distinction between the ends and the means of art.[18] We are likely to misread him if we fail to make this distinction which we also found implied in Dürer. A passage not specifically adduced by Mrs. Alpers defines this aim succinctly. It occurs in the vita of Berna da Siena where Vasari describes a fresco, now lost, of that trecento painter whom he much admired:

There was a painting representing a young man being led to execution, as well done as can possibly be imagined, since one saw in his features the pallour and fear of death quite like the truth, which merited the highest praise. Beside that young man there was a monk to comfort him, very well conceived in his attitude, in short everything in that work was so vividly painted that it was clear that Berna had pictured to himself that most horrible episode as it must occur, full of the most bitter and harsh terror, and that therefore he portrayed it so well with his brush that the real event itself would not arouse more emotion.[19]

In thus describing a work which obviously lacked both correct perspective and a mastery of anatomy, Vasari showed that he was well able to enter into the mind of earlier periods which, in Dürer's words, were still 'blind' and could not see their own mistakes.

Admittedly where he discusses the development of means Vasari sometimes allows his hindsight to obtrude. There are passages where he attributes to earlier artists a dissatisfaction with their own technique which they could only have felt if they could have foreseen coming improvements. His account of the invention of oil painting in the vita of Antonello da Messina is a case in point which is all the more instructive since it deals with a technological problem Vasari considered part of art:

When I reflect on the diverse character of benefits and useful inventions

[18] S. L. Alpers, "Ekphrasis and Aesthetic Attitudes in Vasari's *Lives*," *Journal of the Warburg and Courtauld Institutes*, XIII (1960).
[19] Giorgio Vasari, *Le vite de' più eccellenti Pittori, Scultori e Architettori*, ed. G. Milanesi (Florence, 1878–85), I, 648.

which many masters who adhered to the "Second Style" have contributed to the art of painting, I am bound, in considering their efforts, to call them truly industrious and excellent, since they tried so much to raise painting to a higher level, regardless of difficulties or expense or of their private advantage. . . .

Adhering thus to the method of using no other medium but tempera on panels and canvasses (a method initiated by Cimabue in 1250 when he worked with those Greeks, and then followed from Giotto onwards up to the time of which we are speaking), the same procedure was always followed. . . . Nevertheless the artists knew quite well that tempera paintings lack the effect of a certain softness and vividness, which, if it could have been achieved, would have imparted to these paintings more grace in design, more beauty in colouring and a greater facility in blending the colours, rather than always applying only the point of the brush. But however many had cudgelled their brains searching for such a thing, nobody had as yet found one which was good. . . . a large number of artists had talked about these matters and had discussed them many times without results. The same desire was shared by many outstanding minds who practised painting outside Italy, such as the painters of France, Spain, Germany and other parts.[20]

Vasari's famous account of the accidental invention of oil painting by Jan van Eyck which then follows need not detain us here.[21] What matters is of course his readiness to project the present into the past, to assume that artists working in tempera were longing for a method they did not know. It is safe to say that this is unlikely. Why should a Fra Angelico have been dissatisfied with a medium that he could use with such mastery for his jewel-like panels? Surely a great artist works within the limitations of his means; in fact, these limitations present both a challenge and an inspiration. We can thus discard Vasari's picture of congresses for the overcoming of these limitations. Even so, it is characteristic of the period that collections of recipes were indeed circulating among artists. We may be sure that there were many eager to learn of new methods, including some which made use of oil (of which there were a good many before van Eyck).

The point is surely that once some improved method is claimed to exist any artist worth his salt will want to try it out, even if he had been quite happy with the older procedure before. And though hindsight distorts our view if we imagine Fra Angelico fretting, because his medium lacked softness and vividness, hindsight shows its power if it makes us pay attention to the reasons which secured such a rapid victory of oil over tempera.

For however much Vasari may have simplified the story of this invention and its spread from Flanders to Italy, it remains demonstrably true

[20] *Ibid.*, II, 563–64.
[21] For the bibliography of this problem see E. Panofsky, *Early Netherlandish Painting* (Cambridge, Mass., 1953), I, 418.

that once oil paintings were known they were in demand.[22] We also know from the reliable pen of Vespasiano da Bisticci that that great patron Federigo da Montefeltre had painters come from the Netherlands because he wanted works in the new technique.[23]

By Vasari's time, of course, oil had been so triumphant that tempera had almost become forgotten, a fact which so characteristically prompted Vasari to try his hand at the old technique.[24]

What is apt to irk us in Vasari's account here is obviously the assumption that any improvement must have been striven for and that those who still lacked it felt its absence as a mark of inferiority. Dürer, to repeat, was wiser here, for he spoke from experience.

But though we must acknowledge these perils of hindsight which have vitiated more recent histories than Vasari's and have tended to reduce the history of art to the story of how the past strove to become the present,[25] we must not allow our awareness of these limitations to dismiss Vasari's whole account as unhistorical.

Take his treatment of Signorelli, whose memory he revered as that of a kinsman who had encouraged him in his youth:

Luca Signorelli, an excellent painter, . . . was considered more famous in Italy in his time and his works were more highly praised than had happened to anyone at any period before, because in his paintings he demonstrated the way of representing nudes which, all-be-it with skill and difficulty, can be made to appear alive. He was the follower and disciple of Piero della Francesca and made a great effort in his youth to imitate and even to surpass his master. . . . In . . . Orvieto . . . he painted all the stories of the end of the world with bizarre and capricious inventions, angels, demons, ruins, earthquakes, miracles of the Antichrist and many other similar things; moreover nudes, foreshortened figures and very beautiful ones, imagining the terror of that last and tremendous day. Hence he aroused the mind of all those who came after him, who therefore found the difficulties of that style easy. I am therefore not surprised that the works of Luca (Signorelli) were always highly praised by Michelangelo, nor that some things in (Michelangelo's) Last Judgement were partly gently lifted out of Luca's in-

---

[22] Jakob Burckhardt, "Die Sammler," *Beiträge zur Kunstgeschichte von Italien, Gesamtausgabe*, XII (Berlin, 1930), 316, 317.

[23] Vespasiano da Bisticci, *Vite di Uomini Illustri*, ed. Paolo d'Ancona and Erhard Aeschlimann (Milan, 1951), p. 209.

[24] *Vite*, VII, 686.

[25] For an important discussion of this problem see James Ackerman in James Ackerman and Rhys Carpenter, *Art and Archaeology* (Englewood Cliffs, 1963), pp. 171ff. See also my review article "Evolution in the Arts," *The British Journal of Aesthetics*, Vol. IV, No. 3 (July, 1964).

ventions, for instance, angels, demons, the order of the heavens and other things in which Michelangelo imitated the procedure of Luca as everyone can see. . . .[26]

We may rebel against this typical assessment of an artist's achievement as merely a stepping stone toward Michelangelo's supreme creation (Fig. 7). We are right in wanting rather to forget "The Last Judgment" of the Sistine Chapel when visiting Orvieto (Fig. 6). But can we be quite sure that Signorelli himself did not acknowledge Michelangelo's superiority? Granted that he never saw "The Last Judgment," we know that he did visit Michelangelo's studio in Rome and was, alas, treated less generously than Vasari's account would have made us hope.[27] The incident itself is irrelevant to our problem. What matters is that Signorelli's younger contemporary Dürer freely acknowledged that there was much still to be learned in the rendering of the beautiful nude and that greater masters would achieve in the future what he could only see in a fleeting vision in his dream. Did Signorelli also have such dreams? Did he find that Michelangelo realized what he had wanted to do? These are of course idle questions, but they allow us at least to do a little more justice to Vasari's picture of the Renaissance that finds its culmination in the grandiose opening of the vita of Michelangelo:

While industrious and excellent minds, aided by the light afforded them by such renowned artists as Giotto and his followers, strove to give proof to the world of their value and of the talent which the favour of the stars and harmonious blending of the humours had given them, and while they were desiring to rival the greatness of nature by the excellence of their skill so as to arrive as close as they could at that highest knowledge which many call a vision of truth, and while all of them laboured in vain, the Ruler of Heaven in His goodness graciously turned His eyes towards the earth and perceiving the infinite futility of all these strivings, the most ardent efforts which remained entirely fruitless, and the presumptuous opinion of men who were still further removed from the truth than light is from darkness, He resolved to liberate us from so many errors by sending down to earth a spirit, who would universally master every skill and every profession and demonstrate by himself alone what is perfection in the art of design. . . .[28]

Obviously we cannot take Vasari's view of history here at its face value. The idea of Fra Angelico trying in vain to paint like Michelangelo must again strike the modern reader as ludicrous. But is the idea of Fra Angelico

[26] *Vite*, III, 683, 690.
[27] For Michelangelo's letter of complaint of May, 1518, against Signorelli as a defaulting creditor see G. Milanesi, *Le Lettere di Michelangelo Buonarroti* (Florence, 1875), No. CCCLIV.
[28] *Vite*, VII, 135.

*Fig. 6.—Luca Signorelli, "The Fall of the Damned." Cathedral, Orvieto.
(Photograph by Anderson.)*

*Fig. 7.—Michelangelo, "The Last Judgment." Sistine Chapel, Rome.
(Photograph by Anderson.)*

16

wishing to master the rendering of the human body as perfectly as Michelangelo equally naïve? It is only naïve because in this form the question does not allow of an answer. To answer it, perhaps, would mean to decide to what extent he still belonged to the medieval tradition and to what to the Renaissance. I do not want to be misunderstood. I do not assert that medieval artists or craftsmen never experienced such "sacred discontent." After all, we have Dante's word for it that every "artista" must fall short of his ultimate aim, (Par.XXX. 31), that matter resists the form (Par.I. 127), and that the hand trembles (Par.XIII.76).[29] But we are less sure whether such dissatisfaction was necessarily caused by what Dürer was later to call "mistakes" in painting. This notion obviously presupposes a definite conception of the problem art must solve in order to fulfill its aim of convincing evocation.

If it is true, as the evidence suggests, that the public, including the learned, was easily ready to accept the conventions of more hieratic styles as serving this aim, we are once more faced with the question of why this style ever changed in favor of greater visual truth. How was the leaven of criticism ever introduced, and who were its carriers? Vasari does not give a complete answer, but he certainly shows himself aware of the importance of the critical milieu. He is sure that it is only where many artists congregate and where rivalry is kindled that the atmosphere will favor the kind of progress he has in mind. It is in this way, for instance, that he accounts for Donatello's return from Padua to Florence:

Since he was considered a miracle there and praised by every knowledgeable person he decided he would return to Florence, for, as he said, if he had stayed longer he would have forgotten all he knew being so much praised by everyone; he willingly returned to his hometown to be constantly decried, for these strictures would make him work harder and thus acquire more glory.[30]

In the life of Perugino, Vasari is even more explicit about this secret of the Florentines: he puts the lesson into the mouth of Perugino's ambitious teacher who tells his pupil

that it was in Florence rather than anywhere else that people perfect in all the arts and specially in painting, since in that city people are spurred on by three things: The first is the abuse which is dispensed so much by many, since the air there favours a natural freedom of the mind which is not generally satisfied with mere mediocrity, and is always more concerned with honouring the good and the beautiful than with respect for person. The second reason was that if one wants

---

[29] For a brief discussion of these passages see Schlosser, *La Letteratura Artistica*, p. 82.
[30] *Vite*, II, 413.

18

to live there, one must be industrious which means to say always use one's mind and judgment, and be ready and quick in one's work, and finally be good at earning money; Florence has no large countryside, abundant of products, which would enable one to live there for little, as happens where there is a surplus of goods. The third thing, possibly not less important than the other two, is a thirst for glory and honour which is to a large extent aroused by the air in men of every profession and which does not allow anyone who has his wits together to let others be on his level, let alone to lag behind those whom they consider to be men like themselves, even though they acknowledge them to be masters. In fact the air frequently prompts them to desire their own greatness so much, that unless they are by nature kindhearted or wise they will become evil tongued, ungrateful and thankless. It is true, that once a man has learned enough there, and does not want to go on living from day to day like an animal, but wants to get rich, he must leave that city and sell the quality of his works and the reputation of that city abroad, as the Doctors do with the reputation of the Florentine University. For Florence treats its artists as Time treats its creations, making and unmaking them and using them up little by little. . . .[31]

To some extent, no doubt, this brilliant sociological analysis is also born from hindsight. It serves in its context to explain Perugino's career, his development into a great artist in the Florentine hothouse and his subsequent preference for less exacting surroundings which allowed him to repeat himself and to exploit the vein he had once developed. In fact Vasari goes on to relate toward the end of Perugino's vita how one of the master's paintings done in his old age for Florence "was much abused by all the new artists" mainly because Perugino had there used figures he had employed before. Even his friends accused him of not having taken trouble, but Perugino replied:

"I placed figures into this picture which you had praised at other times and which then pleased you enormously. If they now displease you and you do not praise them, what can I do about it?" But they continued to harass him with sonnets and public insults.[32]

The fact that Perugino was criticized in old age for repeating himself is confirmed by Paolo Giovio.[33] And yet the answer which Vasari puts into his mouth is a perfectly rational one. Judged by the explicit standards of Renaissance art such as they are formulated in Dürer's writings there is no reason why an excellent solution should not be repeated. After all, Perugino's altarpieces were not intended to be all seen side by side, and since

[31] *Ibid.*, III, 567–68.   [32] *Ibid.*, III, 586–87.
[33] P. Giovio's biographies are printed in Tiraboschi, *Storia della Letteratura Italiana* (Florence, 1712), VI, 116.

they contained no mistakes and embodied much beauty, he may well have felt aggrieved when the critical Florentines suddenly produced an additional standard of excellence, that of originality.

Vasari also tells us, it will be remembered, who the people were who raised this criticism. It was the "new artists," critics in other words, who were less concerned with the question of whether a particular altar painting served its purpose well in a particular setting, but who had a professional interest in problem solutions. Artists, we may supplement Vasari's account, did see many of Perugino's works, if not side by side at least in succession, because they were interested in art. Hence their dissatisfaction with a master who repeated himself. He had begun to bore them.

But if we can believe Vasari's simplified model situation, the new artists also carried the public with them. They had fresh and more convincing solutions to offer which made the earlier ones suddenly look inadequate. Even though Perugino's public, like Wolgemut's, had been "blind" and had shared the artist's satisfaction with his fine style, the new masters had shown it up as less than perfect.

Describing the new sweetness and color harmonies of Francesco Francia and Perugino, Vasari writes with his customary zest:

Seeing it, people ran like mad to that new and more vivid beauty, since it seemed to them absolutely sure that it would never be possible to do better. But the error of those people was clearly shown up by the words of Leonardo da Vinci. . . . (Preface to Part III)[34]

It fits in well that Leonardo is here singled out as the artist whose works revealed the limitations of Perugino's achievement. For it was Leonardo who wrote in the *Trattato della Pittura*

Painter, if you please the first painters you will produce good paintings, for these alone can really judge you; but if you want to please those who are not masters, your paintings will contain few foreshortenings and little relief or vivid movement. . . .[35]

IV. THE SOLUTION AND CREATION OF ARTISTIC PROBLEMS

It is here that the argument links up with the previous paper on the Renaissance conception of artistic progress cited above, in which I also

[34] *Vite*, IV, 11.
[35] *Cod. Urb.*, ed. A. Philip McMahon (Princeton, 1956), fol. 33v.

referred to Leonardo's remarks. The paper was concerned with the idea of *dimostrazione*, the display of ingenuity characteristic of Renaissance art. But ingenuity itself may always have had its admirers, and there are few artistic traditions which do not pay tribute to a display of skill. In many societies, however, this skill lies in manual dexterity, in feats of patience and precision easily intelligible to the layman; we all know the intricate fretwork, the minute execution of detail, that leads the guide in the country house to say that it is all made by hand and every one motif is different from the other. Often the admiration of craftsmanship here merges with awe of the rarity of the material, the pearls, corals, or precious stones that characterize the *Kunstschrank* and its contents. We can, if we choose, also speak of problem solutions in such contexts—it is certainly a problem how to acquire a steady hand and a greater one still how to acquire gold and ivory. But obviously *dimostrazioni* which count as artistic progress are of a very different kind. They are admired by the connoisseur who can appreciate their difficulty and their relevance to the overriding problems of art. Problems, one might vary a famous adage, should not only be solved; they should be seen to be solved. Indeed some works of art derive their main interest and their main social function from their character of such demonstrations. Brunelleschi's lost panels demonstrating the laws of perspective provide perhaps the most famous instance.[36] They may be described as an experiment in applied geometry designed to convince artists of the validity of Brunelleschi's method of construction. It is perhaps an open question whether or not we should describe these panels as works of art. But Dürer's testimony alone would suffice to show that this demonstration was felt to be relevant to art. Any problem solved, any difficulty overcome, lent interest to a work in the eyes of a public that watched the progress of the artist's skill in rendering nature as our contemporaries watch the progress of space travel.

To quote a text not adduced in the previous paper: describing the competition for the first Baptistery Doors (Figs. 8 and 9), the anonymous biographer of Brunelleschi who wrote around 1470 takes it for granted that the jury was influenced by such considerations—or that at any rate it should have been:

They marvelled at the difficulties he had confronted, such as the attitude of Abraham with his finger under the chin, his energy, his drapery and the way the body of his son was treated and finished, and the same for the drapery and the posture of the angel, and how he seizes Abraham's hand, and the attitude and treatment and finish of the one who pulls a thorn from his foot, and the same of

[36] A. Manetti, *Vita di Filippo di Ser Brunellesco*, ed. E. Toesca (Florence, 1927). A translation is in Elizabeth G. Holt, *A Documentary History of Art* (New York, 1957), I, 171–72.

21

*Fig. 8.—Filippo Brunelleschi, "The Sacrifice of Isaac." Bargello, Florence.
(Photograph by Alinari.)*

Fig. 9.—Lorenzo Ghiberti, "The Sacrifice of Isaac." Bargello, Florence.
(Photograph by Alinari.)

23

that who drinks bending down, and how many difficulties there are in these figures and how well they perform their roles so that there is not a single limb devoid of life and the same for the action and finish of the animals represented, and every other thing just as the whole of the narrative taken together.[37]

One wonders how far this particular assessment may reflect the intentions of Brunelleschi himself. That he wanted his figures to be lifelike and dramatic is certain. Whether he also wanted to display the difficulties he had overcome by introducing the figures of the *spinario* and the drinking man is a very different question. Maybe the problem here lay in the narrative he had been asked to illustrate. On receiving the dreadful summons Abraham had "saddled his ass and taken two young men with him, and Isaac his son." Seeing the assigned place from afar on the third day he asked the two servants to stay with the ass while he proceeded to the sacrifice. They were therefore to be included in the story, but played no part in the action. The real difficulty was how to represent them within the quadrifoil without their appearing to be witnesses of the main event. Ghiberti's solution here is perhaps the better one, though it is also more traditional. He sets a stage prop between the servants and the main action and uses in addition the device, so frequently found in Giotto, of two bystanders looking into each other's eyes as if to find out their companion's reactions. Brunelleschi apparently spurned this traditional device, and so he had to account for his figures not seeing the main scene, busying one of them with a thorn in his foot and making the other bend down to drink.

Be that as it may, Manetti's praise for the difficult postures chosen by his hero certainly reflects a critical standard that had emerged in the quattrocento. Its formulation, however, is not new. It derives from antiquity. It was Quintilian who passed on to the Renaissance the criterion of variety and difficulty in postures, while advising the orator on effective gesticulation.

It is often useful to depart a little from established and traditional order, and it is also sometimes correct to do so, just as we see that in sculptures and paintings the dress, the face and the attitude is often varied. For there is very little grace in a straight upright body; neither should the face look straight in front, nor the arms hang by the side and the feet joined, to make the whole work appear stiff from top to bottom. A certain bending, I might almost say movement, gives the feeling of action and emotion. And so, neither the hands should all be modelled on one formula, and there should be a thousand different types of faces. Some run and are in violent movement, others sit or lie down; this one is naked, those are draped, some partly the one, partly the other. What can

[37] Manetti, *Vita di Brunellesco*, p. 16.

be more twisted and elaborate than that discobolus by Myron? Yet, if someone were to disapprove of it because it was not straight, would he not lack any understanding of an art in which this very novelty and difficulty is particularly praiseworthy?[38]

The criterion of the "distortum et elaboratum" is more usually today associated with the period of Mannerism, and we shall see that this is not without justification. But it is all the more important to stress that even in Mannerism the critical standards of the Renaissance had not lost their force when it came to assess the validity of certain problem solutions. "Novitas et difficultas" were praiseworthy only when these paramount standards had been met. The most characteristic episode here is the story of Baccio Bandinelli's "Hercules and Cacus" (Fig. 10). It illustrates in all its aspects the spirit of ambition and the craving for fame that the Renaissance bequeathed to the sixteenth century.[39] Bandinelli's intrigues to get hold of a marble originally assigned to Michelangelo, his boast that in carrying out the group of Hercules and Cacus Michelangelo had planned he would surpass the "David," his failure in devising a model that would fit the block, and the hostile reception the finished work was accorded by the Florentines when it was revealed in 1534 are well told by Vasari. It was Benvenuto Cellini, however, the sworn enemy of Bandinelli, who preserved for us in one of the showpieces of his autobiography the criticism leveled against this boastful group:

This talented school says that if one were to shave the hair off Hercules, there would not remain noddle sufficient to contain his brain; and that as regards that face of his one would not know whether it was the countenance of a man or of a lion-ox: and that he is not paying any attention to what he is doing: and that it is badly attached to its neck, with so little skill and with so bad a grace, that one has never seen anything worse: and that those two ugly shoulders of his resemble the two pommels of an ass's packsaddle; and that his breast and the rest of his muscles are not copied from those of a man, but are drawn from an old sack full of melons, which has been set upright propped against a wall. Also the loins seem to be copied from a sack full of long gourds: one does not know by what method the two legs are attached to that ugly body; for one does not know upon which leg he is standing, or upon which he is making any display of pressure: still less does he appear to be resting upon both, (as it is customary sometimes for those masters who know something about the representation of figures). It is easy to see that he is falling forward more than a third of a *braccio*;

[38] Quintilian *Inst. Orat. II.* xiii. 10. The identification of Myron's statue in Roman copies only dates from the eighteenth century.
[39] For the history of this statue see John Pope-Hennessy, *Italian High Renaissance and Baroque Sculpture* (London, 1963), Catalogue Volume, pp. 63f.

Fig. 10.—Baccio Bandinelli, "Hercules and Cacus." Piazza della Signoria, Florence. (Photograph by Alinari.)

for this alone is the greatest and most intolerable fault that those wretched masters of the common herd commit. Of the arms they say that they are both stretched downwards without any grace: nor is there any artistic sense to be perceived in them, as if you had never seen living nudes: and that the right leg of Hercules and that of Cacus make a mixture in the calves of their legs; so that if one of the two were removed from the other, not only one of them, but rather both would remain without calves at that point where they touch: and they say that one of the feet of Hercules is buried and the other appears to have fire under it.[40]

Our age which has rediscovered Mannerism and demoted the standards of classical art has questioned the justice of these strictures. It can be claimed, after all, that Bandinelli never aimed at presenting a naturalistic likeness of a muscular man and that popular criticism therefore missed the point of his work, as it was later to miss the point of Epstein or of Picasso. It is certainly true that Bandinelli had a grievance against Cellini. His famous diatribe is obviously malicious. And yet it is important from the point of view of methodology not to dismiss it too lightly. It is always tempting to argue that an artist did not want to do what he is found not to have done. For the problem he has failed to solve we can always invent one, by hindsight, which he might be said to have solved. If somebody misses the bull's-eye, we can always attribute to him the conscious or unconscious desire to miss it, and we may even be right. But the more sophisticated we become in this technique, the less will it yield in real insights. A sixteenth-century sculptor who aimed at surpassing Michelangelo certainly did not declare his indifference to the accurate rendering of the human body. In seeking an expression of superhuman power and energy at the expense of anatomical plausibility, Bandinelli could in fact be accused of not playing the game. However much the emphasis may have shifted in the appreciation of certain artistic qualities, correct design was still a standard to which the artist was expected to submit.

In this respect the interpretation of Mannerism as an "anti-classical" style has tended sometimes to obscure the continuity of artistic standards which alone accounts for Vasari's profound understanding of the Renaissance.

For Vasari the painter had actually been asked to display a solution of the central problem of the Renaissance at the very period when he put the final touches to the first edition of his literary masterpiece. The document that tells us of this challenge to the painter is worth quoting in full, despite its length, because it not only illustrates the artistic ideals of Vasari's

[40] *The Life of Benvenuto Cellini*, trans. Robert H. Hobart Cust (London, 1927), II, 300–2. I have used this translation.

27

circle but also confirms the importance of criticism as a spur of which, as we remember, Vasari wrote so eloquently. He wrote from experience, for Annibale Caro makes it clear in his letter of 1548 that Vasari's recent work had created a mixed impression which he would do well to wipe out. The artist had just won fame and courted criticism by completing a series of frescoes in the *Cancellaria* in Rome in one hundred days.[41] The opening paragraph of Caro's letter alludes to this situation; the final remarks refer to the *Vite*, in which Caro had some share as a consultant and stylist:[42]

Annibale Caro from Rome to Giorgio Vasari in Florence, 10th May, 1548.

It is my desire to possess a notable work of your hand both for the sake of your reputation and for my own satisfaction; because I want to be able to show it to certain people who know you better as a quick painter than as an excellent one. I discussed this matter with Botto since I do not want to trouble you while you are encumbered with major commissions. But since you yourself have just offered to do it now, you can imagine how great my pleasure is. As to doing it quickly or slowly I leave that to you, because I think that it is also possible to do things quickly and well when the frenzy seizes you as happens in painting which in this respect as in all others is very much like poetry. It is certainly true that the world believes that if you worked less quickly you would do better, but this is a probability rather than a certainty. It might even be said that it is the works which are strained and not finished and carried forward with the same fervour with which they were begun that turn out worse. Nor would I want you to think that my desire to possess one of your works was so lukewarm that I do not wait for it with impatience. Even so I would want you to know that I say "take your time", use deliberation and application, but not too much application, as it is said of that other man among you who was unable to take his hand from the picture. But here I console myself that even the slowest movement you make will be completed before the fastest movement of the others; I am sure that you will satisfy me by any method, for apart from the fact that you are you, I know well that you are well disposed towards me and I see with what zest you turn to this particular enterprise. From your readiness to do the work I have already come to expect the great perfection of the work, so that you should do it whenever and however it comes easily to you.

And as to the invention of the subject matter, I also leave this to you, remembering another similarity that painting has with poetry; all the more so since you are both a poet and a painter, and since in each of these pursuits one tends to express one's own ideas and conceptions with more passion and zeal than those of any other person. Provided there are two nude figures, a male and a female (which are the most worthy subjects of your art) you can compose any story and any attitudes you like. Apart from these two protagonists I do not

---

[41] *Vite*, VII, 680.
[42] W. Kallab, *Vasaristudien* (Vienna, 1908).

mind whether there are many other figures provided they are small and far away; for it seems to me that a good deal of landscape adds grace and creates a feeling of depth. Should you want to know my own inclination I would think that Adonis and Venus would form an arrangement of the two most beautiful bodies you could make, even though this has been done before. And as to this point it would be good if you kept as closely as possible to the description in Theocritus. But since all of the figures he mentions would result in too intricate a group (which, as I said before would not please me) I would only do Adonis embraced and contemplated by Venus with that emotion with which we watch the death of the dearest; let him be placed on a purple garment, with a wound in his thigh with certain streaks of blood on his body with the implements of the hunt scattered about and with one or two beautiful dogs, provided this does not take up too much space. I would leave out the nymphs, the Fates and the Graces which in Theocritus weep over him, and also those love-gods attending him, washing him and making shade with their wings, and would only put in those other love-gods in the distance who drag the boar out of the wood, one of them striking him with the bow, the other pricking him with an arrow and the third pulling him by a cord to take him to Venus. And if possible I should indicate that out of this blood are born the roses and out of the tears the poppies. This or a similar invention, I have in my mind because apart from the beauty it would need emotion without which the figures lack life.

Should you not want to do more than one figure, the Leda, particularly the one by Michelangelo pleases me beyond measure. Also that Venus which that other worthy man painted as she rose from the sea would, I imagine, be a beautiful sight. Even so (as I said before) I am satisfied to leave the choice to you. As to the material I want it to be a canvas, five palms wide and three palms high. About your other work I need not say anything more since you have decided that we should look through it together. Meanwhile complete it entirely as you think for I am sure there will be little else to do but to praise it. Remain in good health.[43]

It may be argued that Annibale Caro's letter was sent to the wrong address and that he should have written to Titian rather than Vasari. Not only would the subject matter have been congenial to the great Venetian; even the distinction with which the letter opens, the discussion of the two modes of performance, one slow and detailed, one rapid and inspired, might have applied to Titian, whose change of manner became a topic of discussion. It has been suggested by Karl Frey that Caro may in fact have had Titian in mind.[44] But the point of his letter is of course precisely that he wants to arouse his friend Vasari to greater efforts by challenging him to create a work that would silence his critics and could be shown in the company of the greatest masterpieces embodying the Renaissance ideal. The

[43] Karl Frey, *Der literarische Nachlass Giorgio Vasaris* (Munich, 1923), I, 220–21.
[44] *Ibid.*, p. 222.

ideal itself is still the one adumbrated by Dürer—the representation of beautiful human bodies in a convincing setting. Of course Caro is less anxious than Dürer was about the possible exploitation of this ability for erotic effects. The power of art over the emotions can be as legitimately tested in the rendering of the alluring as in the evocation of tragedy.[45] Caro's principal suggestion, the death of Adonis, cunningly combines both elements, "for apart from the beauty it would need emotion, without which the figures lack life." He could be sure that there was common ground between him and Vasari if he asked him to strain his poetic faculty in following Theocritus' evocation of that episode. But he also knew that the painter cannot slavishly illustrate the poet. In fact he comes close to Lessing's Laokoon when he reminds the painter of the difference between poetic and pictorial narrative.

We do not know how far Vasari rose to the challenge. The painting of Venus and Adonis he painted for Caro had disappeared in France even by the time Vasari published his autobiography at the end of the second edition of the *Vite*.[46] It is unlikely that this is a very grievous loss. Vasari was not a great artist. Even so it may be argued that he has had rather a raw deal from the art historians. His autobiography, in particular, has suffered from the contrast with Benvenuto Cellini's and from the lack of zest with which Vasari turns from the biography of others to the account of his own career. And yet his own vita offers valuable testimonies about the way a highly intelligent and highly successful artist right in the center of things looked upon his own art. A full analysis of this testimony lies beyond the scope of this paper, but it is clear from the first that this arch-mannerist had meant to practice what he preached and that for him, too, the problem of dramatic evocation was paramount. To the modern observer the drawing for another lost painting by Vasari (Fig. 11), the "Martyrdom of St. Sigismund," which he painted for a chapel of San Lorenzo in Florence, looks like the typical mannerist showpiece of contorted nudes. But it is clear from Vasari's own description that he saw himself pursuing the same aim he had attributed to Berna da Siena:

I painted a panel ten braccia wide and thirteen high, with the story or rather the martyrdom of St. Sigismund, the King, that is when he, his wife and two sons were thrown into a well by another King or rather tyrant, and I arranged it so that the decoration of the chapel which forms an apse should serve me as the frame of the rustica porch of a large palace through which one could see a square courtyard supported by pilasters and doric columns; and I made it

---

[45] For this concept see Leonardo's *Trattato della Pittura*, f.13 and 13ᵛ.
[46] See Paola Barocchi, *Vasari Pittore* (Milan, 1964).

Fig. 11.—Giorgio Vasari, "The Martyrdom of St. Sigismund," drawing.
Musée Wicar, Lille. (Photograph by Giraudon.)

31

appear that through this opening one could see an octagonal well in the centre, surrounded by steps, which servants were mounting with the two naked sons they were throwing into the well. And standing around in the *loggie* I painted on one side people who stand and watch that horrible spectacle, and on the other, the left hand side, I painted a few armed men who, having fiercely seized the King's wife, carry her towards the well to kill her. And under the main gate I painted a group of soldiers who tie up St. Sigismund who, by his yielding and patient attitude, shows that he willingly suffers this death and martyrdom, and gazes at four angels in the air, who show him the palms and crowns of martyrdom for himself, his wife and his sons, which appears wholly to comfort and to console him. I tried equally to show the cruelty and fierceness of the vile tyrant who stands on the upper storey of the courtyard to watch his revenge and the death of St. Sigismund.

In short I made every effort to ensure that all the figures should be animated as far as possible by the right affects and showed the appropriate attitudes, fierceness, and all that is required. Whether I succeeded or not, it is for others to say, but I am entitled to add that I put into this as much study, effort and industry as I was capable of.[47]

There is perhaps more in these repeated protestations than mere conventional modesty. Conventional modesty never imposed similar inhibitions on Benvenuto Cellini. Like Dürer, Vasari is supremely aware of the existence of objective standards, and he feels and says where he has fallen short of them. He knows his weakness and also his strength:

But since art is essentially difficult one must not expect any more from anyone than he can give. Yet I would say, for this I can say truthfully, that I always made my paintings, inventions and drawings of whatever kind if not with utmost speed, at least with incredible ease and without strain.[48]

Thinking of art as a means to an end, facility in invention and execution was certainly an asset. Of the dangers of this kind of productivity to which Annibale Caro had gently drawn his attention Vasari was certainly aware, but the courtier in him made him often yield to pressures in order to please his patrons.

Moreover Vasari was not conscious of having shirked the other requirement of Renaissance aesthetics, the search for "novelty and difficulty." He tells us that he thought out fresh problems which challenged his ingenuity as an artist. In his account of another lost work also known only from a drawing (Fig. 12), his frescoes in Bologna, we read:

[47] *Vite*, VII, 691–92.   [48] *Ibid.*, VII, 669–70.

Fig. 12.—Giorgio Vasari, "Abraham and the Angels," drawing. Musée Wicar, Lille. (Photograph by Gerondal.)

I painted the three angels (an idea that came to me I do not know how) in a heavenly light that can be shown to issue from them while the rays of the sun surround them in a cloud. Old Abraham adores one of these angels though he sees three, while Sarah stands laughingly thinking how their promise could possibly come true, and Hagar with Ismail in her arm leaves the house. The same light also illuminates the servants who are making ready, among them some who are unable to bear the splendour and shield their eyes with their hands; since strong shadow and bright lights impart strength to paintings, this variety of things gave it greater relief than the two preceding compositions and having varied the color, made a very different effect. Had I only always known how to realize my ideas! For then, and later, I always went on with new inventions and imaginings, searching out the difficulties and obstacles of art.[49]

The emphasis on light effects as the problem Vasari had set himself is significant. It was a problem where much remained to be done. The paramount problems of art mentioned by Dürer, perspective and the rendering of the nude, had been solved. Nobody could surpass Michelangelo in the rendering of the human body. But Mrs. Alpers has shown that we misread Vasari if we interpret his glorification of this victory as the feeling of an epigone who believes that nothing remains to be done any more. In one of his most famous critical discussions, inserted in the life of Raphael, Vasari is eager to remind artists that there are other problems which remain to be explored.[50] He mentions the rendering of light among them. No wonder therefore that in his autobiography he frequently lingers on his attempts to render light effects. It was a problem that had interested him since the time when he had struggled pathetically with the rendering of the sheen on Alessandro de Medici's armor.[51] Perhaps his training and his gifts were really not sufficient for him to make this novel contribution which he seems to have envisaged for himself. Perhaps also the weight of the Florentine tradition was too strong for him, which saw in invention and design the main problems worthy of an artist's aspirations. Color and light are less easily subject to rational standards and rational criticism, and hence progress was much less measurable there than in the mastery of perspective and the nude.

It is perhaps significant that it was ultimately an artist from the north who put this critical bias of the Florentines for problem solutions in this field to the test, and won through. In the famous story of Giovanni da Bologna's response to criticism the theme of this paper comes to its climax

[49] *Ibid.*, VII, 665–66.
[50] *Ibid.*, IV, 376. For a quotation and discussion of this passage see my papers "Mannerism: The Historiographic Background," a contribution to the XXth Congress of the History of Art, New York, 1961, now reprinted in *Norm and Form.*
[51] *Vite*, VII, 657.

(Figs. 13 and 14). It is a well-documented story, since Raffaello Borghini's account in *Il Riposo* of 1584 was published less than two years after the event.

When Giambologna had produced many figures of bronze, both large and small and an infinite number of little models he had so well demonstrated his mastery of his art that even certain envious artists could no longer deny that he succeeded uncommonly well with things of this kind. They admitted he was very good in devising appealing figurines and statuettes in the most various attitudes which had a certain beauty. But they added that he would not have succeeded in the creation of large marble figures which after all constituted the true task of sculpture. It was for this reason that Giambologna pricked by the spur of ambition resolved to show to the world that not only he knew how to make ordinary marble statues but also several together and that the most difficult that could be made which would display all the skill in creating nude figures (showing the decline of old age, the strength of youth and the delicacy of womanhood). And thus he fashioned for no other reason than to demonstrate the excellence of his skill and without proposing to himself any particular subject matter a group of a proud youth who snatched a most beautiful girl from a weak old man; and when he had nearly completed that marvellous work it was seen by His Highness Francesco de Medici, our Grand Duke, who so admired its beauty that he decided it should be placed where it is still to be seen. But to avoid the group having to be shown without any title he urged Giambologna to find a subject. He was told, I do not know by whom, that it would be a good thing to link it up with Cellini's story of Perseus, to say that the raped girl was intended to be Andromeda, the wife of Perseus, the abductor Phineus, her uncle, and the old man Capheus her father.

But when one day Raffaello Borghini happened to find himself in Giambologna's studio, and having looked at this beautiful group of figures with much pleasure he showed surprise on hearing the story it was alleged to represent. On noticing this, Giambologna asked him urgently to tell him his views, and he made it clear that he should on no account give that name to those statues; the Rape of the Sabines would fit better and that story being judged suitable lent indeed a name to that work.[52]

Here, then, is a work of art which every visitor to Florence has seen, which owes its existence entirely to the desire of its maker to show that he could solve an artistic problem. The theme, as he himself put it in a letter, was chosen "to give scope to the understanding and the mastery of art."[53] The interest in problem solutions had ultimately led to a shift in the very aim of art. For Dürer, as we have seen, the aims of art were still unequivocally religious: the new means were to serve the elimination of mistakes;

[52] Pope-Hennessy, *Renaissance and Baroque Sculpture*, Catalogue Volume, pp. 82–83.
[53] Letter of June 13, 1579, quoted *ibid.*

Fig. 13.—Giovanni da Bologna, small bronze, representing a rape. Museo di Capodimonte, Naples. (Photograph by Soprintendenza alle Gallerie, Napoli.)

36

Fig. 14.—Giovanni da Bologna, "The Rape of the Sabine Women." Loggia dei Lanzi, Florence. (Photograph by Alinari.)

even Vasari still implied the primacy of the narrative function to which the means were subordinate. In Giambologna's group the display of virtuosity as such has gained priority over the subject matter. The dramatic story is selected (not without qualms) after the group is finished, though the artist then obliged and placed a relief of the "Rape of the Sabine Women" on the pedestal to serve as a label (Fig. 15).

Seen through the eyes of hindsight, then, the development of the *dimostrazione*, the display of problem solutions in art really falls into its place from the vantage point of Giambologna's challenge to his critics. We also know from Borghini's account who these critics were. Not, to be sure, the *letterati*, but his fellow artists who had sought to damage his reputation by insinuating that his skill, however appealing, was limited to small-scale work. It so happens that Vasari tells of a similar criticism and of a similar reaction in the vita of the painter Giovan Francesco Caroto:

It is true that either through envy or for any other reason he was labelled as a painter who could only do small figures; in doing the painting for the chapel of the Madonna in S. Fermo . . . he therefore wanted to show that he had been groundlessly slandered and made the figures more than lifesize and so well, that they were the best he had ever done . . . and truly this work is never considered anything but good by the artists. . . .[54]

But let us remember that Caroto's work was an altar painting; he had had to wait for his demonstration till he received a commission. The monks hardly worried about the size of the figures which he enlarged to show his fellow artists what he could do. Giambologna's case was different here. He could embark on a large marble group of no specified subject or purpose, sure that it would attract admiration and find a buyer. He could be sure because the setting of the event was Florence. In his native Boulogne such a gesture would have been as futile as it might have been even in Siena or Naples. But in Florence a public had grown up in the course of almost 300 years which would appreciate the point and hail the problem solution as a feat deserving of fame and support.

It would be tempting to end this survey with this triumphant manifestation of the creative power of criticism, but it would also be a little misleading. For what really mattered in this acceptance of rational standards was not the external display of virtuosity, but the internalization of these demands, that "sacred discontent" that makes criticism creative.

Perhaps the most beautiful description of the workings of the artistic conscience is the report of Titian's working methods which the master's

[54] *Vite*, V, 284.

Fig. 15.—Giovanni da Bologna, "Rape of the Sabine Women," relief. Loggia
dei Lanzi, Florence. (Photograph by Alinari.)

faithful pupil Palma il Giovine passed on to Marco Boschini, who tells how Titian would lay in his paintings rapidly with a few brushstrokes.

Having laid these precious foundations he turned the paintings to the wall, and there he sometimes left them for several months without looking at them. And when he then wanted again to apply the brush, he scrutinized them with rigorous attention as if they had been his deadly enemies, to see if he could find a fault in them. And having discovered any feature that did not correspond to his sensitive understanding, he would, like a beneficent surgeon treat the invalid, if necessary remove some excrescence or superfluity of flesh, or straighten an arm, if the shape was not adjusted to the bone structure; or if a foot had taken up an inconsistent attitude he set it right without paying heed to the pain, and so in all things of this kind.

Thus shaping and revising the figures, he reduced them to the most perfect symmetry that could represent the Beauty of Nature and Art. . . .[55]

We do not usually connect Titian's art with this concern for correct representation, and it is quite possible that Palma-Boschini stressed this element to counter the criticism of Titian's *disegno* Vasari had attributed to Michelangelo.[56] But whatever Titian's main emphasis may have been in these revisions, we need not doubt that note of self-criticism that marks the great master, the hostile scrutiny of his own work, the awareness of his limitations and the wish to transcend them. Here our account may link up with its beginning and return in conclusion to the testimony of Albrecht Dürer.

Eighteen years after Dürer's death Philip Melanchthon recalled in a letter what Dürer once said to him:

When he was a young man he had loved florid paintings with a maximum of variety, and had admired his own work, being mightily pleased when he contemplated that variety in any of his own paintings. But when he had grown old, later, he had begun to see nature and had attempted to perceive her real face; then he had understood that this simplicity was the greatest ornament of art. Since he could not achieve this, he said he no longer was an admirer of his own works as before, but often sighed when looking at his pictures reflecting on his weakness.[57]

When Melanchthon wrote down this memory, in 1546, German art had entered into a decline from which it was not to recover for centuries. It is usual to blame the Reformation for this loss of creative power, but the lack

---

[55] Marco Boschini, *Le Ricche Minere della Pittura* (Venice, 1674), 2d ed., fol. b4–b5.
[56] *Vite*, VII, 447.
[57] Letter to Georg von Anhalt, Dec. 17, 1546, in *Philippi Melanchthonis Opera*, ed. C. G. Bretschneider (Halle, 1839), VI, 322.

of ecclesiastical patronage did not have similar disastrous effects on the arts of the Netherlands. Somehow, in Germany the new standards had killed the vigor and enjoyment of art, the Gothic pleasure in richness and variety, without leading to a more than skin-deep adoption of the new critical tradition. Perhaps there was too much in this tradition that was implicit rather than explicit, and these implications were lost in teaching.

For obvious reasons this paper has been concerned with the explicit and formulated standards that made it possible for a new critical attitude to emerge. But though these standards were rigorous in establishing priorities, they were only conceived as a framework, as certain specifications that a work of art had to fulfill if it was to be acceptable. What the Renaissance never formulated was the crucial experience that for every problem solved, a new one could be created,[58] that the mastery of perspective and the nude had destroyed certain assets of the medieval tradition which had to be compensated for in new solutions of calculated composition and conscious color harmonies, in short in all those specifically artistic problem solutions which the modern art historian attempts to reconstruct from hindsight.[59] Among these potent but less articulate ideals, that of "simplicity" which Melanchthon remembers Dürer to have mentioned certainly stands out in what we call the "classic" art of the Renaissance. The history of this ideal is not yet written. There is little doubt, though, that for Dürer the recognition of the profound simplicity of nature was the consequence of a grasp of her mathematical structure. It was the simplicity we admire in Piero della Francesca and in Mantegna, the mastery of essentials which stands in no need of the subsidiary values of richness and variety. It may be claimed that the generation of Vasari had lost sight of this ideal and had replaced it by its deceptive double, the ideal of ease of execution which mattered so much to Vasari.[60]

[58] For a general formulation of this principle see K. R. Popper, *Of Clouds and Clocks*, The Arthur Holly Compton Memorial Lecture (St. Louis, 1966), pp. 22–25.

[59] For the fresh problems raised by the application of perspective and the consequent impairment of the compositional pattern see i.a. Otto Pächt, "Zur deutschen Bildauffassung der Spätgotik und Renaissance," (originally presented to Fritz Saxl in 1937, *Alte und neue Kunst*, I) (1952), 2, (where our Fig. 1 is discussed from this point of view); R. W. Valentiner, "Donatello and the Medieval Front Plane Relief," *The Art Quarterly*, II (1939) and in *Studies of Renaissance Sculpture* (London, 1950); my book *The Story of Art* (London, 1950), especially pp. 190–91, and 200–1. For a more detailed and very subtle discussion see John White, *The Birth and Rebirth of Pictorial Space* (London, 1957). For the problem of reconciling conflicting demands see also my papers on "Raphael's Madonna della Sedia" and "Norm and Form," the title essay of the volume cited above. The lecture to the Humanities Seminar at The Johns Hopkins University on which this paper is based was intended as an introduction to the discussion, in two subsequent lectures, of various examples of such problems and their solutions.

[60] For a discussion of this ideal see Craig Hugh Smyth, *Mannerism and Maniera* (New York, 1962), p. 9 and n. 50, giving further literature.

41

But the time was to come when simplicity was to be exalted into a value higher even than "perspective and the nude." When this shift was complete and the "primitives" were rediscovered, the rational ideals of progress collapsed and a new era had begun. The rebellion against the standards of the Renaissance now acted as a critical ferment. Once again, as in the time of Dürer's youth, apprentices produced their work "powerfully, but without forethought, simply as it pleased them." The artist and the professional faultfinder had dissolved their partnership.

PARAGONE: ASPECTS OF THE RELATIONSHIP BETWEEN SCULPTURE AND PAINTING

✃ JOHN WHITE

or those who are not themselves involved, there is probably nothing duller than a certain kind of academic argument, and the Paragone, the discussion as to primacy among the arts—which was the better, which more fit to elbow its way into the charmed, closed circle of the Liberal Arts—was an academic argument par excellence. The major impulse had been provided by Leonardo da Vinci,[1] and by 1546, when Benedetto Varchi's question about the relative values of sculpture and painting was directed to the five sculptors; Michelangelo, Cellini, Tribolo, Tasso, and Francesco da Sangallo; and to the three painters; Pontormo, Bronzino, and Vasari; both Michelangelo and Pontormo clearly showed what they thought of the whole business.[2]

Michelangelo, in by far the shortest and, not unexpectedly, the most intelligent reply, wished amongst other things that "since the one and the other stem from the same understanding, that is to say sculpture and painting, they could make a fine peace together and give up so many quarrels, for they take more time than making figures." Pontormo, for his part, mockingly concluded that the two arts "are like clothing; sculpture a fine cloth which lasts longer and costs more, and painting a cloth that is cottoned in hell which lasts little, and costs less, for once the nap is worn off it is no longer of any account."

Among the most frequently repeated themes were those of the relative cost and durability of the two arts, and of whether or not drawing was fundamental to them. Another problem was whether there was a valid contrast to be made between the additive methods of painting, which allow of constant correction, and the subtractive quality of sculpture, in which what has once been chiseled away can never be replaced. There was much discussion of the degrees of difficulty involved in presenting a single view on a flat surface or of producing a multiplicity of views in a single object

---

[1] See I. A. Richter, *Paragone* (Oxford, 1949), in which the opinions of Michelangelo and Pontormo quoted below are given in full on pp. 89ff.
[2] See M. G. Bottari and S. Ticozzi, *Raccolta di lettere sulla pittura, scultura ed architettura* (Milan, 1822), I, 17ff.

in the round. The greater descriptive range of painting was opposed to the greater truthfulness of sculpture in those things which it did represent. The sculptors praised sculpture for being what it is, instead of being merely an illusion. The painters lauded painting because the three-dimensionality of sculpture is, like the hardness of marble, a function of nature not of art, while the illusions of painting, being the creations of intellect and science, are truly artistic. Finally the physical fatigues of sculpture were pitted against the mental agonies of painting.

Although these oft-repeated arguments were almost wholly futile in their ostensible, immediate aims, they are highly revealing in psychological and sociological terms and in relation to the history of ideas and of aesthetics. How important a part the simple facts which underlie these seemingly platitudinous statements play in the actual history of the two arts, and above all in their constantly changing interrelationships, is often forgotten or underestimated. This dialogue is nowhere more significant than in the art of the Renaissance and in that leitmotiv of central Italian sculpture and painting, man's exploration of himself and of his immediate surroundings. This investigation of the actual appearances of things proved to be the preliminary, and then a major initial motive, for the investigation of their inner structure and for the development of experimental science. Moreover, it is in this area, in which the goals of the two arts overlap and coincide to the maximum, that the history of their interrelationships provides implicit answers to the problem of relative difficulty with which the disputants were so concerned.

In the early stages of this exploration, motivated by new concepts both of the relationship of the created world to its Creator and of the function of religious art, the primacy of sculpture is clear-cut. Throughout the first half of the thirteenth century in France there is nothing in manuscript illumination or stained glass to match the solidity and sense of structure, the straightforward naturalism in the description of the fall and fold of drapery, or the subtle humanity and range of expression in the heads, which are to be found in the finest of the sculpture on the north portal at Chartres, at Notre Dame in Paris, or on the west portals at Reims and Amiens. A similarly sweeping generalization holds good for the Italy of the 1260's.

It is not merely that in terms of the anatomical description of the nude human figure, of a sense of muscle and sinew over bone, Nicola Pisano's "Hercules" on his Pisa pulpit of 1260 (Fig. 1) has no contemporary pictorial counterpart. With the notable exception of the minuscule direct imitations of sculpture which occur in the fresco of the "Trial by Fire" in the Upper Church of S. Francesco at Assisi (Fig. 3), probably dating from the mid-1290's, there seems to be nothing remotely comparable in painting

Fig. 1.—Nicola Pisano, "Hercules." Baptistry, Pisa. (Photograph by Anderson.)

45

*Fig. 2.—Master of St. Francis Cycle, "St. Francis Renouncing his Patrimony,"
detail. S. Francesco, Assisi. (Photograph by Bencini and Sansoni.)*

Fig. 3.—Master of St. Francis Cycle, "Trial by Fire," detail. S. Francesco, Assisi. (Photograph by Bencini and Sansoni.)

before the second quarter of the fifteenth century. It is a recurrent phenomenon in the history both of painting and of sculpture that such extraordinary adventures and achievements in observation and representation should occur in very small-scale works, in the minor details of major compositions, and in such peripheral features as predella panels or manuscript borders, and that they should then remain without full-scale progeny for decades or even centuries. In terms of major figure-painting, in which the bounds of pre-existing mental schemata are so much stronger, the formal problems more obtrusive, and the pressures of public taste and opinion more directly brought to bear, the stage reached by the mid-1290's is shown by the torso of the saint in the fresco of "St. Francis Renouncing his Patrimony," which is also in the Upper Church at Assisi (Fig. 2).

Here it is very noticeable that the relatively rarely tackled problem of the profile nude body has led to a thoroughly wholehearted, if, by Nicola Pisano's standards, only partially successful attempt to depict the layers of flesh upon the rib cage and over and beneath the shoulder blade as it is pulled forward by the raising of the saint's arm in prayer. As soon as the less unusual anatomical situations and, in terms of pictorial practice, more familiar parts of the body are involved, the situation, instead of improving as might be expected, rapidly deteriorates. Immediately the lower arm and the hand are reached there is a swift return to relatively fleshless, stiff, and wooden forms. Even for so adventurous an artist as the Master of the S. Francesco Cycle the strain involved in breaking new ground is such that pre-existing formulae always tend to retain their grip to a much greater extent where the artist has not involved himself in a problem for which no ready-made solution is to hand.[3]

A major factor in the pressures to which late thirteenth-century painters subjected themselves in struggling to create a more realistic art was naturally the difficulty so frequently referred to by the painters concerned in the arguments of the Renaissance Paragone; that of rediscovering or recreating visual formulae which would give the appearance of three-dimensionality to a flat surface. There can be little doubt that the primary cause of the vast gap so rapidly opened up by French and Italian sculptors as soon as they turned to the problem of the realistic representation of the human form is the fact, repeatedly observed by the Renaissance disputants, that the sculptor has solidity as of right, his very materials being three-dimensional, whereas

---

[3] This phenomenon is discussed, in particular connection with the marked contrast between Giovanni dei Grassi's relatively traditional and schematic approach to the drawing of iconographically familiar animals, and his freshness of observation and originality of draftsmanship when faced by wholly unfamiliar creatures such as ostriches, in J. White, *Art and Architecture in Italy 1250–1400* (Harmondsworth and Baltimore, 1966), p. 384.

the painter has to fight for its appearance with insubstantial pigments on a two-dimensional surface. For decades and even centuries after the sculptor had found himself free to concentrate on the particularities of bone and muscle structures, and of the personalities that inhabit them, devising subtle and complex formulae for dealing with nuances of form or expression, the painter was still struggling to make the simplest represented solid seem to be just that—a solid. By and large therefore, it was only by a prolonged struggle with the simplest formal analogies for essentially complex anatomical structures that the painter could proceed at all.

The fundamental, unavoidable problem was greatly aggravated by the nature of the artistic heritage on which the thirteenth-century artist was so largely dependent. In Italy in particular, but up to a point in France also, the sculptor could make use of antique carvings. Whether he intended to work in the round or in any particular kind of relief, these provided him with complete solutions to the basic formal problems. Both the materials and the tools in use had remained substantially unchanged. Only the urge to use what was so readily to hand was needed. For the painter outside Rome, however, virtually nothing from the seemingly short-lived period of three-dimensionally realistic Roman painting seems to have survived above ground. Even in Rome itself and its immediate surroundings, the visible remains equivalent to the then unknown Second, Third, and Fourth Pompeian Styles were almost certainly both few and fragmentary. Finally, even where they did survive, the thirteenth-century painter was confronted by the barriers of a lost technique and unknown medium. The painters therefore largely depended on the relatively pale echoes of antique three-dimensionality and anatomical realism which survived in Early Christian painting. Again, outside the field of manuscript illumination, Rome was the only place in which, despite substantial loss and overlay, enough Early Christian painting survived to provide the basis for a new art when, in the final quarter of the thirteenth century, the exigencies of Papal policy turned attention to the refurbishing of the great basilican pictorial cycles. The existence of such a basis was the prerequisite for any such swift revolution in pictorial style as in fact occurred. The thirteenth-century artists were conditioned by centuries-old traditions of substantially non-realistic art and of a very high degree of dependence on the repetition of existing patterns. Consequently they would also have lacked the techniques required for any kind of accurate life-drawing, had such an activity been conceivable to them. Without the Early Christian prototypes even the most daring spirits would have been unable to break through the barriers of existing style with a rapidity which, if slow and tentative in comparison with the sculptors' pace, was revolutionary when seen within the context of its own inherent problems and matched against preceding rates of change.

None of this is a denial of the extent to which, even in the realm of anatomical description, Italian sculptors in the mid- and later thirteenth century used pictorial prototypes. It is simply that whenever they did draw on the immediately preceding or the contemporary pictorial tradition, the intensification of the degree of descriptive realism was extreme. It does not matter whether Nicola Pisano's crucified Christ on the Pisa pulpit of 1260 is compared with Giunta Pisano's crucifix of about 1250–55 in S. Domenico at Bologna, or the head of Pietro Oderisi's Clement IV in S. Francesco at Viterbo of 1271–74 with that of the crucifix attributable to Coppo di Marco-valdo of about ten years earlier, now in the Pinacoteca Civica at S. Gimignano.[4] In the first instance almost all the crisp linear stylizations on which the painter depends have been abolished in favor of a determined attempt at a softly modeled fleshiness characterized by continuous surface movement and by subtlety of transition. In the second (Figs. 4 and 5), nearly all the painter's linear stylizations and sharp definitions of change of plane and of the boundaries of individual anatomical forms have been retained by the sculptor. In spite of this much closer cleaving to the painter's detailed pattern, the latter's inability to suggest the three-dimensional rotundity of the skull with its facial structure, or his disinterest in doing so, is such that the sculptor's evocation of the actuality of the figure he portrays is by far the more compelling. Whenever a major sculptor is involved and the borrowing occurs in the reverse direction, the element of realism is invariably diminished.

The situation is not substantially changed in the early years of the fourteenth century. In terms of anatomical conviction and of structural certainty Giovanni Pisano's "Hercules" on his Pisa pulpit of between 1302 and 1310, indebted as it is in all probability to French models, and his neighboring figures of "Fortitude" and "Prudence" (Fig. 15), are hardly less in advance of Giotto's Christ in the Arena Chapel Baptism of between 1304 and 1313. Indeed, the finest of Giotto's Paduan figures hardly match the angle figures of the "Virgin and Child" on Nicola Pisano's Siena pulpit of 1265–68 (Fig. 6) in matters of volume and of structural conviction. Both in the latter carving and in such figures as that of "Hope" on this same pulpit, Nicola had been able to draw upon French sculptural patterns in his search for a truly Christian humanity with which to breathe life into the cold pagan prototypes on which he had depended for the structural basis of his figures on the Pisa pulpit of 1260. This step-by-step attack on anatomical and structural fundamentals, and then on spiritual and emotional realism,

[4] See *ibid.*, pp. 57–58, where the question of the portrait element in the effigy of Clement IV is discussed.

Fig. 4.—Coppo di Marcovaldo, Head of Christ, detail of Crucifix. Pinacoteca, S. Gimignano. (Photograph by Soprintendenza alle Gallerie, Florence.)

51

Fig. 5.—Pietro Oderisi, Head of Clement IV, detail of tomb. S. Francesco, Viterbo. (Photograph by Gabinetto Fotografico Nazionale.)

Fig. 6.—Nicola Pisano, "Virgin and Child," detail of pulpit. Duomo, Siena. (Photograph by Anderson.)

which seems to provide the explanation for Nicola's seemingly sudden change from Classical to Gothic models, already places him upon a level of descriptive development which Giotto was, for all his genius, at first unable to achieve. A comparison of the head of Nicola's Virgin (Fig. 6) with that of Giotto's similar figure from the "Last Judgment" at Padua (Fig. 7) shows the painter still to be engaged, despite the grandeur of his image, in tackling problems which the sculptor of the 1260's had already left behind him. Not only are the Virgin's draperies still substantially displayed on the plane surface, but the rotundity of the magnificent head is only attainable through its reduction to the simplest ovoid forms. The contrast with the sense of soft and living flesh already within Nicola Pisano's grasp by 1268 is notable. Moreover, on the upper side walls of the Arena Chapel it is clear that even in such heads as that of the remarkable sleeping figure of Joachim in "Joachim's Dream," the firmness of the more fully differentiated and boldly foreshortened structure is only achieved at the expense of the fleshy quality so frequently observable among the marble figures of Nicola's Siena pulpit. On the other hand, the speed of Giotto's development as he wrestles with his problems is no less obvious. In the lowest frescoes on the side walls, whether in the head of Pilate on the right of the "Mocking of Christ," or in the gentle "Hope" or plump-faced "Fortitude," Giotto attains solutions analogous to those of the sculptor. The same approach to descriptive parity in the rotundity and complexity of fold fall in the draped figures is visible in images such as those of "Temperance" and "Faith."[5] Indeed, although it is probably true to say that in descriptive complexity, in richness of pose and fold form, and in subtlety of spiritual and psychological insight, there is no individual figure in the entire range of Giotto's surviving production to match the finest of the Sibyls of Giovanni Pisano's Pistoia pulpit of 1301, it is at this point that a radical change in the relative positions of the two arts begins to take place.

The underlying reasons for this are seemingly not far to seek. In Italian art throughout the period the whole emphasis, in painting and sculpture alike, was placed on the dramatic narrative. The driving force behind the growing interest in nature was apparently the wish to give immediacy to the sacred stories and to make them a living reality. There was a new theological emphasis on the intimacy of the relationship between the Creator and His creation, and a new stress on the innocence and suffering of God made man. In the visual arts these new insights were primarily expressed in terms of human action. All the greatest artists and their hosts of lesser

---

[5] The back-turned seated figures in the "Lamentation" are more than a match for such earlier sculptural essays in the same genre as the thirsting figure from Arnolfo di Cambio's dismembered Perugia fountain.

Fig. 7.—Giotto, Virgin, detail of "Last Judgment." Arena Chapel, Padua. (Photograph by Alinari.)

followers were each in his own manner concerned firstly with finding ways of bringing out the dramatic essentials of the sacred stories, and secondly with fleshing out the often bare dramatic bones with every kind of realistic touch in order to drive the actuality of the event home to the onlooker. For enterprises such as these the descriptive range of paint and the monumental scale of fresco represented the ideal medium.

In northern Europe different climatic and cultural conditions had, both directly and by virtue of the architectural types inspired by those conditions, always inhibited a growth of mural painting comparable in strength and quantity to that in Italy. There, on the other hand, the ingrained humanism of the Graeco-Roman world had inhibited the subjection of the human form to the distorting demands of architectural ornament and structure. In the North the small scale of manuscript illumination and the physically uncompromising two-dimensionality of the parchment page, and in stained glass the no less insistent immateriality, together with the formal fragmentation and linear emphasis arising from the demands of the leading, constantly militated against the development of narrative realism. In the South the total lack of architectural function similarly inhibited the development of the carved single figure in an age when large-scale sculpture was only conceivable in an architectural context.

Italian Romanesque sculpture was largely confined to the application of fundamentally pictorial forms in low or moderate relief to flat architectural surfaces.[6] It had little or no structural or architectonic function, and nothing occurred in thirteenth-century Italy to introduce any substantial change in this respect. Giovanni Pisano's use of the façade, which he in all probability himself designed (about 1285–97) for the new Duomo at Siena as a stage for figure drama, specifically does not disturb this pattern. His figures of prophets and sibyls, often in rapid motion, are never used in the French manner to replace or even to accentuate a specific architectural member. There is a similar lack of specifically architectural function in Arnolfo di Cambio's projected sculptural decoration of his façade for the Duomo at Florence about 1294, although the figures in the round were evidently much more carefully contained in niches than they had been in Giovanni's work. There is also a substantial return towards the Romanesque precedents in the reintroduction of extensive areas of relief carving in and around the portals.

The nature of these large-scale architectural complexes is largely conditioned by the fact that in late thirteenth-century Italy architecture and sculpture primarily came together on the fundamentally sculptural scale

[6] See G. H. Crichton, *Romanesque Sculpture in Italy* (London, 1954). S. Zeno, Verona; S. Michele, Pavia; and the Baptistry at Parma provide outstanding examples.

of pulpit, tomb, and altar canopy. If the four Pisano pulpits are considered as a group, it is clear that the pictorial demands which were paramount in the Tuscan Romanesque examples, insofar as sculpture was not confined to a purely decorative role, rapidly reasserted themselves. There seems to be little doubt that it was an urge for increasingly vivid narration of the sacred stories that controlled Nicola Pisano's move from the relatively architectonic and concentrated narrative relief of the Pisa Baptistry pulpit to the much more pictorial, figure-swarming relief style of the Siena pulpit with its proliferation of descriptive detail and narrative incident. The same urge and the same scale of values seem to underlie the way in which the sculptural forms on the body of the Siena pulpit flow over the previously distinct architectural framework. It has already been noted that emotional and psychological realism is the goal towards which Nicola Pisano moves immediately he has conquered the basic problems of physical description, and it is as much with the emotions as with a continued increase in accuracy of physical description that Giovanni is concerned in his Pistoia pulpit of 1301.

In this connection there are seemingly two remarkable developments in Giovanni Pisano's sculpture.[7] The first is a positive exploitation of anatomical distortion for dramatic purposes. Figures are stretched beyond the limits of the anatomically possible in ways which greatly increase the dramatic impact of the scenes in which they are involved. The forward-thrusting neck of the angel Gabriel in the "Annunciation" on the Pistoia pulpit is an excellent example. The second is that when dealing with inherently violent subject matter such as the "Massacre of the Innocents," the chisel itself is handled with an expressive directness that is in complete contrast to the normal high degree of finish. In the "Massacre" the children are reduced to crude, blocked-in forms, wholly distinct from the beautifully detailed naturalism of the infant Christ in the "Nativity." The furrows left by the chisel are virtually undisguised. There seems to be no question here of studio intervention on a massive scale, and the selective use of this expressionist technique seems to argue for its purposive character. About its effectiveness in raising the emotional temperature there is no doubt at all. Even in sculpture there is hardly any subsequent parallel for Giovanni's achievement before the late works of Donatello are reached. In painting, the painstaking struggle, brush-point stroke by brush-point stroke, to build up satisfactorily descriptive forms upon the flat surface is such that there is no counterpart of any kind before the sixteenth century.

The pressure of the revolution in narrative description, as well as the

[7] For more detailed discussion of these points, see White, *Art and Architecture*, pp. 77–78. The view of Nicola Pisano's artistic development reflected earlier is discussed on pp. 43 and 48–49.

proof that Giovanni was, like his father before him, thoroughly involved in a struggle to increase the human realism of his sculpture, can be seen in the Pisa pulpit of 1302–10. Whether the Nativity (Fig. 8) or the Adoration panels are compared with their forerunners on the Pistoia, Siena, or Pisa Baptistry pulpits, it is clear that another stride towards the development of a fully pictorial relief style has taken place. Never before had the complete, if sloping, ground plane, on which the animals and figures stand and out of which the bushes grow, been laid out for inspection in the manner common in the pictorial prototypes. The descriptive complexity of the somewhat differently organized Crucifixion relief in which all attention is concentrated on the crowd of figures is, indeed, such that there are few parallels for it in contemporary painting. Giovanni's success in the Pisa pulpit, in which for the first time the figure sculpture is extensively used for purposes of architectural support, is such that it marks the end of the road for just over a hundred years. By this time Giovanni is working so completely on the painter's ground that he is fully committed to a competition which he cannot possibly win.

The fundamental handicap of the relatively tiny scale on which, in the context of pulpit sculpture and the like, Giovanni Pisano was obliged to work, in contrast to the broad sweep of the fresco-painter's field, was not the only limiting factor as soon as sculpture became committed to a pictorial approach to dramatic narrative. Without a fully developed perspective system; without a complete realization of the possibility of treating the relief plane as a pictorial intersection, as a window through into an extension of the spectator's world; without a complete abandonment of the conception of the relief figure as a solid sculptural form, no further advance was possible along the road on which the Pisani had so consistently been traveling. In the absence of such changes there was no hope of creating a sculptural counterpart for the descriptive range which was being achieved, in their very different ways, by painters such as Giotto and Duccio. In his work Giovanni Pisano is, like Giotto, like Masaccio and Donatello after him, like Raphael and Michelangelo, a prime illustration of what might be called the principle of the unavoidable pause. He has advanced so far in a particular direction, and in so doing has uncovered problems of such magnitude and intensity standing in the way of further travel, that nothing more is possible until sufficient time has elapsed for a fundamental change to take place in the whole approach to artistic production. Before this can happen innumerable experiments and moves in other directions are normally required within the field of art itself. Almost invariably, substantial changes in the social function of art and in the patterns of patronage are also needed before the frame of reference is sufficiently transformed to allow of the solution of what were

Fig. 8.—Giovanni Pisano, "Nativity," detail of pulpit. Duomo, Pisa. Photograph by Alinari.)

once apparently insoluble problems. It is not a question of a conscious return to unsolved problems of one kind or another, nor is it a matter of some rigid and repetitive cycle of development with a historically determined outcome. It is simply that one of the prime, recurrent concerns of late and post-medieval Western art, and not merely of Florentine or Tuscan art in the thirteenth, fourteenth, and fifteenth centuries, has been man's exploration of himself and of his environment. The result has tended to be that as soon as a sufficiently radical shift in the general frame of reference has taken place a new art form has emerged, and this has incidentally led to the solution of the problems which had previously blocked a certain line of development.

The relationship of Baroque art to High Renaissance art does not repeat that between the High Renaissance and the Early Renaissance or that between the Early Renaissance and the art of the late thirteenth and early fourteenth centuries. None of the major styles involved were predictable. Yet after the event the nature of the pauses which in fact occurred in certain central streams of development, and the contributions made in the interim to the eventual resumption of movement, are clear. The pressures which have introduced this particular element into the pattern of art historical change during the last seven hundred years or so of Western history, though varying in each particular instance, are in their own way as evident as, for example, is that more constant, underlying pressure exerted by the flatness of the representational surface on all pictorial styles which involved the creation upon it of any kind of three-dimensional space.

The reality of the late thirteenth- and early fourteenth-century sculptural impasse which has been discussed is confirmed by Maitani's carvings of between 1310 and 1330 on the Orvieto façade and by Andrea Pisano's work on the bronze doors of the Baptistry at Florence between 1330 and 1336.[8] In Andrea's case it was only by a seemingly carefully calculated avoidance of any possibility of competition with Giotto's pictorial prototypes in the Peruzzi Chapel in S. Croce that he felt himself able to proceed. The break-up of the rectangular and potentially pictorial subdivisions, inherited from Bonannus' Pisan doors, by Gothic quatrefoils enabled him to achieve an even more tightly disciplined exploitation of Giotto's principles of dramatic concentration than had been attained by Giotto himself. It also rendered wholly irrelevant any attempt to compete with the kind of architectural and spatial description which is one of the most revolutionary elements of Giotto's late style. Invidious comparisons were neutralized before they could be made.

[8] Lorenzo Maitani and Andrea Pisano are discussed in some detail in *ibid.*, pp. 291ff. and 303ff.

The first main characteristic of Lorenzo Maitani's façade at Orvieto is that it is even more pictorial than either Giovanni Pisano's Siena façade or, insofar as it can be reliably reconstructed, Arnolfo's façade for the Duomo in Florence. It is not merely that figure sculpture in the round or near-round plays no architectonic role. Even allowing for the four symbols of the evangelists and for the central, canopied Maestà, set on its own carefully bounded stage above the central doorway, its total contribution is radically reduced. Instead, the whole sculptural scheme is based on the pictorial relief and on pictorial mosaic.

In terms of lowness of relief and of the melting of the figures into the wide, almost atmospheric areas of smooth ground, Maitani does indeed move further than Giovanni in the direction of a fully pictorial relief style. The larger scale also allows of a certain increase in the descriptive realism of landscape details such as trees. In anatomical richness and articulation certain of the figures in the "Last Judgment" undoubtedly compare with anything in Giovanni Pisano's work, and in sheer grace and delicacy some of the figures in the Creation scenes on the first pier are unsurpassed (Fig. 9). Nevertheless, in terms of the realistic, total description of a dramatic event, of the descriptive setting of a single scene and of the spatial articulation of a crowd, there is, significantly, nothing that seems to move beyond the point of furthest advance marked out in the reliefs of Giovanni Pisano's Pisa pulpit. This is not surprising since, apart from the lack of a perspective theory to provide the framework for a complete pictorialization of the relief surface, sculpture, by virtue of the natural, material solidity so heavily emphasized in the arguments of the Paragone, accentuates one of the major intellectual problems faced by all the late thirteenth- and early fourteenth-century artists. Indeed the rediscovery of descriptive naturalism in the late thirteenth century underlines a phenomenon that is earlier visible in the first great age of realism in the ancient world.[9] In the work of the Greek vase painters or in that of the archaic Greek sculptors, as in the arts of primitive peoples, the first aspect of the visible world to be encompassed is invariably the single, isolated, solid object. It is only by the slowest and most painstaking processes, extending over long periods of time, that coherent linkages between the previously isolated objects of attention are established and a feeling for the reality and continuity of the intervening space is achieved.

In the thirteenth and early fourteenth centuries the process of rediscovery naturally moves more quickly. Nevertheless, the creative struggle that separates the self-isolating architectural blocks of "St. Francis Renouncing his Patrimony" in the Upper Church at Assisi from Giotto's final

[9] See J. White, *The Birth and Rebirth of Pictorial Space* (London, 1957), pp. 34ff. and 236ff. and elsewhere.

61

Fig. 9.—Lorenzo Maitani (?), "Creation of Eve," detail of façade. Duomo, Orvieto. (Photograph by Raffaelli, Armoni, and Moretti.)

and most notable achievement in this respect, the "Raising of Drusiana" in the Peruzzi Chapel in S. Croce, is very clear. It is therefore no surprise at all to find that in the sculpture on the Orvieto façade there is a constant tendency to turn back from the dissolution of the solid object into its pictorial equivalent towards the creation of a solid, three-dimensional counterpart in stone for the solid three-dimensional reality in nature. Leaving aside the extreme illustration of this tendency in the carvings by the second major atelier, concentrated on the lower part of the two central piers, the bell tower in the scene of "Tubalcain" at the very top of the first pier, and therefore one of the latest elements in its design to be completed, provides an excellent case in point (Fig. 10). Despite the bold attempt to foreshorten the bell wheel and despite the wide range of relief that is called into play, it demonstrates the insistence of the pull towards truly sculptural and actually three-dimensional, as opposed to fully pictorial, virtually two-dimensional forms.

The extent to which the leadership in the visual arts had, by this time, been taken over by the painters can be seen by turning to Giotto's work in the Arena Chapel. Already in the second register the solid-object-dominated starting point in the "Refusal of Joachim's Offering" has been left far behind by the spatial subtleties of the "Teaching in the Temple" or the "Feast at Cana." By the time the Passion scenes in the third and lowest narrative zone have been reached, the compositional and psychological complexities of the "Lamentation" or the "Mocking of Christ" and the grasp of descriptive fundamentals, from the expression on an individual face to the interaction of numerous figures ranged throughout a convincing space, make the most ambitious relief designs of Giovanni Pisano, of Lorenzo Maitani, of Andrea Pisano, or indeed of any sculptor active before the 1420's appear to be extremely restricted in narrative scope, however beautiful and effective they may be as works of art. If Giotto's later works such as the "Apparition at Arles" in the Bardi Chapel or the "Raising of Drusiana" and the "Dance of Salome" in the Peruzzi Chapel in S. Croce are considered, the gap becomes correspondingly greater. By the late 1330's and early 1340's the fresco painters' exploitation of the large scale of their medium, and of the descriptive range which was so interminably emphasized by Leonardo and by all the subsequent protagonists of painting in the debates of the Paragone, is such that any kind of detailed comparison is hardly meaningful. The range of Ambrogio Lorenzetti's town and landscape compositions in the Palazzo Pubblico in Siena, with their combination of detailed observation and panoramic sweep, is ample confirmation of the nature of the changed relationship already apparent during the first decade of the fourteenth century.

If sculpture no longer held the primacy among the visual arts in the

Fig. 10.—Lorenzo Maitani (?), "Tubalcain," detail of façade. Duomo, Orvieto. (Photograph by Raffaelli, Armoni, and Moretti.)

years of bold experiment and of pictorial exploration which preceded the Black Death, there was nothing in the religious or in the economic climate of the years that followed to restore the balance. It was a world in which visionary intensity rather than descriptive logic was the primary answer to the spiritual needs of the times, and one in which the psychological tensions, founded as much on social and economic chaos and uncertainty as upon religious feelings of guilt and fear, found their expression in a return toward non-realistic aesthetic ideals.[10] In such a world the sculptors were if anything less qualified than ever to recapture their lost lead. By and large they remained quietly to one side of the newly turbulent main stream of development. Gentle, sweetly smiling Madonnas in sinuously flowing robes, and quietly representational reliefs, are, with a few exceptions, characteristic of the period in which artists such as Francesco Traini, Giovanni da Milano, Giovanni del Biondo, Bartolo di Fredi, Paolo di Giovanni Fei, and Agnolo Gaddi were in their different ways giving direct expression to the emotional tensions of the times.

The essential problem is summed up in the career of Andrea Orcagna, the first major Italian painter-sculptor of whom we have record. In his Strozzi altarpiece (Fig. 11) the inherent contradiction in the late fourteenth-century use of the realistic representational vocabulary, evolved in the preceding half of the century, as a means of creating an essentially non-realistic visual language is exploited with majestic coherence. Such conflicts as those between emphatic contour and enclosed bulk, or between sculptural solidity and non-existent space, are securely based on the fundamental pictorial paradox involved in any representation of three-dimensional forms on a two-dimensional surface. The element of contradiction is consistently exploited at every level.

In Orcagna's corresponding relief of the "Assumption" (Fig. 12) on his tabernacle of 1359 in Orsanmichele, the very fact that his figures do have actual bulk and that the enclosing mandorla is a solid sculptural form seemingly leads to a correspondingly increased emphasis on decorative inlay in order to redress the balance. Unfortunately, the pictorial processes cannot successfully be applied with such directness to the sculptural problem. The result is no longer the creation and negation of space and the achievement of an intensely unreal reality. The effect of the inlaid pattern of the sky is not to counter or negate the materiality of the angels. It simply adds to solid angels solid sky. Similarly the pattern element in the mandorla, far from reducing its solidity, adds definition to its material structure. Despite his evident disinterest in representational realism as such, Orcagna was, like all his contemporaries, evidently unable, when thinking as a sculptor, to

[10] See M. Meiss, *Painting in Florence and Siena after the Black Death* (Princeton, 1951), pp. 44ff.

*Fig. 11.—Orcagna, Strozzi Altarpiece. S. Maria Novella, Florence.
(Photograph by Alinari.)*

*Fig. 12.—Orcagna, "Assumption," detail of tabernacle. Orsanmichele, Florence.
(Photograph by Alinari.)*

transcend the concept of sculpture as an art in which solid objects are carved out of a solid material. A process analogous to that which in painting allows of a grand intensification of inherent paradox leads in sculpture to a largely self-defeating accumulation of detail. The design as a whole remains quite obstinately solid sculpture, and no visionary super-reality is achieved. In spite of this the "Assumption" remains an extremely influential design and one which bore important fruit in the changed artistic and intellectual climate of the early fifteenth century.[11] Orcagna's stature is, indeed, confirmed by his very willingness to tackle sculptural problems which were quite beyond the scope of his contemporaries. Furthermore, the very factors which prevented him from expressing his ideas as impressively in sculpture as in painting ensured that sculpture should once more be in the van as soon as interest in the exploration of the physical actualities of the human form revived. There is no contemporary counterpart in paint for the vivid corporeality and sense of human movement, of structure and of volume, of sinew, flesh, and muscle over bone, that is found in Jacopo della Quercia's ruined relief of the "Expulsion" on the Fonte Gaia, which was probably carried out between 1414 and 1419 (Fig. 13).

The impressive line of figures carved by Donatello and Nanni di Banco from about 1408 onwards shows that the primacy of sculpture is not a chance event. This is confirmed by the extent and nature of Masaccio's use of sculptural precedents as a basis for the pictorial revolution which he was almost singlehandedly consolidating. There is a long-familiar, close connection between the Eve of Masaccio's "Expulsion" in the Brancacci Chapel (Fig. 14) and Giovanni Pisano's "Prudence" at the base of his pulpit in Pisa Cathedral (Fig. 15). Recently attention has also been drawn to the relationship between the Virgins both in the center panel and in the predella of the Pisa polyptych and the Virgin carved in Giovanni's shop for the exterior of the Baptistry at Pisa, and also between two prophets likewise carved for the Baptistry in Nicola and Giovanni Pisano's workshops respectively and the figures of St. Paul and St. Andrew from the polyptych.[12] Such borrowings from late thirteenth- and early fourteenth-century sculpture are, of course, more than matched in the Brancacci Chapel by Masaccio's debt to Giotto's work in S. Croce. When it comes to his own contemporaries, however, it is the great series of standing figures by Donatello, beginning with the "St. Mark" of 1411–12 and continuing with two unnamed prophets for the

[11] Nanni di Banco's relief of the "Assumption" on the Porta della Mandorla of the Duomo in Florence is perhaps the most noteworthy of the works indebted to Orcagna's example.

[12] See E. Borsook, Note on Masaccio in Pisa, *The Burlington Magazine*, CIII (1961), 212–15.

Fig. 13.—Jacopo della Quercia, "Expulsion," detail of Fonte Gaia. Palazzo Pubblico, Siena. (Photograph by Alinari.)

Fig. 15.—Giovanni Pisano, "Fortitude and Prudence," detail of pulpit. Duomo, Pisa. (Photograph by Anderson.)

Fig. 14.—Masaccio, "Expulsion." Brancacci Chapel, S. Maria del Carmine, Florence. (Photograph by Alinari.)

Campanile (Figs. 17 and 18), carried out in 1415–18 and 1418–20, and the "Jeremiah" of 1423–26 (Fig. 20), which foreshadows his own achievement in paint.[13] In volume and in grandeur of form; in the blocking out of the large masses of such heads as that of St. John in the "Tribute Money" (Fig. 16); in the soft and massy freedom of certain of the draperies; and in the sheer physical weight and presence of such nude and semi-nude figures as those in the scene of baptism, the general debt to Donatello's prior achievements seems to be as clear as direct borrowings in terms of details are hard to find.

As far as the treatment of the individual figure is concerned, the natural advantages of his medium have once again allowed the sculptor to outdistance the painter in certain important respects. Inevitably Masaccio is largely concerned to use drapery as a means of creating volume on the flat surface of the wall. He is, therefore, nowhere able to indulge in rich complexities of fold like those to be found in the bearded, unnamed "Prophet" (Fig. 18) and in the "Jeremiah" (Fig. 20) or, for that matter, in the "Habakkuk" or "Zuccone" (Fig. 21) if the latter, instead of being dated 1527–36, is interchanged with the "Jeremiah" and connected with the documents of 1523–26.[14] It is very noticeable that in Donatello's own low reliefs, such as the "St. George and the Dragon" of 1416–20, the "Dance of Salome" of 1423–27, the "Donation and Ascension" of about 1427–30, and the Bronze Doors of the 1430's for the Old Sacristy of S. Lorenzo, where he is, to a greater or lesser degree, faced by the problem of creating an impression of three-dimensionality in forms that do not actually exist in the round, drapery

[13] The identification and relative dating of the "Jeremiah" and the "Habakkuk" have been the subject of some controversy. Stylistically it seems to me that the "Jeremiah" represents an accentuation of tendencies in earlier works by Donatello and that its connections with the output of Nanni di Banco and Jacopo della Quercia are strong enough to justify connecting it with the earlier documents. The "Habakkuk" seems to be more radical in its stylistic innovations. The inscription on the "Jeremiah," though challenged by H. W. Janson, *The Sculpture of Donatello* (Princeton, 1963), pp. 39–40, seems to carry a good deal of weight, and Janson's argument from the fact that the word *Gëmia* is on the verso of the scroll which the prophet holds in such a way that he can declaim his message from the recto seems to be two-edged. The earlier, beardless prophet pointing to his message follows the Gothic prototypes in so doing. In the face of this tradition the arrangement in the later work could well represent a significant and typically Donatellesque rethinking of old practices, instead of a mere misattribution by a later hand. J. Pope-Hennessy, *Italian Renaissance Sculpture* (London, 1958), p. 277, acknowledges the arguments of Kauffmann and Janson, but decides in favor of the view implicit in the works of Lanyi, Planiscig, and Paatz, stressing that the inscription in question evidently dates from before 1464 and that confusion was unlikely at so early a time. It must, however, be emphasized that the arguments put forward in the present context are, as will be seen, of such a kind that they are substantially unaffected by an interchange in date between the two statues.

[14] See preceding note.

is used in quite a different way from that found in his objectively three-dimensional, standing figures. It clings far more closely to the body and becomes much more linear. Its main formal function is to use this emphatic, linear element to indicate the solidity of the underlying bodies and to create a sense of movement.[15] Indeed, in the narrative reliefs, expressive movement, expressive grouping, and expressive interaction between figures are the primary means of releasing the spiritual and psychological forces with which, since they lie at the dramatic heart of the story, Donatello is finally concerned.

When, on the other hand, the sculptor is concentrating on the single figure in the round, he has both the natural advantage that his material is truly solid and the natural restriction that, like the painter tackling a similar problem, he has at his disposal far more limited means for suggesting psychological and spiritual character. Expression, pose, and, potentially, the handling of drapery not merely for descriptive but for dramatic purposes, are the sole tools at his command. In Masaccio's "Tribute Money," for example, the heads of St. Peter and of St. John (Fig. 16) create a vivid sense of the totally contrasted character and psychological make-up of the two apostles. Their draperies, on the other hand, are wholly noncommital in this respect. There is nothing in those of St. Peter to strengthen the impression of his individual personality and to contrast it with that of St. John. The same is generally true of the remainder of this fresco and of all of Masaccio's narrative designs as well as of his single figures of saints and apostles. The draperies may be disturbed by action or hang differently on a fat saint and on a thin one, but they are not directly manipulated to reinforce the distinction between one personality and another. In this respect Masaccio merely follows in the footsteps of all his pictorial and sculptural predecessors right back to the mid-thirteenth century and beyond. The one and only exception is Donatello. This is not to suggest that painters' and sculptors' styles in draperies, as in everything else, were not continually liable to change. It is simply that within the general style of a given moment or period the possibility of detailed variation for purposes of individual characterization was evidently not considered. The general style could be calm and monumental and structurally emphatic, as in the case of Giotto, or rhythmically sinuous, as in that of Duccio. It could be highly agitated in a rhythmic sense, as in the case of Lorenzo Monaco and of many International Gothic artists. It could take on the rumpled and excited forms of Jacopo della Quercia. In the latter's Trenta altarpiece of 1416–22 (Fig. 22), for example, the drapery of all the figures, male and female alike, shares the excitement

[15] Andrea Pisano's use of linear elements in drapery for very subtle compositional and iconographic purposes is discussed in White, *Art and Architecture*, p. 306.

Fig. 16.—Masaccio, Sts. Peter and John, detail of "Tribute Money." Brancacci Chapel, S. Maria del Carmine, Florence. (Photograph by Alinari.)

of their architectural surroundings without the least distinction between them. This shows that the agitation in his drapery style, which so increases the emotional temperature of his figures, is a generic rather than a particular or selective characteristic. The same is true of the prophets on the Siena Font and of the reliefs framing the main doorway of S. Petronio in Bologna, where a great variety of subject matter is likewise accompanied by a uniform agitation in the draperies. Although the style is vastly different and its development more marked, the situation is in principle no different in the case of Ghiberti.

The contrast in the case of Donatello's work of the second and third decades of the fifteenth century is striking. The armored figure of "St. George" is a special case, but the "St. Matthew," the two unnamed prophets for the Campanile (Figs. 17 and 18), the "Jeremiah" (Fig. 20), and the "Habakkuk" (Fig. 21) each have their own individual psychological character which seems to be reinforced by a distinctive treatment of the draperies. The thin, grey, beardless prophet (Fig. 17) has thin, liquescent draperies. The fold forms seemingly flow downwards, echoing the sad face with its down-turned mouth and the hands and scroll held down against the stomach as if in resignation at God-given messages too long unheeded. Not only does the closed configuration of the arms express the pensive quality of the heavy, bearded prophet (Fig. 18), but the mass and static weight of draperies enlivened by a heavy play of folds creates a visual analogy for the heavy convolutions of his inward thoughts. Such mass and complexity of form is only made possible by Donatello's feeling for the underlying structure, and by the objective solidity of the marble statue. None of the painters of the period who were primarily interested in presenting the body as a solid structure could attempt such draperies without endangering their prime objective.[16] Still less could they, or did they, attempt any equivalent of the abruptly broken, deep-cut folds to be seen in Donatello's "Jeremiah" (Fig. 20). It is evidently not until Piero della Francesca's Borgo S. Sepolcro altarpiece, probably of 1445 (Fig. 23), that any painter felt able to incorporate broken, deeply cut fold patterns of this kind into his figures without endangering their underlying structure. The folds of the "Jeremiah" establish a harsh and jagged linear pattern and a boldly broken and contrasted play of light and shade. These in their turn create a visual tension that exactly matches, and therefore intensifies, the sense of physical strain betrayed by the tensed wrist forcing the hand against the thigh, by the taut tendons of the neck,

---

[16] The International Gothic painters, sculptors, and illuminators in northern Europe, and men such as Ghiberti in Italy, during his Gothic phase, were not, of course, at all hesitant in such matters, since for the most part they were only minimally, if at all, concerned with underlying structure. They were, therefore, free to create drapery patterns of the greatest complexity in pursuit of their decorative and emotive aims.

Fig. 17.—Donatello, "Prophet." Museo dell'Opera del Duomo, Florence. (Photograph by Alinari.)

Fig. 18.—Donatello, "Prophet." Museo dell'Opera del Duomo, Florence.
(Photograph by Alinari.)

and by the compression of the upper lip. The drapery, like every physical detail of the body, is used to heighten the emotion seething in this great doom-saying figure. Its every aspect adds conviction to the belief that this is indeed the Jeremiah of the Jeremiads.

Everything that has been said of the "Jeremiah" as regards the sculptor's relative freedom to create grand masses of drapery which add to the mass and stature of the figure without endangering the sense of underlying structure is also true of the "Habakkuk" (Fig. 21). Again, the drapery is distinct in type and treatment. Again, it intensifies the austere grandeur of a supremely individual personality as "the burden which Habakkuk the prophet did see" (Hab. 1:1) finds its material counterpart.

In Donatello's case, the different types and treatments of drapery clearly do not reflect a simple chronological development. This is confirmed by the "St. Louis" of 1423–25 (Fig. 19). Its draperies, like those of the "Jeremiah," have much in common with the convoluted forms developed by Jacopo della Quercia (Fig. 22). This similarity only serves to underline the distinctive use that Donatello makes of the system on each occasion.[17]

The "St. Louis" (Fig. 19) has never been particularly popular and this is to some extent because of a misunderstanding of Donatello's purpose. The question of subject matter is often as important when undertaking the stylistic analysis of a single figure as it is when considering a narrative scene. It is not mere fancy to try to relate drapery pattern to personality in Donatello's figures, and where, as in this instance, the identification is certain, the effect may be to transform the observer's attitude to the visual facts. Vasari merely tells a little tale to give historical substance to his evident opinion that this was a bad statue of a stupid saint.[18] He attributes to Donatello mere insensitivity to the underlying meaning of the life of a saint, who, though he became the patron of the Parte Guelfa, was also the patron of the strict Observants in the Franciscan Order. For himself St. Louis' only ambition was the humble life of the Franciscan Friar, and the Bishopric which was thrust upon him because of his Angevin connections was almost as distasteful to him as the earthly kingdom which he eschewed. Already in 1317 in the great cult-image designed to confirm the legitimacy of his brother's power, Simone Martini had weighed down the impassive saint with

[17] In variety of drapery style Nanni di Banco provides a partial parallel to Donatello, by whom he was strongly influenced, but the range does not seem to be so great nor the variation so meaningful. Nevertheless, the convoluted style of the Porta della Mandorla relief seems certainly to be connected to the subject matter, and to have been adopted for emotive reasons, as well as in order to create a sense of movement.

[18] Giorgio Vasari, *Le vite de' più eccellenti pittori, scultori e architettori*, ed. G. Milanesi (Florence, 1878–85), II, 416.

Fig. 19.—Donatello, "St. Louis." Museo dell'Opera di S. Croce, shown in original setting, Orsanmichele, Florence. (Photograph by Courtauld Institute, London.)

a burden of episcopal insignia that all but hides the Franciscan habit that St. Louis sought and loved. Now, for all its over-lifesize actual dimensions, Donatello's gentle, resigned figure is almost smothered in its heavy cloak. The miter (Fig. 27) is no slim and graceful form, as it might well have been, and as it is in Nanni di Banco's anything but graceful "St. Eligius," which was likewise commissioned for Orsanmichele. Indeed, it is a massive object pressing down over the ears and forehead. Like the enormous but now truncated crozier, it adds to the observer's sense of the oppressive weight of the insignia of St. Louis' unwanted office. It is not by chance that Donatello chose to transmute the broken folds of the "Jeremiah" (Fig. 20) into these slow, smothering, velvet convolutions, when he came to design this particular figure. It is an attempt in the true Tuscan manner to compel mere matter, whether paint or stone or metal, to provide a counterpart for the intangibilities of a character and a state of mind. Donatello's "St. Louis" looks as it does by deliberate intention and for the very best of reasons.

Probably because of the factors which have already been discussed, Donatello did not attempt to extend this revolutionary concept of the function of drapery to his reliefs. These, like the single figures themselves, are, however, revolutionary in many other respects. The invention of the new artificial perspective gave scientific, logical validity and precise theoretical formulation to the long-established empirical practice of treating the picture as a view through the surface of the wall or panel into a new, three-dimensional world beyond it. In doing so it provided the sculptor with precisely what was needed to enable him to escape from the overriding tendency to see a work of sculpture as a solid object. The way which had been closed to Giovanni Pisano and his immediate contemporaries and successors was thrown open. Although nothing that could be accomplished within the relatively small compass of a monochrome or near-monochrome relief could hope to rival the totality of what Masaccio could achieve in fresco, Donatello was immediately able to explore certain fundamental aspects of the new perspective system as thoroughly as any painter. In the process he created a new type of relief. It is arguable that the mottled, grey-white marble of the pictorial, stiacciato relief of the "Ascension and Donation" betrays a greater sense of atmospheric perspective than any painting of the first half of the fifteenth century. Certainly Donatello's exploration of the visual ambivalences which could be exploited with the aid of the new system of linear perspective is more thoroughgoing than that of any painter of his generation. In the end, however, neither his system of calculated ambiguities nor the piling up of surface-stressing and depth-stressing counter-indications evolved by artists such as Filippo Lippi provided the final so-

Fig. 20.—Donatello, "Jeremiah." Museo dell'Opera del Duomo, Florence. (Photograph by Alinari.)

lution espoused by the painters of the High Renaissance.[19] Similarly, Michelangelo, precisely because he was heir to the innumerable experiments carried out by sculptors and painters during the fifteenth century, no longer needed to follow Donatello's lead in using the expressive potential of drapery for the creation of character.

The gilt-bronze bust of "S. Rossore" (Figs. 24 and 25), which Donatello completed sometime between 1422 and 1427, is one often underrated milestone on the road to the mastery of the human form which freed men such as Michelangelo of the need to use devices which were revolutionary advances at the time of their creation. Apart from the probability that the soft collar overlapping the armor and pressing up under the beard is a later addition,[20] several things have contributed to a misunderstanding of the work. They have led to doubts, now generally resolved, that Donatello is the author, and even to denials that the work dates from the fifteenth century at all. The most important contributory factor is that, unlike all the figures so far considered, this reliquary bust was not associated with an architectural setting and was not designed to be seen from a distance. In the heads of figures such as the "Jeremiah" (Fig. 28), intended for positions far above the spectator's head, the need to compete with the surrounding architectural detail is intimately associated with Donatello's concentration on the massive basic structure. The broad almost geometric planes of the underlying form are clearly visible. Even allowing for the effect of weathering, the surface treatment is summary, and there is little incidental detail, a characteristic which is still more noticeable in the smoothly modeled heads of the "St. George" (Fig. 26) and the "St. Louis" (Fig. 27). The whole physical form of the "Jeremiah" (Fig. 20), of which the head is the focus, expresses a similar breadth and strength both in structure and in emotional content.

The characteristics of the bust of "S. Rossore" (Figs. 24 and 25) are very different. The concentration of the features is greatly increased by the ab-

---

[19] See White, *Birth and Rebirth*, pp. 148ff. and 170ff. The High Renaissance solution is primarily in terms of balance and economy.

[20] See J. Lanyi, "Problemi della critica Donatelliana," *Critica d' Arte,* IV–V, Parte Seconda, 1939–40, pp. 9–23, in what is the key article on the bust, and Janson, *Donatello,* pp. 56–59. Lanyi, p. 14, n. 7, refers to the Trecentesque quality of the goatee. Almost exactly this type of beard recurs contemporaneously with the S. Rossore in the figure of Theophilus in Masaccio's "Raising of the King's Son" in the Brancacci Chapel. L. Berti, *Masaccio* (Milan, 1964), pp. 24ff., and nn. 43 and 259, refers to the convincing identification of this figure as a portrait of Gian Galleazzo Visconti (d.1402) in P. Meller, "La Cappella Brancacci: Problemi ritrattistici e iconografici," *Acropoli,* III–IV (1961), 186ff. and 273ff. However, the same type of beard is also used by Masolino in the figure of the emperor in the fresco of "St. Catherine Exhorting the Emperor" at Castiglione d'Olona. There is, therefore, no reason to think that such a beard and mustache would be an anachronism in a work of art executed during the 1420's.

Fig. 21.—Donatello, "Habakkuk." Museo dell'Opera del Duomo, Florence. (Photograph by Alinari.)

Fig. 22.—*Jacopo della Quercia, Trenta Altarpiece. S. Frediano, Lucca.*
(Photograph by Alinari.)

Fig. 23.—*Piero della Francesca, "St. John the Baptist," detail of altarpiece.*
Palazzo Comunale, Borgo San Sepolcro. (Photograph by Anderson.)

Fig. 24.—Donatello, "S. Rossore." Museo Nazionale di S. Matteo, Pisa. (Photograph by Alinari.)

*Fig. 25.—Donatello, "S. Rossore," detail. Museo Nazionale di S. Matteo, Pisa.
(Photograph by Alinari.)*

Fig. 26.—Donatello, "St. George," detail. Museo Nazionale, Florence. (Photograph by Soprintendenza alle Gallerie, Florence.)

Fig. 27.—Donatello, "St. Louis," detail. Museo dell'Opera di S. Croce, Florence. (Photograph by Courtauld Institute, London.)

Fig. 28.—Donatello, "Jeremiah," detail. Museo dell'Opera del Duomo, Florence. (Photograph by Alinari.)

sence of the body as a whole. The possibilities of characterization through pose are restricted to a subtly pensive forward tilt of the head itself. At every point the work depends on delicacy of detail. The intricacies of drapery and armor harmonically intensify the visual vibrations and the impression of psychological subtlety and of complexity of thought set up by the sensitivity of surface with which the head itself is modeled. The fine underlying structure has now become the basis for an endless play of light and shade across innumerable variations of hollow and projection. No other work by Donatello himself or by any earlier or contemporary artist gives so vivid and complete an impression of the infinite subtleties, and of the continuous change of plane and configuration, which enliven every square centimeter of a human head. No other work creates so detailed an interplay between the underlying structure and its thin and mobile overlay of flesh. Whatever view is chosen, each contour, every silhouette, is composed of an infinity of subtle changes of direction.

There is a further reason why this at first sight almost anachronistic achievement in anatomical description should have taken this particular form. The human head is, quite simply, the one point in the human frame in which almost the whole of the underlying bone structure is close to the surface and can be experienced without need for dissection, since the overlay of muscle tends to be a relatively thin skin, not a many-layered volume. Whenever the achievements in terms of the human nude of men such as della Quercia and Masaccio, or of Donatello himself, are considered, the corollary to the revolutionary aspects of their work is the patently severe limitations of their anatomical knowledge. The magnificently effective skeleton lying beneath Masaccio's "Trinity" in S. Maria Novella shows how little he knew of actual bone structure, not how much.[21] The legs, ankles, and feet of Adam and Eve in his "Expulsion" (Fig. 14) tell much the same story in this respect as those of Giovanni Pisano's "Prudence" (Fig. 15) of over a century earlier. The splendidly evocative lines of strain in the forward leg of the angel as he hurls the sinners from the garden (Fig. 13) shows the extent of della Quercia's ignorance of the actual interplay of muscle, bone,

[21] Illus. C. de Tolnay, "Renaissance d'une fresque," *L'Oeil*, Numéro 37 (1958), p. 41.

There is extreme inaccuracy in the number, individual form, and general distribution of the ribs, as well as in the relationship between the rib cage and the pelvis. The shape of the pelvis itself, and the form and relationship of the bones of the lower arm, are also far from correct, to speak only of the most obvious features of the substantially undamaged parts of the design. To speak, as does de Tolnay, of "the first known representation of a scientifically exact human skeleton half a century before the anatomical drawings of Leonardo" in a phrase referred to approvingly by W. Schlegel, "Observations of Masaccio's Trinity Fresco in Santa Maria Novella, *Art Bulletin*, XLV (1963), 33, n. 104, is only to obscure the true historical development.

and tendon in the human leg. Like Masaccio in the "Baptism" in the Brancacci Chapel, della Quercia is most successful where he is able to confine himself to a generalized impression of well-padded muscular or fleshy overlay as in his figure of "Acca Larentia" for the Fonte Gaia. When Donatello does essay a complete adolescent nude in his bronze "David," dating from the thirties at the earliest, it is no surprise that to the modern eye it is notable not for the detailed anatomical truth to nature which Vasari singled out for routine praise but for its extreme idealization and generalization of form (Fig. 37). The almost abstract geometry of intersecting planes with which the three-dimensional implications of the pose are articulated is what now catches the eye.

A final factor bearing on the isolation which the "S. Rossore" shares with many revolutionary works that seemingly appear before their time is that there was no contemporary method of drawing that was capable of encompassing a comparable, detailed sensitivity of form. At the level of generalized formal description current in the late thirteenth and early fourteenth centuries, just such a fundamental unity of style between drawing, painting, and sculpture had existed. The continuous, even, form-following curves used in their paintings by such men as Cavallini, the Isaac Master, the Master of the St. Francis Cycle (Fig. 2),[22] or Giotto himself, are the exact equivalent of the careful linear system used by the Giottesque master who made the pen-drawing (Fig. 29), now in the Uffizi, which corresponds to the fresco of the "Visitation" in the right transept of the Lower Church at Assisi. The brush-point strokes of the painters and the even parallel strokes of the pen are as close to each other as both are to the form-following parallel striations of the claw-chisels which give a corduroy-like surface to the penultimate main stage of the carving of the reliefs on the façade at Orvieto carried out by Maitani and his workshop. As far as sculptors were concerned, however, this unity of the arts was certainly only valid in relationship to men such as Maitani who were dealing in highly generalized, ideal forms. The differential rate of representational advance in the two arts meant, as has already been emphasized, that there was as yet no pictorial or graphic equivalent for Nicola Pisano's more particularized adventures in anatomical description. By the 1540's there was unity again, but on a quite new level. In answering Varchi's questionnaire, Michelangelo was able to assume that painting and sculpture had similar objectives. Both Pontormo and Vasari were, moreover, particularly insistent that drawing was the foundation both

[22] The Master of the St. Francis Cycle is, in this respect, the least regular and least controlled of the painters mentioned. The unusually complex problems faced in this torso, and discussed on p. 48 above, at several points disturb the even flow of strokes observable in the head and lower shoulder.

Fig. 29.—Follower of Giotto, "Visitation," detail. Uffizi, Florence.
(Photograph by Soprientendenza alle Gallerie, Florence.)

of sculpture and of painting, and this was an argument that Sangallo was careful to counter.

The situation was very different in the 1420's. As in the later fourteenth century, there were still two basic types of approach to draftsmanship, and this is true both of northern and of central Italy. The first approach is exemplified by the drawing of the "Prophet David," attributed to Fra Angelico and now in the British Museum (Fig. 30). It depends on a smooth idealization or generalization of the forms, whether of drapery or of anatomy, and is characterized by an even-running, continuous line. A steady pressure is maintained upon the pen, and any minor breaks in continuity or direction are fortuitous. There is a self-effacing quality, a feeling almost of anonymity, in the handling of the pen, and because the resultant contour is so thoroughly idealized it is quite unable of itself to suggest the solidity of the forms which it contains. As a result contour definition and internal modeling constitute two completely distinct functions. In this particular drawing the modeling happens to be carried out by washes applied with a brush. In other cases, ranging from the early fourteenth-century Giottesque drawing in the Uffizi (Fig. 29) to drawings attributed to Lorenzo Monaco, Parri Spinelli, Benozzo Gozzoli, Jacopo della Quercia, and other Tuscan artists working in the early fifteenth century, various types of internal hatching are used. Long or short parallel strokes, sometimes of hair-fine delicacy, may be used.[23] Sometimes, as notably in drawings attributed to della Quercia and Lorenzo Monaco, this may be combined with a limited use of rather loosely organized cross-hatching.[24] Occasionally, as in the work of Parri Spinelli, cross-hatching of this kind may be quite extensively employed.[25] In every case, however, whether the drawing is precise and highly finished or relatively free and sketchlike, a glance at any area which is free of whatever system of internal modeling is being used immediately reveals that the contour line itself is powerless to suggest the solidity of the contained form. The head of the Virgin in the early fourteenth-century "Visitation" (Fig. 29) sinks into the paper and becomes wholly flat and insubstantial above her splendidly bulky body, while the folds upon the forward knee of the "Prophet David" (Fig. 30) become paper thin.

[23] A particularly fine example is the "Head of a Youth" formerly given by Berenson to Fra Angelico and confirmed as Benozzo in A. E. Popham and J. Wilde, *Italian Drawings at Windsor Castle* (London, 1949), p. 172 (No. 12812 recto), Pl. 3, which is also illustrated in L. Grassi, *I disegni Italiani del trecento e quattrocento* (Venice, 1961), Pl. 24.

[24] See the study for the Fonte Gaia, Victoria and Albert Museum, Dyce 181, illus. *Italian Drawings Exibited at the Royal Academy London 1930* (London, 1931), Pl. 7, as possibly studio, and the sheet with "Six Saints" (Uffizi, N.11.E recto), illus. Grassi, *I disegni Italiani*, Pl. 11.

[25] Among many examples are the drawings of "Fortitude" (Uffizi, N.35.E recto) and the "Madonna and Child" (Uffizi, N.35.E verso), illus. Grassi, *I disegni Italiani*, Pls. 12 and 13.

Fig. 30.—Fra Angelico, "Prophet David." British Museum, London.
(Photograph by the Museum.)

*Fig. 31.—Stefano da Verona, "Prophet." British Museum, London.
(Photograph by the Museum.)*

*Fig. 32.—Pisanello, "Costume Studies." Ashmolean Museum, Oxford.
(Photograph by the Museum.)*

The situation is no different if attention is turned to northern Italy. Time and again Pisanello's drawings demonstrate the importance of the generalized contour in suggesting bulk without the use of wash or of the hair-fine hatching systems which he favored. In the breath-takingly beautiful sheet of costume studies with pale magenta, green, blue, and yellow washes, now in the Ashmolean Museum (Fig. 32), the figures strut and glide like peacocks on the page. Even a cursory examination shows, however, that only the short parallel hatching strokes, moving timidly in from the generalized contour, give any hint of solidity to the matchstick legs of the male figure on the right. The limitations of the contour are no less clear in the central female head wherever, as in the forehead, there is no internal modeling. The most startling effect of all is in the head of the figure on the left where there is no internal hatching of any kind. The splendid presence suddenly becomes a wisp, a ghost upon the page, so insubstantial that to call it flat gives far too great a flavor of material substance. It fully confirms that in the central head it is only the delicate hatching strokes in the angle of the neck and chin, and hints of shading round the eye and between the nose and cheek, that lend corporeality to the whole.

The other pole in early fifteenth-century draftsmanship is represented by the drawing of a "Prophet" in the British Museum (Fig. 31), attributed to Stefano da Verona, and by that in the Albertina, attributed to Lorenzo Ghiberti.[26] Here there is no technical dualism and no separation of function. There is no distinction between contour lines and hatching or modeling lines. Particularly in the da Verona drawing there are great liveliness, freedom, and character in the individual, discontinuous pen stroke. There is no attempt to maintain a controlled and even pressure on the pen. The corollary is that except in the splendid barrel of a sleeve, the individual mark or stroke gives hardly any indication of the nature of the particular feature that is being described. This is very evident in the whole of the trunk and lower body and in the drapery around the legs. The forms are built up by a cumulative process, and if the particular stroke concerned happens to be a contour, it is no more, and indeed in most cases, rather less, able than the carefully generalized or idealized type of contour to suggest the existence of a contained volume. The outcome gives rise to a high emotional charge. Particular details of form are only very partially or summarily described, and the sense of life and movement becomes more important than the carrying out of precise descriptive tasks. Although the approach is very different from that exemplified in the "Prophet David," it is also very much a process of generalization. As far as the individual mark upon the paper is

[26] Illus. *ibid.*, Pl. 19.

98

Fig. 33.—Antonio Pollaiuolo, "Hercules and Hydra." British Museum, London. (Photograph by the Museum.)

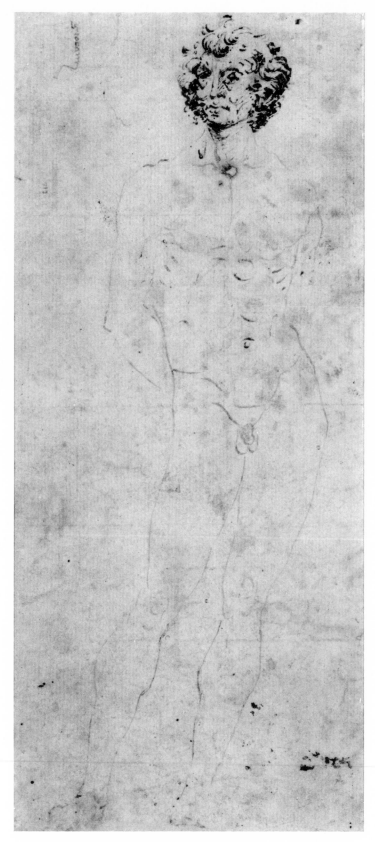

Fig. 34.—Antonio Pollaiuolo, "St. Sebastian." Staatliche Museen, Berlin-Dahlem. (Photograph by the Museum.)

concerned, and this is the foundation of all draftsmanship, the choice in the early fifteenth century clearly lies between greater energy or more control.

Between the two poles of continuous contour plus internal modeling and the free, all-purpose stroke there were naturally many variations and intermixtures. In particular, contour and modeling systems were used with greater or lesser degrees of sketchiness. This led to a greater or lesser continuity in the contour and to a tendency to move in the direction of the undifferentiated, all-purpose stroke without in any way affecting the essential descriptive limitations which have been discussed.[27] About 1460, however, Antonio Pollaiuolo achieved a major expansion in the possibilities of the drawn line and in the significance of the individual stroke or mark in describing the human form.

The nature of this revolution at the height of its development can best be seen in the "Study for a St. Sebastian," now in Berlin (Fig. 34), which Pollaiuolo probably carried out in the mid-seventies. It is clearly a revolution based upon new knowledge, and Benvenuto Cellini stresses Antonio Pollaiuolo's importance as a draftsman in one of the most eloquent passages in the introduction to the *Trattati dell'oreficeria e della scultura.* He says of him that "he was such a great draughtsman that not merely did all the goldsmiths use his most beautiful drawings, which were of such excellence, but also many sculptors and painters, and I speak of the most accomplished in these arts, also used his designs, and through them achieved the greatest honour. This man did few other things, but only drew most wonderfully, and always concentrated on this *gran disegno.*"[28] Vasari indicates the foundation of Pollaiuolo's achievement in draftsmanship in the passage in which he says that "Antonio's treatment of the nude is more modern than that of any of the masters who preceded him, and he dissected many bodies to examine their anatomy, being the first to represent the proper action of the muscles."[29] Like the invention of a scientifically accurate mathematical perspective system, the advent of carefully observed dissection marks a great stride forward in the history of art as exploration and as an investigation of the nature of man and of the world which he inhabits. With Antonio Pollaiuolo comes the realization that the artist can go no further in his art by simple observation of the fleshy envelope and external surface of the human form. The needs of art are calling science into being, and a new,

[27] This is particularly true in the case of the *sinopie* or brushpoint sketches on the under plaster which prepared the way for fresco painting. For extensive illustration of characteristic *sinopie* see P. Sanpaolesi, M. Bucci, and L. Bertolini, *Camposanto Monumentale di Pisa: affreschi e sinopie* (Pisa, 1960), and U. Procacci, *Sinopie e affreschi* (Milan, 1961).
[28] B. Cellini, *I Trattati dell'oreficeria e della scultura,* ed. L. De-Mauri (Milan, 1927), pp. 4–5.
[29] Vasari, *Vite,* III, 295.

detailed, scientific knowledge needs a new technique of drawing for its exploitation.

The most striking thing about Pollaiuolo's "St. Sebastian" (Fig. 34) is the way in which the contours are, unaided, able to create a powerful impression of the solidity of the enclosed forms. There is a minimum of intervening modeling at any point. In the entire length of the legs the only internal forms that are actually indicated are the kneecaps. Nevertheless, the whole of the volume and structure of these same legs seems to be instantly appreciable. The reasons for this, though often merely characterized by references to "vigorous" or "sensitive" draftsmanship, are quite precise.

The full descriptive range of any line which is used to describe a continuous form and which is drawn with a single instrument using a single homogeneous material, liquid or otherwise, may readily be summarized. They are its continuity or discontinuity in relation to the total length of the form described; its direction at any given point and its change of direction from point to point; its breadth and its change of breadth; its color; and finally its tone and change of tone. Not all of these characteristics need appear as positive features of any given line. In a straight line, for example, the change of direction is nil. It is, however, most important in analyzing the factors bearing on the appearance and descriptive meaning of any individual line or stroke that the element of change, or its absence, should be considered. Its significance is somewhat analogous to that of acceleration in relation to velocity in physics.

Pollaiuolo's handling of contour in the "St. Sebastian" and in other similar drawings marks the opening of a new era in that every one of these basic possibilities of variation appears to be put to positive use. His line is discontinuous, and the discontinuities, particularly where the end of one stroke runs inside the beginning of the next, nearly always indicate points of strain or places at which there is a particularly noticeable interaction between the underlying bone and tendon structure and the overlying muscle. Because the discontinuities are descriptively significant, and not a mere mannerism or trick of draftsmanship, the lengths of continuous line are extremely varied. This in itself adds to the liveliness of the contour system.

The second major feature of these contours is that even within a general curve there is almost invariably a constant series of minor and irregular changes of direction. These changes also are controlled by the nature of what is being described and not by rhythmic or decorative considerations. In comparison with this constant variation both in direction and in rate of change of direction, the individual stroke in Fra Angelico's "Prophet David" (Fig. 30) appears to be extremely smooth and regular. The changes of direction are far fewer and much more even in their acceleration and deceleration.

102

Fig. 35.—Antonio Pollaiuolo, "Prisoners before a Judge," detail. British Museum, London. (Photograph by the Museum.)

Such breaks in continuity as do occur seem most often to be connected with the accidents of movement of the pen over the surface. They certainly add variety within the over-all uniformity of the line, but they seldom appear to be informative either in intention or in effect. In Stefano da Verona's vigorous pen work (Fig. 31), on the other hand, where there is nothing which can strictly be called a contour line, the length of any stroke is seldom the outcome of descriptive necessity. It does not matter whether the innumerable, slightly curved single sweeps of line or the hook- or loop-shaped strokes are examined. With a few rare exceptions the impression given in terms of change of direction is of smooth accelerations caused by single swift gestures, rather than of constant meaningful change within the individual stroke.

The final, linked considerations in Pollaiuolo's line are the result of constant changes in his pressure on the pen. These give rise to an ever-changing breadth of line and depth of tone as a greater or lesser quantity of bister is deposited on the paper for any given length of stroke. Once again, the constant swelling and darkening, thinning and fading of the line are descriptively significant at every point. There are occasions on which the change in thickness and tone reflects no more than the recharging of the pen, but in a very high proportion of cases the thinning and paling of the line reflect a concave form or a relaxation of anatomical stress. A calf muscle swelling as it contracts; the hip bone of the standing leg forced outwards, causing a sudden change of linear direction in the process—it is at such points that the pressure on the pen likewise increases and the line moves swiftly up from thin and faint to thick and dark and back again. In none of these respects is there in Fra Angelico or Stefano da Verona, or in the work of any of Pollaiuolo's predecessors in the surviving history of human draftsmanship, anything comparable in its range, in its constant variation, and in the descriptive meaning with which the unprecedented richness of the line is charged.

The extent to which this new kind of drawing is based on anatomical knowledge which is at once precise and so ingrained as to have become instinctive can be seen in the composition sketch in the British Museum (Fig. 33). This is presumably for Pollaiuolo's early "Hercules and the Hydra" and is therefore to be dated about 1460. In this swift, small-scale sketch Antonio was seemingly intent on capturing a vigorous pose. It is unlikely in the extreme that as he drew the forward-stretching arm or the contracting muscle of the forward leg he was specifically considering anatomical problems. It is simply that, for the first time since antiquity at least, precise anatomical understanding has become so much a part of an artist's subconscious store of knowledge that it pours forth instinctively, however

Fig. 36.—Antonio Pollaiuolo, "Dancing Figures," detail. Villa Torre del Gallo, Florence. (Photograph by Brogi.)

swiftly pen is put to paper and for whatever reason. Here again the discontinuous line, and the constant changes in the pressure on the pen and in the direction of its movement, impart a wealth of accurate information in a form which matches the vigor of the visual presentation to the vigor of the subject matter. It is this kind of drawing which alone makes possible some of the liveliest of Leonardo's sketches, such as that of the "Virgin and Child with a Bowl of Cherries" in the Louvre.[30] Only on the basis of a similar or even greater anatomical knowledge, and by exploiting the full range of possible linear variation, could two such distant, swiftly sketched-in contours be made to swell the chubby infant flesh of the Christ Child on the page without the help of any system of internal modeling whatsoever.

The effect of Pollaiuolo's increased flexibility of line when considerable amounts of hatching and other indications of internal form are added to the contours can be seen in the "Study for a St. John the Baptist" in the Uffizi.[31] Even in the rather more careful drawing of an "Athlete" in the Musée Bonnat at Bayonne[32] or the "Adam and the Eve" in the Uffizi,[33] in which the internal modeling of the musculature is washed in and the contours are rather more continuous and less bold in their variations of breadth and tone, the effectiveness of Pollaiuolo's new use of line is hardly less evident. Its efficacy when the factors of breadth and tone are almost wholly eliminated and his new understanding of the human form can only be expressed in the subtleties of constant change of direction is evident in the drawing of "Prisoners before a Judge" in the British Museum (Fig. 35) and in the ruined frescoes of "Dancing Figures" in the Villa of Torre del Gallo on the outskirts of Florence (Fig. 36). By intention in the one case and through the accidents of time in the other, the figures are very largely presented in terms of rippling, form-following, form-creating silhouettes. These distilled essences of the continuous contour are so effective in creating an impression of contained volume that time and again a mere discoloration on the wall or on the surface of the paper is taken by the eye as an indication of internal modeling. The discoloration near the buttocks of the two male figures in the detail of the drawing illustrated (Fig. 35) and those on the legs of the dancing figures on the right of the detail of the fresco (Fig. 36) demonstrate that in

[30] Louvre, R. F. 486, illus. A. E. Popham, *The Drawings of Leonardo da Vinci* (London, 1946), Pl. 25.
[31] Uffizi, N.399.E recto, illus. Grassi, *I disegni Italiani*, Pl. 41.
[32] No. 1269, illus. B. Berenson, *The Drawings of the Florentine Painters* (Chicago, 1938), Vol. III, Fig. 82, and J. Bean, *Les dessins Italiens de la Collection Bonnat*, Inventaire Général des Dessins des Musées de Province, 4, Paris, 1960, No. 120. Prior to the latter publication this drawing has seldom been well treated in reproduction. Berenson's near-rejection as autograph is not perhaps altogether justified.
[33] N.95.F, N.97.F, illus. Grassi, *I disegni Italiani*, Pls. 42 and 43.

*Fig. 37.—Donatello, "David," detail of back. Museo Nazionale, Florence.
(Photograph by Courtauld Institute, London.)*

the presence of such outlines every visual cue and every unintentional, chance mark tend to be read in only one way. Even when the most extreme foreshortenings are involved, it is seen in terms of volume and of structure.

The part played by Pollaiuolo's draftsmanship in finished paintings such as the "Hercules and the Hydra" and "Hercules and Antaeus," the "Apollo and Daphne," or the large-scale "Martyrdom of St. Sebastian" needs no emphasis. The small bronze "Hercules and Antaeus" (Fig. 38) shows that Pollaiuolo's graphic and pictorial means of describing form are directly translatable into sculptural terms and back again. Here once again are the expressive, broken contours, the muscles bulging in the taut, angular forms of strain. In contrast to the sculpture of the first half of the century the entire surface is set alight or set in motion by the rippling highlights on the bronze. Nevertheless, as the back view of Hercules' shoulders demonstrates, this is still no matter of mere reproductive naturalism. The rocky simplifications and stylizations that replace the smooth, ideal geometry of Donatello's "David" (Fig. 37) represent at a new level a similar transmutation of knowledge into art.[34] Indeed, one should say into the arts, for it is Pollaiuolo's prime achievement that he not only united in his own practice the major visual arts, as Andrea del Orcagna had before him, but was also a revolutionary innovator in drawing, as in painting and in sculpture. His achievements open the way for the technical and artistic developments in all three media at the hands of men such as Leonardo and Michelangelo. Pollaiuolo is the immediate precursor of the universal artist of the High Renaissance who looms behind the arguments of the Paragone. His control of line sets the stage for the even closer and more wide-ranging unification of stroke in pen and brush and chisel at the hands of Michelangelo. For the first time in the history of the Renaissance and of the period leading up to it the arts are fully in step at the very highest level of achievement. In artistic, if not in sociological or philosophical, terms, the ground has been cut from beneath the feet of the disputants before the discussions of the Paragone have properly begun. As Michelangelo himself implies, the theorizings of the Paragone are precisely of the kind which follow rather than precede creation.

[34] Pollaiuolo's frescoes and drawings show that the greater liveliness of surface in the "Hercules and Antaeus" is not merely a reflection of its small scale. In the case of Ghiberti's torso of *Isaac* in the competition relief for the first bronze doors, comparison with larger works seems to show that the small scale is indeed a factor influencing the unusual liveliness of surface. Among Ghiberti's life-size figures and among the small heads on the two bronze doors a smooth, generalized form is the normal rule. Only the "Prophet" on the North Door, illustrated in R. Krautheimer, *Lorenzo Ghiberti* (Princeton, 1956), Pl. 67, and the "Self-Portrait" on the Paradise Doors (illus. Krautheimer, Pls. 135A and 136B) seem to approach the complexities of structure and surface visible in the "S. Rossore."

Fig. 38.—Antonio Pollaiuolo, "Hercules and Antaeus." Museo Nazionale, Florence. (Photograph by Alinari.)

PREFATORY NOTE. To a large extent, this paper makes use of my previously published research. While it appeared useful to prepare a lecture for the Humanities Seminar on the musical innovations of the Renaissance, with which I have been concerned for many years, it seemed questionable whether such a lecture should be published. However, Charles Singleton kindly and persuasively insisted on including my lecture in the present volume, and two considerations seemed to favor this step: whereas in my previous publications I had used the magnifying glass of the scholar in scrutinizing new phenomena, here it was possible to catch a glimpse of a vast development from one single viewpoint. A second argument was that there is a time and a place where the specialist should communicate the results of his research to the larger community of scholars.

The lecture was admirably illustrated by performances of the compositions by the Bach Society of Baltimore under the direction of George R. Woodhead. I had hoped to be able to include a recording of the works sung, but due to the many technical difficulties involved it proved impossible.

E.E.L.

THE MUSICAL AVANT-GARDE OF THE RENAISSANCE OR: THE PERIL AND PROFIT OF FORESIGHT ᴥ EDWARD E. LOWINSKY

hen we speak of avant-garde in the arts, we have in mind something more than a new style or new techniques; we think of an artistic attitude of bold, even reckless, experimentation and a deliberate flaunting of tradition. Such an attitude we have come to expect in an age of confusion, anxiety, alienation, and revolutionary upheaval such as the twentieth century, but not in the Renaissance, one of the great periods of civilization. Yet the Renaissance, with the introduction of the Spanish Inquisition, the expulsion of the Jews from Spain, the clash between Savonarola and Pope Alexander VI culminating in Savonarola's execution; with the shattering events of the Reformation, the Sack of Rome, the wars of religion in the sixteenth century; and with the collision between paganism and Christianity, the revival of ancient Greek philosophy and Scholasticism, and the rise of a spirit of free confession and free inquiry in constant life-and-death struggle against the persecution of free thought and free conscience —this age was as much one of anxiety and confusion, of alienation and revolutionary upheaval, as is our present period.

Art always reflects the society that creates it, although often we may not know enough of the art or the society to read the reflected image. The art of the Renaissance expresses not only the few great moments in which a certain equilibrium was achieved, the moments of a Raphael, a Palladio, a Palestrina, but also its moments of confusion and distress, of struggle and upheaval, the moments of Michelangelo's "Last Judgment" or his "Captives," of El Greco's tortured visions, of Gesualdo's pained outcry in the madrigals of his late years.

It is not easy to sketch for non-musicians the evolution of avant-gardism in the music of the Renaissance. Music, to be understood in its structural aspects, needs a terminology which is often foreign to the non-specialist. Since avant-garde art, to deserve the name, must overthrow the whole structure of the preceding artistic tradition, we shall first sketch out the nature and structure of that tradition and explain as concisely as possible the terms entering into such a description.

Music until 1500 was governed by the system of the church modes, it was firmly based on the diatonic system, it was rooted in the Pythagorean tuning system. Western music includes twelve tones within the octave; but each church mode, as well as the later major and minor modes, operates with only seven of these twelve tones. The seven tones, consisting of five whole-tone and two half-tone steps (the white keys on the piano), are what we call the diatonic system. The chromatic twelve-tone scale consists of these seven tones plus the five "chromatic" tones (the black keys on the piano). The Pythagorean tuning system, as Pythagorean philosophy in general, is based on mathematical proportion. If we divide a string in the middle and cause that half-length to vibrate, we obtain a tone an octave higher than the tone emitted by the undivided string. The proportion of a tone to its octave is therefore 1:2, the proportion of a tone to its fifth is 2:3. The Pythagorean tuning system works with these two proportions. If we superimpose the interval of the fifth twelve times (C to G to D to A, etc.—which is what we call the circle of fifths) we obtain the twelve notes of our tone system. But the strange, irrational fact is that with the proportion of 2:3, on the thirteenth fifth, we come to a tone approximately one-eighth higher than the tone of departure. The farther we go, therefore, in the circle of fifths, the more out of tune the intervals and harmonies will be.

The medieval tone system was incapable of producing chromatic music; it could neither conceptualize melody in successive half-tone steps nor produce chords on all twelve tones; it was also incapable of modulation going systematically from one key to another in the circle of fifths. The number of "keys" for composers and players was limited. This does not mean that the medieval tone system employed no chromatic tones. This system, dating back in its theoretical formulation to about 1000 and attributed to Guido of Arezzo, contained one chromatic tone, the B-flat. Other chromatic tones could be obtained by notating accidentals, so called because they were treated as accidental rather than essential to the tone system, or by *musica ficta*, "feigned music." The system of *musica ficta* was a set of rules that allowed the singer to sing an accidental where none was written, according to two principles, that of necessity and that of beauty (*necessitatis causa* and *pulchritudinis causa*). The principle of necessity concerned the raising or lowering of a tone to obviate imperfect and prohibited intervals; the principle of beauty concerned the introduction of chromatic tones for the sake of sonority and harmonic color. The very fact that chromatic notes were often introduced by the singer and instrumentalist in performance but left unnotated[1] betrays the disinclination of the medieval musician to accord the chromatic notes the status enjoyed by the fundamental notes of the modal system.

---

[1] These chromatic notes that do not appear in the original notation but must be introduced according to the rules of *musica ficta* appear above the notes in our scores, in harmony with present-day practice.

The mathematical basis of the medieval tone system was responsible for a number of propositions such as that the tone was divided into major and minor semitones or that thirds and sixths were dissonances, although the ear judged differently—propositions that were, as we shall see, gradually and more or less generally rejected by the musicians of the Renaissance.

The musicians of the avant-garde of the Renaissance felt that medieval diatonicism, the system of church modes, and the Pythagorean tuning system were too confining. They experimented with chromaticism, with modulation, with new sounds and new harmonies, and, to achieve all this, they explored new tuning systems in which the differential of one-eighth tone was distributed in some fashion over the twelve tones so as to be absorbed by them in a more or less equal proportion. If the distribution proceeds in an equal proportion, we speak of an "equal" temperament; if not, of an "unequal" temperament.

If one were to look for the first musician who initiated the musical avant-garde of the Renaissance, the choice would have to be the Spanish theorist and composer Bartolomeo Ramos, who left Salamanca in the 1470's and went to Bologna, where he, in 1482, published a treatise, *Musica practica*, that was to arouse the ire of generations of theorists. Ramos replaced Guido's tone system and solmization theory based on six tones with one based on the octave; he recognized thirds and sixths as consonances; he introduced a complete chromatic scale; he originated significant changes in tuning; he advocated a rich use of chromatic notes; and he wrote a motet, *Tu lumen, tu splendor patris*, in which he demonstrated for the first time in the history of Western music the use of the chromatic and enharmonic genders. Unfortunately, this composition is lost, as is almost every other composition of the Spanish master. Likewise lost is a Spanish treatise of his; the Latin treatise of Bologna is preserved only in two copies. Surely the poor preservation of the works of this Spanish radical was due to the fact that he was clamorously opposed by the most illustrious authorities of his day. Nicolaus Burtius of Bologna, Franchino Gafurius of Milan, John Hothby, the English Carmelite monk residing in Italy—they all took up their pens in a vigorous defense of the old Guidonian system. Ramos had to leave Bologna and never published another word. He experienced the full impact of the peril of foresight: isolation and ostracism were his fate. Yet, all that he foresaw came to fruition in the further evolution of music.[2]

One of the earliest experiments with modulation from one key to another in the circle of fifths is preserved in the first print of polyphonic music, the *Odhecaton*, the Hundred Song Book, published by Ottaviano

[2] Edward E. Lowinsky, "Music of the Renaissance as Viewed by Renaissance Musicians," in *The Renaissance Image of Man and the World* (Columbus, Ohio, 1965), pp. 129–77; 156–60.

Petrucci in Venice in 1501.[3] It is the *Fortuna d'un gran tempo*, by the great Fleming Josquin des Prez, who spent most of his life in Italy. Although making use of a popular melody, this is an instrumental composition for three voices, each one written in a different key, the highest voice, as we would say, in C major, the middle voice in F major, the lowest voice in B-flat major. Obviously, a composition written in three different keys must lead to clashes between the three voices that can be adjusted only through judicious use of *musica ficta*. Such adjustment leads to further modulation to G minor, F minor, and B-flat minor, involving E-flat, A-flat, and D-flat. Unless these unwritten chromatic tones are introduced, the composition sounds poor and dissonant. With them the piece, conceived in an animated tempo, becomes one of the most graceful and elegant works of the master. We quote the beginning of the composition.[4] (See Ex. 1.)

Although "conflicting signatures"[5] occur often in the music of the time, a composition with three different key signatures indicating three different keys is something extraordinary for the time of 1500—in fact, it remains an extraordinary procedure right down to the twentieth century and the arrival of polytonal music, that is, music for parts written in different keys. Also, a modulation as extensive as the one in Josquin's *Fortuna d'un gran tempo* was novel at the time.

Now, Fortune's outstanding trait is her capriciousness; she is the goddess of change. Modulation is change; its name in Renaissance terminology was the same term used for the activities of Fortuna: *mutatio*. Boethius, in his *Consolatio philosophiae*, has Fortune say, "This continuous play we are playing: we turn the wheel in hasty circle and find pleasure in changing low to high and high to low"; the original Latin is *"infima summis, summa infimis mutare gaudemus."*[6]

The Renaissance composers were motivated by the desire to create an image in tones of the Goddess Fortuna:

Medieval and Renaissance thinking was permeated by so strong a belief in Fortune's power that it almost amounted to a second religion. This strange phenomenon can perhaps be explained, in part at least, by the prevailing character of medieval and Renaissance government. In non-democratic societies, and

---

[3] *Harmonice Musices Odhecaton A*, ed. Helen Hewitt (Cambridge, Mass.: The Mediaeval Academy of America, 1942), No. 74.

[4] For the complete score see Edward E. Lowinsky, "The Goddess Fortuna in Music," *Musical Quarterly*, XXIX (1943), 45–77; 51–53. A recording of the composition based on our interpretation of it was issued by Elaine Music Shop (EMS 213), performed by Safford Cape's Pro Musica Antiqua.

[5] See Edward E. Lowinsky, "The Function of Conflicting Signatures in Early Polyphonic Music," *Musical Quarterly*, XXXI (1945), 227–60.

[6] See Lowinsky, *Musical Quarterly*, XXIX, 66.

Ex. 1.   Fortuna d'un gran tempo   Josquin des Prez

especially in those founded on arbitrarily seized power as well as on intrigue and conspiracy, the change of government is usually accompanied by the catastrophic downfall of those expelled from positions of power. The most widespread symbol of Fortuna shows the goddess turning the spokes of the wheel on top of which sits the crowned king with scepter and sphere; he looks to his predecessor on the right tumbling down headlong who in turn beholds the one who is at the bottom of the wheel, whereas on the left the sitting figure clinging to the spokes of the wheel, but lifting eyes and left hand longingly upward, is on the ascent to the seat of power.[7]

Fortunately, Josquin composed another work of an even more advanced character that proves, through its explicit and careful notation, that his harmonic imagination indeed went far beyond the conventional framework of contemporaneous music. In his four-part motet *Absalon fili mi*, the setting of David's passionate plaint on the death of his son Absalom, a work for men's voices with a key signature of two flats in the upper two parts, three flats in the tenor, and four in the bass, Josquin penetrates, in the final section, to the regions of E-flat, A-flat, D-flat, and G-flat, clearly marking each chromatic note either as an accidental or in the key signature or both. The final words on which this great modulation occurs are, *"Descendam in infernum plorans"* ("I shall descend into my grave, weeping"). The work is written in a slow tempo adapted to the character of mourning.[8] (See Ex. 2.)

In Josquin's *Fortuna* and plaint of David two motivating forces behind the progressive movement of Renaissance music reveal themselves: the animated, elegant *Fortuna* is intended as the musical symbol of the pagan goddess whose domination of the popular imagination of the time was contested only by that of the Virgin Mary; it also reveals the role of the enigmatic that was destined to play an outstanding part in the further evolution of musical innovation in the Renaissance. The slow, deeply affective plaint of David bares the roots of the new musical idiom in the realm of human emotion.

Josquin's scheme of modulations was as yet limited to four or five steps in the circle of twelve fifths. It could still be performed within the confines of Pythagorean tuning. It was another Flemish musician, Adrian Willaert, likewise living in Italy, who wrote, twenty years later, the first composition going through the whole cycle of fifths, the famous *Quid non ebrietas*.

This piece, bold and enigmatic as its predecessor, Josquin's *Fortuna*, but technically much more radical, was set for four voices, of which only three survive. Fortunately, however, the artifice involved can be demon-

---

[7] Edward E. Lowinsky, "Matthaeus Greiter's *Fortuna*: An Experiment in Chromaticism and in Musical Iconography," *Musical Quarterly*, XLII (1956), 500–19; 515–16; and XLIII (1957), 68–85.

[8] For the complete score, see Helmuth Osthoff, *Josquin Desprez* (2 vols.; Tutzing, 1962–65), II, 382–84.

118

Ex. 2. Absalon fili mi, m. 77 to end   Josquin des Prez

Ex. 5.

strated in the two parts of soprano and tenor, and it is as a "duo" that the work has been transmitted in treatises of music until late in the seventeenth century.

Like Josquin, Willaert shows the way by notating some accidentals—indeed Willaert goes as far as G-flat and C-flat—but then leaves the singer alone to find the unprecedented F-flat, B-double flat, E-double flat, and A-double flat needed to go around in the complete circle of fifths. We present the modulation in a four-part score (bass added by the present writer) complete with a keyboard reduction for easier reading. (See Ex. 3.)

Unlike Josquin, Willaert restricts the modulation to one single voice part, the tenor, while the other voices stay in the same key. This leads to unavoidable temporary clashes between the chromatic and non-chromatic parts and produces intervals prohibited in the contemporary books on counterpoint and composition. These clashes, however, stay within the limits of the tolerable. But, if the F-flat and the various double flats which are not written down were left out and the tones were sung as written, the work would sound intolerable and would end, against all possibilities in Renaissance music, not on an octave, but on a seventh.

Willaert's work is the manifesto of the musical avant-garde of the Renaissance—Josquin's works were harbingers—and indeed it was the most hotly debated work of the Renaissance.[9] Here is the first work in the history of Western music that cannot be performed in the Pythagorean tuning system of the Middle Ages. The question arises in what tuning system the work was intended to be executed.

Of course, the ancient Greeks had been the masters of musical mathematics, the originators of a whole number of tuning systems, of which the Pythagorean was only one, albeit the dominating one. The great antagonist of Pythagoras was Aristoxenus, disciple of Aristotle; he held that not mathematics but the human ear was the decisive factor in determining consonance and dissonance. He set forth two propositions that were anathema to the Pythagoreans. Instead of dividing the whole tone into a major and minor semitone, he proposed dividing it into equal semitones; he further proposed that the octave could be divided into six whole tones, whereas the Pythagoreans divided the octave into five whole tones and two unequal semitones. The theory of equal temperament throughout the Middle Ages was known under these two propositions of Aristoxenus—known, I should say, and derided. From Boethius' fifth-century *De Musica* on, Aristoxenus was held up for contempt as the man who trusted the senses more than

[9] For publication and translation of a number of erudite contemporaneous letters on Willaert's composition, see my study on "Adrian Willaert's Chromatic 'Duo' Reexamined," *Tijdschrift voor Muziekwetenschap*, XVIII (1956), 1–36.

Ex. 3. Quid non ebrietas, mm. 10–25   Adrian Willaert

(Ex. cont.)

reason. And nothing could provoke a medieval thinker's contempt more than so naïve an attitude. It is significant that humanists, not musicians, revived the teachings of Aristoxenus at the end of the fifteenth and at the beginning of the sixteenth centuries.[10]

Now it should be mentioned that, whereas the keyboard instruments of the Renaissance were tuned according to various systems preserving a division of the tone into a major and a minor semitone, on some string instruments of the Renaissance, the lute and the viols, the strings were divided into equal semitones by frets, which allowed performance of works that could not be executed on the organ or the harpsichord. Of course, the human voice was not tied to any specific tuning system. Contemporary reports tell us that Willaert sent his composition to the musicians of Pope Leo X, a music enthusiast, and that they tried it on viols and for voices, but had an exceedingly hard time of it.

Willaert had, of course, an old score to settle with the singers of the Papal Chapel. When he first went to Rome, he heard the Papal Chapel sing a work of his which the singers thought was Josquin's. Informed that it was his, the Roman singers dropped the piece from their repertory. Now Willaert showed them that while it was in their power to discard a work of his, it was in his power to write one that they could not place in their repertory even if they wanted to. But this was not the reason that Willaert wrote the composition.

As in the works of Josquin, the text will aid us in understanding the artist's intention. Willaert took the words from Horace's fifth epistle (Book I). He chose the following lines:

Quid non ebrietas dissignat? operta recludit,
Spes iubet esse ratas, ad proelia trudit inertem,
Sollicitis animis onus eximit, addocet artes.
Fecundi calices quem non fecere disertum?

(What a miracle cannot the winecup work. It unlocks secrets, bids hopes to be fulfilled, thrusts the coward into the field, the load from anxious hearts, teaches new arts. The flowing bowl—whom has it not made eloquent?)[11]

Although the connection between the winecup and its power to unlock secrets and teach new arts is as suggestive as it is humorous, it fails to disclose an adequate reason for Willaert's extraordinary undertaking. The deeper cause must be sought in the intention to revive the Aristoxenian

[10] *Ibid.*, pp. 7–13.
[11] Horace, *Satires, Epistles, and Ars poetica*, trans. H. Rushton Fairclough (London, 1926), pp. 280–83.

philosophy of music, the primacy of the ear over mathematics,[12] the recognition of the musical realities contested by the mathematical doctrines of the Pythagoreans: the acceptance of thirds and sixths as consonances, the divisibility of the whole tone into two equal semitones, the possibility of dividing the octave into six equal whole tones and hence into twelve equal semitones, and, with these, equal temperament that frees music to circle back and forth from one key to another in the newly conquered and unified harmonic space.[13]

It took a full generation until Willaert's chromatic experiment was followed by a work of similar daring, this time, contrary to all expectation, by a German composer. Matthaeus Greiter's *Fortuna*, written in or before 1550, the year of his death, appears as a strange and powerful confluence of Josquin's *Fortuna* and Willaert's *Quid non ebrietas*.[14] From Josquin Greiter took the Fortuna motif and the idea of modulation; from Willaert, the bold extension of the modulatory system that required equal temperament for its performance.

Greiter chose his text from Ovid's *Tristia* (v.8.15), written by the poet at a time when he himself had experienced the turn of Fortune's wheel, having been banished by Emperor Augustus from Rome to desolate Tomi, where he died in exile. These are the lines of Greiter's composition:

Passibus ambiguis Fortuna volubilis errat
Et manet in nullo certa tenaxque loco.

Fortune doth wander and flit and her step one can never determine
Nor doth she stay in one place, fixed and reluctant to go.
                                        (Konrad Gries)

The text, although accompanying the music, cannot be sung to it: the two words at the beginning, *Passibus ambiguis*, for example, while clearly set so as to depict Fortuna's desultory manner, go on for over seventeen measures, stretching over more than sixty notes. The text, it seems, served as a program for the composition to be performed by a quartet of viola da gamba instruments. In this four-part work the tenor repeats in *ostinato* fashion the initial phrase of a famous Italian song, *Fortuna desperata*. This motif, played in the slow tempo of a *cantus firmus*, is transposed, step for step, from F throughout the circle of fifths, beyond the modest modulation of Josquin's

---

[12] On this point see my paper in *Renaissance Image*, pp. 136–38, where it is shown that Tinctoris, half a century before Willaert, established the ear as superior to mathematics in musical thought and practice of the Renaissance.

[13] See my paper on "The Concept of Physical and Musical Space in the Renaissance," *Papers of the American Musicological Society for 1941* (New York, 1946), pp. 57–84.

[14] See my study in *Musical Quarterly*, XLII and XLIII.

Fortuna, to D-flat, G-flat, C-flat, and finally F-flat, involving B-double flat. At this point the piece has fallen one semitone below its original level, and there it ends.

The fundamental difference between Willaert's and Greiter's compositions is that in the former only one voice, the tenor, goes through the modulations; the others accommodate the modulatory part as best they can, producing a number of inescapable dissonances, whereas in Greiter's *Fortuna* all four voices share in the modulatory process. One generation after Willaert, his daring came to fruition in a work of rare beauty and undisturbed harmoniousness.

This composition is one of the most astonishing works of art of the Renaissance. Published in the Latin treatise of Gregorius Faber in Basel in the year 1553,[15] the work is distinguished from similar experiments by its careful notation. Every new chromatic note is clearly marked by flats. Nothing is left to speculation. It was written by a man who was a monk and chorister at the cathedral in Strasbourg, but who left the monastery, turned Protestant, and married in 1524. His output is limited to about one dozen German songs and about twenty German chorales. Nothing in the whole history of German music, from which chromatic experiments are conspicuously absent, nothing in Greiter's own output, prepares us for such an extraordinary work. There is no adequate explanation for an artistic event of this order. But there are two circumstances, one political, the other artistic, that render the event, if not less marvelous, at any rate more comprehensible.

The political circumstance attaches to the fact that the work was dedicated to Albrecht von Brandenburg.[16] Now Albrecht, first Duke of Prussia and, like Greiter, an adherent of the Reformation, was a great friend and patron of music; he engaged in correspondence with the leading German composers of the time. He shared the humanistic tendencies of the prominent Italian and German courts of his time. If Greiter intended a homage to so exalted a patron, it is understandable that he should have undertaken so unusual an enterprise. Greiter may have written the work in the hope of receiving a court appointment, a hope conceivably shattered only by his early death in 1550, for the work cannot have been written long before that.

The second, artistic, circumstance I refer to is a technique of composition cultivated by the leading composers of the North, the Netherlanders, in works of religious music that I have described as "secret chromatic art."[17] The technique can be defined as the art of unwritten chromatic modulations hinted at by notational and compositional means based on the time-honored

---

[15] *Musices practicae erotematum libri II autore M. Gregorio Fabro Luscensi, in Academia Tubingensi Musices Professore ordinario* (Basel, 1553), pp. 140–51.
[16] See the author's article in *Musical Quarterly,* XLIII, 82ff.
[17] *Secret Chromatic Art in the Netherlands Motet* (New York, 1946; reissued, New York, 1967).

rules of *musica ficta*. In ordinary *musica ficta* one note is changed; in the secret chromatic art the first note to be changed is placed into such a melodic, harmonic, and contrapuntal context that a whole series of notes in the circle of fifths follows from it. It is a musical version of the domino theory: if the first tone goes the way of *musica ficta*, the whole phrase follows. In the small circle of about one dozen works in which these secret chromatic modulations occur, we deal with motets on texts of a mystical nature such as the Annunciation, the Passion, the Resurrection, Baptism through the Holy Ghost; or of an emotional nature such as plaints of David, of Rachel, of Job, and of certain fervent psalm passages.

A representative example is a motet of about 1550 by the famous Flemish composer Clemens non Papa, the text of which treats Lazarus' resurrection. It is a work of rich sonority set for six voices. Throughout the work the second soprano sings in slow notes the call with which the whole work ends in the second part, Jesus' command, *"Lazare, veni foras"* ("Lazarus, rise!"). The first part of the work has two passages containing secret chromatic modulations. The first one occurs close to the beginning on the words *Fremuit spiritu Jesu*, which reflect Jesus' shock at the news of Lazarus' death; the second, at the words *et lachrimatus est Jesus* ("and Jesus wept"). Again, the modulations are so natural and flow with such ease and logic that, to modern ears, they are hardly perceived as extraordinary.[18] (See Ex. 4.)

In musical design and expression, Clemens follows in the footsteps of Josquin. Comparison of Examples 2 and 4 reveals the closeness of the two great modulatory passages, one expressing David's mourning of Absalom, the other Jesus' reaction to Lazarus' death. It is part of the secret chromatic technique that it allows two readings: one, as written, although awkward and technically deficient, conforms to the conventional style of the period; the other, the secret chromatic reading, daring and beautiful, nevertheless convinces through its modulatory logic, its conformance to the emotional meaning of the text, and its flawless return to the diatonic mainstream of the music.

If the musical avant-garde in Italy was motivated by the humanists' revival of ancient art and letters, in the Netherlands the new artistic means were used in the service of a deep religious emotion. Characteristically, the texts used by the Netherlandish composers interested in the secret chromatic art treat topics close to the religious thought of the Reformers. This is not surprising if we keep in mind that in the year 1568 the Netherlanders re-

---

[18] Mr. Woodhead's ensemble, after a performance of the whole work, gave a dual illustration of the two passages, in which the sound of the chromatic interpretation was contrasted with the sound of the passage as written.

Ex. 4.   Fremuit spiritu Jesu, Part I, mm. 1–14   Jacobus Clemens non Papa

(Ex. cont.)

mm. 63–74

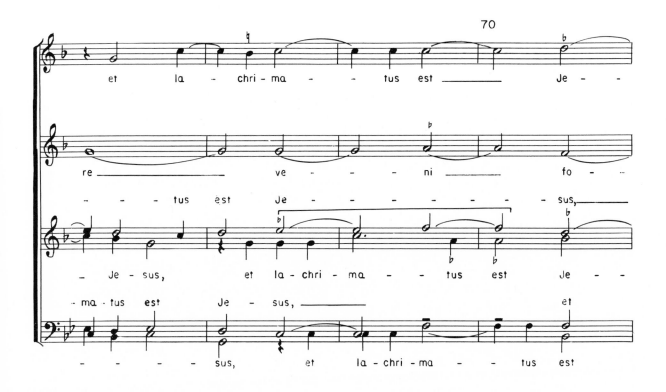

volted against the Spanish regime, driven by their distaste for political tyranny and their search for freedom of conscience in matters of religion.

Nor should we forget that chromaticism, dear to the humanists who knew of the chromatic music of the Greeks, was frowned upon by the Church. This anti-chromatic attitude started with the old Church Fathers who identified chromaticism with paganism; it continued throughout the Middle Ages, where the theorists, while dutifully referring to the three *genera* of the Greeks because they were treated by Boethius, remind the reader that *Mater Ecclesia ex his tribus [generibus] dyatonicum elegit.*[19] Moreover, the medieval tone system, as suggested above, did not allow for the use of chromaticism.

Finally, the political-intellectual climate of the Netherlands, where Charles V had introduced, in 1527, the Spanish Inquisition, explains why an open embracement of chromaticism must have seemed risky to Netherlandish composers, who, unlike their Italian confreres, were mostly not in the employ of princes but of local churches. Whereas the enigmatic in the art of the avant-garde musicians practicing in Italy was a part of the intellectual attitude toward art—the traditional love of guild secret, the humanistic delight in mystery and concealed meaning—in the Netherlands there was added to these elements the fear of appearing in any way unorthodox. Innovators of all kinds were suspect.[20] This may well have been the over-riding reason for the secrecy of the Netherlandish use of chromaticism. However, there is a deeper layer of meaning in the Netherlandish attempt at a secret musical technique; it is an idea pervading all manifestations of the time—literary, artistic, philosophical—that the world offers two faces to man: one esoteric, full of significance and profound truth, accessible only to the initiated; the other exoteric, moving on a level of common understanding, open to the many, the vulgar.[21]

So far we have dealt more with avant-garde music than with avant-garde musicians, for although the single works studied, and in particular the epoch-making *Quid non ebrietas* by Willaert, were genuine monuments of avant-garde music, they constituted only a tiny fraction of the composer's output and were in fact isolated works, forays into unexplored territory: "experiments." Only in the "secret chromatic art" do we encounter a whole circle of works, but even this is small in numbers.

It was in Italy, in the mid-sixteenth century, that musical avant-gardism turned into a full-fledged movement supported by powerful patrons. There we meet, in the person of Nicola Vicentino, a man who devoted his

[19] Lowinsky, *Secret Chromatic Art,* pp. 111–12.
[20] See the chapter on "The Religious Background," *ibid.,* pp. 111–34.
[21] See the final chapter of *Secret Chromatic Art,* "The Meaning of Double Meaning in the Sixteenth Century."

130

whole life to the cause of radical modernism. Vicentino was, as we would say today, a professional revolutionary. A disciple of Adrian Willaert, he was determined to draw the full consequences of his master's first tentative experiments. His whole work, theoretical and practical, was dedicated to the overthrow of the old order and the introduction of the most extreme and systematic use of chromatic and enharmonic music. To this end he built the *archicembalo*, a fantastic instrument with six rows of keys, containing, instead of twelve, no fewer than thirty-one tones for each octave; to promulgate the new gospel, he wrote a comprehensive and brilliant treatise whose title reveals his program: *L'antica musica ridotta alla moderna prattica*, or "How to apply the chromatic and enharmonic tone systems of the Greeks to modern music." In it he published a series of motets and madrigals to illustrate the use of the Greek genders in contemporary music. Around his portrait (see Fig. 1) we read: "Incerta et occulta scientiae tuae manifestasti mihi — Archicymbali Divisionis Chromaticique ac Enharmonici Generis Practicae Inventor — Nicolas Vicentinus Anno Aetatis Suae XXXXIIII" ("You have revealed to me the uncertain and the occult of your science [Ps. 50:8] — Inventor of the practical division of the chromatic and enharmonic genders — Nicolas Vicentinus at the age of forty-four").

It was in Rome, the seat of the Papal Chapel, center of traditionalism, that the greatest and most articulate representative of the musical avant-garde discussed his ideas and later published his treatise—not without getting into trouble (like his master, Adrian Willaert) with the singers of the Papal Chapel. One day in the year 1551, Vicentino met with some Papal singers. He fell into an argument with the Spanish singer and composer Vincentio Lusitano over a certain composition on the Marian antiphon *Regina caeli*. Vicentino claimed that "nowadays no composer knows in what gender he writes"—which was his way of saying that the days of writing pure diatonic music were over and that actually most composers were treating the church modes with so much liberty that they were unintentionally using the old chromatic gender of the Greeks. But the Spanish composer, quick on the uptake, wagered two gold ducats that Vicentino could not prove his statement that no living composer knew in what mode he was writing, whereas he, Lusitano, would prove the opposite to be true.

The public disputation took place with pomp and circumstance on June 11, 1551, in the presence of the whole Papal Chapel and of a number of princes of the Church, notably Vicentino's patron, Cardinal Ippolito d'Este. The arbiters were without exception Papal singers. The verdict went unanimously against Vicentino—an outcome that could have astonished only its victim, who underestimated the peril of foresight in an environment dedicated to the cultivation of tradition. But Vicentino determined to

131

Fig. 1.—Nicola Vicentino, L'antica musica ridotta alla moderna prattica *(Rome, 1555), reverse of title page. (Courtesy of The Newberry Library.)*

vindicate himself by writing his great treatise, which appeared four years later in Rome in the same year in which Palestrina became a singer at the Papal Chapel. In one of those double counterpoints, of which history is rich, the greatest traditionalist and the greatest revolutionary of the sixteenth century worked at the same time (and, for periods, at the same place) with equal dedication and in completely opposite directions, and, which is stranger still, for the same patron, Cardinal Ippolito d'Este, although at different periods of time.[22]

If Vicentino was capable of publishing his revolutionary theses in the seat of musical conservatism, it must be attributed to the patronage and protection of the powerful Cardinal Ippolito d'Este of Ferrara, who, in 1553, after the death of Pope Paul III, came close to being elected Pope himself. Vicentino's treatise was dedicated to the Cardinal, who most likely bore the expense of publishing the large and handsome work.[23]

Astonishing is the radicalism with which Vicentino converted his theories into practice. To demonstrate the use of the chromatic gender of the Greeks, he set, for example, a text from the Lamentations, *Jerusalem convertere*, for which he used the chromatic tetrachord of the Greek tone system in all five voices, ascending and descending.[24] (See Ex. 5, p. 119.) The harmonic result is a succession of major and minor chords that was a complete *novum* and must have startled the ears of the musicians and listeners alike. This is followed by a free and bold modulatory section that distinguishes itself from preceding modulatory experiments which pursued

[22] Palestrina was in Ippolito d'Este's service from 1567 to 1571.

[23] The treatise has been made available in a facsimile edition by the present writer in the series *Documenta Musicologica, Erste Reihe*, XVII (Kassel, 1959). For a full-length treatment of Vicentino see H. W. Kaufmann, *The Life and Works of Nicola Vicentino (1511–c. 1576)* (Rome: American Institute of Musicology, 1966). For previous accounts of the significance and the impact of Vicentino's treatise, see Theodor Kroyer, *Die Anfänge der Chromatik im italienischen Madrigal* (Leipzig, 1902); Hugo Riemann, *Geschichte der Musiktheorie* (2d ed.; Berlin, 1920), pp. 367–77 (in the English translation of R. Haggh [Lincoln, Nebraska, 1962], pp. 311–20); H. Zenck, "Nicola Vicentino's 'L'antica musica' (1555)," *Theodor Kroyer Festschrift* (Regensburg, 1933), 86–101; the Postface of the facsimile edition by the present writer; and the same writer's *Secret Chromatic Art*, pp. 88ff, and "Music in the Culture of the Renaissance," *Journal of the History of Ideas*, XV (1954), 509–53; 553.

[24] For a transcription of this piece see the writer's *Tonality and Atonality in Sixteenth-Century Music* (2d rev. printing; Berkeley, 1962), p. 42, or Kaufmann, *Life and Works of Vicentino*, pp. 139–40. Unfortunately, Kaufmann's practice of placing the natural sign of the initial motif above the notes instead of into the score creates the misleading impression that these chromatic steps are editorial interpolations. Since Vicentino consistently places accidentals only before those notes that he wishes to lower or to raise, a B without accidental in his notation is a B-natural and must be notated as such in the score to make it clear that it constitutes the unmistakable intention of the composer.

LIBRO TERZO

Vattro compositioni differenti di Musica uariata sono dimostre . hora mi ritrouo alcuni essempi fatti iquali sono in questa opera et gli ho impressi per non stàr à farne de tutti Cromatici, & de tutti Enarmonici, per ch'io non hò tempo à me basta ch'io habbi dato le regole & il modo di comporre detti Generi partati, & il discepolo li còporrà ad ogni suo piacere; Io so che il Lettore haurà piu piacere ch'io habbia seguito l'opera, che poi ch'io nò hauesse finita essa opera, per qualche disturbo, & per caggione di perder tempo in far questi essempi così giusti; perche il tempò è breue & nostre uoglie sono lunghe & l'opera, perche ua fatta latina & uolgare porta assai tempo con seco, per hora hò deliberato di porre quelli essempi ch'io mi ritrouo fatti, & ritornarò al quarto essempio ch'io ho dato, che è misto, di tutte tre le sorti di spetie de i tre Generi, accio il cantante, odi che effetti fanno le diuersità de i gradi, & per dimostrare in una sola compositione le spetie de i tre Generi per fare il paragone di una & di l'altra spetie, porrò qui sotto un essempio à quattro uoci di tre uersi Latini; & il primo sarà composio nelle spetie della Musica Diatonica. Il secondo sarà cantato con le spetie della Musica Cromatica. Il Terzo sarà dimostro con le spetie Enarmoniche, come di sopra hò detto, & questi tre uersi Latini sono stati fatti per honorare il mio signore & Patrone, lo Illustrissimo & Reuerendissimo Cardinale HYPPOLITO DA ESTE. hora il sù detto essempio con la diuersità di tre ordini di Musica qui presente è notato.

Musica prisca caput   tenebris modo sustulit al tis

Musica prisca caput tenebris modo sustulit al tis Dulcibus ut nume-

ris dulcibus   ut numeris priscis certantia factis   dulcibus ut numeris

priscis certantia fac tis   facta tua Hyp polite   facta tua Hyppo-

lite excelsum super aethera mittat facta tua Hyppolite   excelsum

super aethera mit tat.

Fig. 2.—Nicola Vicentino, L'antica musica ridotta alla moderna prattica *(Rome, 1555), fol. 69ᵛ. (Courtesy of* The Newberry Library.*)*

a logical stepwise path in the circle of fifths by its combination of stepwise procedure with chromatic shifts to third-related keys (from B-flat major to G major; from D to B major). The over-all result of these new techniques I have described as varying from "floating tonality" to "triadic atonality."[25]

To illustrate, "in one single composition," the effects of the three genders, Vicentino wrote (according to his own statement)[26] "an example for four voices of three Latin verses: the first will be composed in the system of diatonic music; the second will be sung in the system of chromatic music; the third will be demonstrated with enharmonic degrees. And these three Latin verses have been written to honor my lord and patron, the most illustrious and reverend Cardinal Ippolito d'Este." (See Fig. 2.) Its text, in translation, runs as follows:

Ancient music has upheld her head, through long obscurity,
Only that she, with the sweetness of the old intervals,
May send high up to the heavens the fame of your heroic deeds, O Hyppolitus!

The opening words, "*Musica prisca caput tenebris modo sustulit altis,*" sung twice, are set in the traditional diatonic gender; the continuation, "*dulcibus ut numeris priscis certantia factis facta tua,*" follows in a highly chromatic style, first melodically chromatic, then harmonically in a sequence of the most astonishing chromatic progressions. (See Ex. 6.) Finally, at the invocation of the Cardinal's name, *Hyppolite,* the enharmonic section of the work begins. The quarter tones in the original are indicated by dots over the notes, each dot raising the note by one quarter tone. We draw the reader's attention to the truly incredible sequence of harmonies in the chromatic section, from A-flat major to A minor, A major, B-flat major, B major, cadencing abruptly on F major (mm. 24–28 in Ex. 6). Nothing that Arnold Schoenberg did in his first atonal works fell with less preparation on contemporaneous ears than these wild measures of Vicentino's written to recall the "sweet" sound of Greek music.

Vicentino's two works are so extraordinary in their harmonic boldness that there is no difficulty in appraising them as manifestations of the musical avant-garde of the Renaissance. At the same time, these samples of avant-garde music are true children of Renaissance humanism, inspired, if not by an exact knowledge of the complexities and mysteries of Greek musical theory, which Vicentino did not possess, nevertheless by the Greek concept of music as based on the word, its meaning and emotion, and the Greek system of chromatic and enharmonic genders. To let the emotion of the

---

[25] *Tonality and Atonality,* Chap. IV.
[26] *L'antica musica,* fol. 69ᵛ. (See Fig. 2.)

Ex. 6. Musica prisca caput

Nicola Vicentino

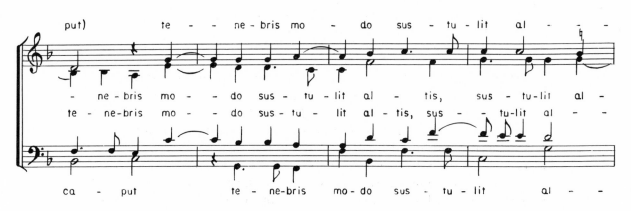

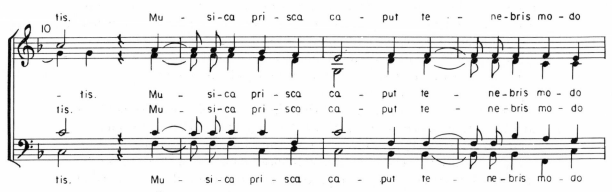

(Ex. cont.)

text shine through in the intense new colors of chromatic and enharmonic harmonies—this indeed appeared to Vicentino the true mission of music.

A culmination point of musical avant-gardism in Italy is 1555. This is the year in which Cipriano de Rore's chromatic Latin ode *Calami sonum ferentes*, by an unknown, probably neo-Latin poet, appeared. Showing its extraordinary character already in its setting for four basses, it begins in all voices with a chromatically ascending line stretching over the four-tone unit of the tetrachord. (See Ex. 7.)

It is odd that we know nothing about the relationship between Vicentino and Rore, notwithstanding the fact that both were connected with the court of Ferrara. The temptation is great to credit Vicentino with providing the original stimulus to radical chromaticism in view of his theoretical work and his compositions; but it should be resisted. A mere comparison of the motives of *Jerusalem convertere* by Vicentino (see Ex. 5, p. 119) and the *Calami sonum* by Rore illustrates a telling difference: Vicentino's motive is taken over from the chromatic tetrachord of the Greeks; Rore's has no such connection. Vicentino is motivated by the desire to demonstrate the revival of ancient Greek music on a new, polyphonic basis; Rore, while unquestionably inspired by the pervading humanistic visions of ancient music, creates his own version of it in a free chromatic style that seems to owe nothing to Vicentino.

Rore's ode was published for the first time in the *opus I* of the young Orlando di Lasso, his *Il primo libro dovesi contengono madrigali, vilanesche, canzoni francesi e motetti a quattro voci*, printed by Tielman Susato in the year 1555 in Antwerp, shortly after Lasso's return from Italy after a stay of over ten years. In this print Lasso himself published a highly modern work, a secular motet, *Alma nemes*, in which he celebrated the *dulce novumque melos* of the avant-garde with a marvelous modulation stretching all the way from F-sharp major to G major, introduced by a chromatic step from D major to F-sharp major and ending in a plagal cadence A minor, E major.

But Orlando di Lasso's most direct contribution to the musical avant-garde of the sixteenth century is his cycle of twelve Sibyl compositions in the advanced chromatic idiom of Vicentino and Rore. This cycle was written by the young Lasso during his years in Italy, where he had been brought from South Flanders at the age of twelve and where he stayed until the age of twenty-three, in 1555. Lasso could hardly have chosen a more fitting subject. The ancient female prophets who, in ecstatic and enigmatic language, foretell the coming of Christ, are made to speak in the musical language of the future which, at the same time, is related to the musical language of the Greek past. Number VIII, the *Sibylla Phrygia*, is representative of the bold style chosen by the young Lasso in his depiction of the

Ex. 7. Calami sonum ferentes

Cipriano de Rore

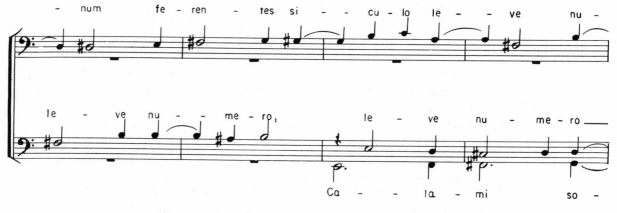

Sibyllinic figures that must have impressed him in his Italian days in the fresco paintings of Raphael, Michelangelo, and Pinturicchio.[27] (See Ex. 8.)

Lasso's modulatory style owes its individuality to the skill and elegance with which he employs three techniques combined in constantly changing patterns: (1) modulation in the circle of fifths, with occasional skips of one step; (2) modulation to third-related degrees through a chromatic step in one and the same voice (for example, B-flat major to G major in measures 3–4, or G major to E major in measure 21) or through false relations between various voices (as from D major to B-flat major and from E-flat major to C major in measures 9–10); and (3) constant harmonic change over wide key areas.

Lasso was intent upon creating a musical likeness of the incoherent and visionary character of prophetic speech. In choosing harmonies foreign to the common style such as B major (mm. 14, 32) or G-sharp minor (Number IV, m. 40) with their unusual D-sharp, or F-sharp major with A-sharp (Number VII, m. 3), he attempted furthermore to echo in sounds the unusual and miraculous character of Sibyllinic utterance.

From 1555 on, the year of Vicentino's epoch-making treatise, Rore's chromatic ode, and Lasso's first publication, avant-garde music is not counted in single and isolated works, but in dozens. Few are the composers who, like Vicentino, wrote almost exclusively avant-garde music, fewer still those who remained aloof from the feverish excitement of new visions of sound. Most composers are affected by the avant-garde and use a harmonic vocabulary of considerably greater freedom and flexibility. Modulation becomes a chief means of musical expression; chromaticism is injected where the text evokes feelings of particular intensity.

The distance between the experimental and the ordinary style, which had been tremendous in Willaert's *oeuvre*, shrinks in the writing of composers like Lasso and Rore and vanishes in the work of the radical modernists such as Vicentino and Gesualdo da Venosa, for the simple reason that the distinction, for them, has lost its meaning. It is the text that determines the means to be chosen. And it is the choice of the text by the composer that reveals his interest in making the experimental an ordinary means of musical expression. To put it differently, the "experimental" becomes the norm.[28]

---

[27] Adolf Sandberger ("Orlando di Lasso und die geistigen Strömungen seiner Zeit" [Munich, 1926], p. 21) believed there was a special connection with the Sybils of Pinturicchio, who is the only painter to have presented twelve Sibyls as did, half a century after him, Lasso.

[28] We must not forget that the early works of both Vicentino and Gesualdo were relatively conservative. Odd is only the fact that Vicentino's first madrigal book of 1546, although in no way unconventional, was advertised by Vicentino as avant-garde music *avant la*

Ex. 8. Sibylla Phrygia

Orlando di Lasso

142

(Ex. cont.)

(after H. J. Therstappen [ed.], Orlando Lasso,"Prophetiae Sibyllarum," Das Chorwerk 48, pp. 18–19)

This is not yet the case in the madrigals of the greatest and most Italian of madrigalists, Luca Marenzio, who became the idol of his countrymen and the rage of Elizabethan England. In his rich output the experimental is still reserved for the extraordinary.

Radical chromaticism is encountered in two famous settings of Petrarchan texts. One is from the ninth book of five-part madrigals (1599), Marenzio's setting of the poet's sonnet

Solo e pensoso i più deserti campi
  Vo misurando a passi tardi e lenti

Petrarch pictures the love-struck wanderer, out in the wilderness, alone with his grief. Gaiety seeks company, melancholy solitude. Marenzio, in an unprecedented inspiration, creates a sound symbol of loneliness by setting the soprano apart from all other voices: while they move in quarter notes, the soprano limps in whole notes. While they take the large steps of a chain of thirds with fourths, fifths, and octaves appended, the soprano takes the smallest melodic steps, going up the chromatic scale in fourteen steps, down in seven. (See Ex. 9.)

If, in this example, the primary inspiration was a melodic one and the harmonies result, more or less, from the melodic, the opposite situation obtains in the second example, a work from Marenzio's second book of madrigals for five voices, published in Venice in the year 1581, "*O voi che sospirate.*" The text is taken from Petrarch's *sestina* on Laura's death, *Mia benigna fortuna.* Marenzio chose the verse in which the poet pleads for an early death. The work is famous for the passage addressed to Death:

Muti una volta quel suo antico stile

(*But let him change for me his ancient song*)

For this line Marenzio changed the *stile antico* of diatonic harmony to a ravishing modulation of the *stile nuovo.* (See Ex. 10.) Starting on G major, we are led through C, F, B-flat, E-flat, A-flat, D-flat, to G-flat major; now, in an "enharmonic modulation"—using the term in the modern sense—G-flat major is turned into F-sharp major, which, in a surprise denouement, resolves to B minor, going on to C major and A minor, and settling down in a cadence on A. While not going entirely through the circle of fifths, this

---

lettre, so to speak. For in the title, in which he calls himself "disciple of the unique Adrian Willaert," he announces: "Madrigals for five voices composed, through theory and practice, in the new style discovered by his celebrated master" ("*Madrigali a cinque voci per theorica et pratica…. composti al nuovo modo dal celeberrimo suo maestro ritrovato*").

Ex. 9. Solo e pensoso

Luca Marenzio

(after L. Torchi, L'Arte musicale in Italia, *II, 228–29)*

Luca Marenzio

(after Einstein, Luca Marenzio Sämtliche Werke, I, 69–70)

passage is certainly the closest approximation to Willaert's enterprise in his Dionysian hymn. But what a difference: whereas Willaert confined the modulation to one voice, Marenzio lets all five voices take part in it. Whereas Willaert wrote an isolated passage in a modal environment, Marenzio's passage is embedded in a free modulatory harmonic style that uses strong and lush chromatic colors throughout. Beginning with a call to all lovers in a harmonic setting oscillating between major and minor, between the tonal centers of A, E, and B, Marenzio continues with an invocation of Death, *"non mi sia più sorda morte"* ("Pray that for me be no delay in death"), on a chromatic motif going through all five parts. This passage concludes with the poet's prayer for an end to his tears, *e fin del pianto*, set by Marenzio in an unusual chromatic cadence going from F major to B-flat major, C minor, then taking the unlikely skip to A major, involving a chromatic step in two voices and a double false relation and settling across the sixth chord of F-sharp minor in G major, the harmonic point of departure for the central event of the composition, its great modulation.

This modulation, on the text *muti una volta quel suo antico stile*, proves that Josquin, with his *Fortuna d'un gran tempo*, created a tone symbol of enduring strength when he used modulation to depict Fortuna's constant mutations. Indeed, in Marenzio's six-part setting of *Spent' eran nel mio cor l'antiche fiamme*, the third stanza of Sannazaro's *sestina*, *O fere stelle omai datemi pace*, almost four generations after Josquin, he writes a modulation stretching from F major to D-flat major on the words *E tu fortuna muta il crudo stile*.[29] (See Ex. 11.)

The harmonic style has grown richer and fuller, the modulatory technique smoother, the tonal definition sharper, but the line from Josquin to Marenzio is unmistakably clear; the unity of the century across its immense variety of genres, styles, and personalities is unbreakably solid. Marenzio is surrounded by kindred spirits such as Gioseppe Caimo, Giaches de Wert, or Luzzasco Luzzaschi, to name only three of the greatest honorary members of the Italian avant-garde, that is, those who mastered all aspects of avant-garde technique but reserved it for special texts calling for extraordinary means.[30] These masters reap the full benefit of the pioneers' foresight; the early experiments now bear the ripe fruit of fully matured works. That among these masters were also an isolated Frenchman and some Englishmen

---

[29] Alfred Einstein, *The Italian Madrigal* (3 vols.; Princeton, 1949), III, No. 80; see Einstein's commentary in II, 666.

[30] See Alfred Einstein's masterly treatment of these three composers in the second volume of his great work on *The Italian Madrigal*. Giaches de Wert's madrigals are now becoming available in the edition by Carol MacClintock (Rome: American Institute of Musicology, 1961–).

Ex. 11.   O fere stelle, mm. 9–19

(Ex. cont.)

(after Einstein, The Italian Madrigal, *III, 253)*

—although the latter used only the most moderate of avant-garde techniques and then very occasionally—has been shown by Kenneth J. Levy[31] and Joseph Kerman.[32]

Carlo Gesualdo, Prince of Venosa and, of all princely composers, in or out of the Renaissance, the most authentic creative musician, made the radical style of the avant-garde his personal idiom. Aside from the first two books, which he composed before his visit to the court of Ferrara,[33] the hotbed of musical radicalism, to which the whole avant-garde remains indebted, Gesualdo's madrigals breathe, in steadily increasing measure, the fierce intensity of an utterly unbridled temperament and a nervously vivid imagination. His inexhaustible harmonic, melodic, and rhythmic inventiveness is fired by the images, the passions, the contrasts expressed in poems that Einstein believes were commissioned by the Prince with the specification "that half the pieces were to contain the word *morte* or *morire*, that the other half was to consist of lively pastoral pieces, and that the whole was to be as epigrammatic and as full of oxymora as possible."[34]

The famous *Tu m'uccidi, o crudele*, from the composer's fifth book, published in Naples in 1611, one generation after Marenzio's setting of Petrarch's *sestina*, shows the avant-garde of the Renaissance in its last stage. The very intensity with which the composer attacks the single conceits of the poem shakes the concept of a coherent organic musical form to its foundations. Whereas the dancelike *balletti* of Giacomo Gastoldi of 1591 celebrate triumphs of early tonality,[35] a work such as the present madrigal shies away from cadence as a means of tonal definition, and, with its unexpected harmonic skips and resolutions and its far-flung modulatory excursions, constantly presses against the confines of a vaguely divined tonal center. The beginning motif could be felt to be in B major, B minor, or E minor. (See Ex. 12.) Phrases following tend toward—but never settle in— A minor–C major (mm. 7–9), F minor–A-flat major (mm. 9–11), D minor (mm. 11–13), A minor (mm. 13–15), B minor (mm. 15–16), C-sharp minor –E major (mm. 17–18), A major (m. 18), A minor (m. 20), E minor–G major (mm. 21–24), A minor (mm. 24–26), and so on. Only in the last phrase does the C major–A minor oscillation prevail and settle down on a final chord of A major in a V–I progression that lacks the subdominant as well as the necessary complement of rhythmic retardation and melodic circumscription to define the A major as a clearly established tonal center.

---

[31] "Costeley's Chromatic Chanson," *Annales musicologiques*, III (1955), 213–63.
[32] *The Elizabethan Madrigal* ("Studies and Documents," No. 4; New York: American Musicological Society, 1962).
[33] See Einstein, *Italian Madrigal*, II, 695.
[34] *Ibid.*, p. 710.
[35] See Lowinsky, *Tonality and Atonality*, Chap. V.

Ex. 12. Tu m'uccidi, O Crudele

Carlo Gesualdo

(Ex. cont.)

(Ex. cont.)

The revolution set in motion by Willaert has now come full circle. The prohibited intervals and dissonances that Willaert used experimentally to solve a technical problem are employed by Gesualdo to achieve the utmost of expression. Indeed, the most vital difference between his style and that of his predecessors such as Marenzio lies precisely in the liberation of dissonance which, instead of following the traditional rules of placement on the weak beat, of preparation and resolution, obeys exclusively the demands of the text. And the text is followed, indeed pursued, by the composer with a ferocious intensity. Words such as *uccidi, crudele, morir,* and *aspro martire* are set in uncompromising dissonance. But melody, rhythm, tempo, texture, counterpoint—they all are used with the same single-minded intent of expressive realism. This gives Gesualdo's writing its aphoristic brevity, its discordant diversity. It lives on the clash of opposites. When Gesualdo, whom Einstein rightly called "the musician of the oxymoron,"[36] thinks of *morire amando* (at the end of the madrigal), he conceives of it not as of one act and one concept. "Woe me, I die" is set in a phrase of biting dissonance, grating chromaticism, and a painfully retarded motion. *Amando,* "loving," is portrayed in the intertwined contrapuntal lines of a fast-moving scale motif of the greatest simplicity. One indeed needs the partial repeat of the last phrase to find one's bearing in the extreme switch of mood and motion.

From Bartolomeo Ramos on, the peril of foresight lay in the clash with tradition, in the risk of isolation. But the rejected pioneers of yesterday become the prophets of tomorrow. Not all they see comes true. Vicentino's music with quarter tones and even his keyboard instrument with multiple division of the tone was tried again and again; yet, it never took hold. Nor is there any telling at what particular point in time a prophecy will be fulfilled. Vicentino's and Gesualdo's experiments in "atonality" came to fruition in the twentieth century. The true profit of foresight lies in the integrity of a new vision which, while alienating the present, may conquer the future.

It is the characteristic mark of the musical avant-garde of the Renaissance that it acquired its impetus and inspiration from classical antiquity as revealed and interpreted by the humanistic movement. In this it constitutes a unique example of a radical avant-garde movement set in motion by study of the past, of generations of musicians held under the spell of scholars, for the revival of the knowledge of Greek music would have been impossible without the aid of the Greek scholars of the Renaissance who rediscovered and translated the old Greek treatises on music. Halfway through the century, however, we observe a significant change. Willaert, Greiter, Cipriano

[36] *Italian Madrigal,* II, 709.

161

de Rore wrote their avant-garde music to texts from Horace, Ovid, or Latin odes of probably neo-Latin origin. Josquin was inspired by the image of a pagan goddess. But in the second half of the sixteenth century avant-garde music increasingly becomes synonymous with the Italian madrigal, with music set to texts written in the native tongue of the composer and his audience. If one were to put it that humanism fathered modern music, then it might well be added that native poetry and the richness of its feeling and imagery played a mother's role. "Modern" music, in the broadest sense of the word, is unthinkable without either of these two great extra-musical forces.

THE MUSIC FOR "QUANT'È BELLA GIOVINEZZA" AND OTHER CARNIVAL SONGS BY LORENZO DE' MEDICI &⁓ WALTER H. RUBSAMEN

he philosophic basis of the Italian Renaissance finds apt expression in Lorenzo the Magnificent's carnival song, the *Trionfo di Bacco ed Arianna*, for in it he appears to reject the medieval Christian doctrine of the necessity for patient suffering during this life in order to merit blessed happiness during the next, and substitutes the *carpe diem*, advice to be joyful and carefree whenever possible, since the future is so uncertain. *Il Magnifico* also contemplates with melancholy the fleeting nature of youth, in retrospect the happiest age of man, and seeks to divert his listeners with the enactment of a popular myth whose characters represent the other facet of Renaissance thought, the world of classical antiquity. Bacchus and Ariadne, nymphs, satyrs, Silenus, and Midas are introduced in turn to the carnival throngs, most being joyful, sensuous creatures who love life and each other and who therefore serve as excellent models for those who wish to take their pleasures where they can find them. In the case of Midas, however, Lorenzo reminds us that the mere possession of worldly goods does not bring contentment and that avarice interferes with the perception of beauty. Having thus returned to the pensive tone of the beginning, the poet urges us to forget our cares as we sing, dance, and make love during the festive season. This pagan attitude, and especially the unabashed hedonism of the section concerned with the deities, contrast markedly with the traditional Christianity of Lorenzo's lauds, written later, perhaps as an act of penance.

The melancholy and nostalgia of "Quant'è bella giovinezza" are easily explainable, for Lorenzo's illnesses of long standing became acute in 1488 and remained so until his death. It is no wonder that he considered youthful bliss to be so transitory, for his beloved wife, Clarice, had died on July 30 of that year, at the relatively tender age of thirty-eight. The ruler of Florence himself had but a few years remaining before the end came on April 8, 1492. For a decade after the revolt of the Pazzi in 1478, during which he was wounded and his brother, Giuliano, killed, Lorenzo was of no mood to organize or to participate in elaborate carnival festivities. Only after the victory of Sarzana in 1487, having secured peace between Florence and her neighbors, did he allow himself to enter the realm of public amusement. As Machiavelli writes of *il Magnifico's* activities during this period of tranquillity: "In these peaceful times he maintained his people in a state of continuous festivity, during which one often saw jousts and representations and *trionfi* of antiquity, and his purpose was to keep the city in abundance, with the people united and the nobility honored. . . . He took marvelous delight in architecture, in music, and in poetry, and many verses came forth, not only composed but also expounded by him."[1]

Machiavelli's use of the term "trionfi antichi" must refer particularly to the *Trionfo di Bacco ed Arianna*, for Lorenzo's other example of this category, "Sette pianeti siamo," is an allegory, to be sure one that expresses the same philosophy:

Il dolce tempo ancor tutti c'invita
lasciare i pensier tristi e van dolori.
Mentre che dura questa brieve vita,
ciascun s'allegri, ciascun s'innamori.

Actually one can distinguish between three types of carnival songs during the age of Lorenzo: the *mascherate*, sung by groups of masked men and boys who serenaded the ladies of Florence during the period preceding Lent or during the Calendimaggio; the *carri*, in which singers representing one of the trades or professions appeared on sumptuously decorated wagons that formed part of the carnival procession; and those *carri* called *trionfi*, often consisting of several elaborate floats on which sat the personifications of ancient deities or of the allegorical virtues. An illustration of such a *carro* drawn by centaurs, depicting a bacchanal, may be found in an engraving after a design by a contemporary artist. (See Fig. 1a and 1b.[2]) One of the centaurs is playing the lyre, the other a pipe, as they pull a four-wheeled wagon in which are seated the intoxicated Bacchus and Ariadne, both young and handsome. Herbert P. Horne at first believed that the artist was Botticelli,[3] then changed his mind and assigned the drawing to Bartolommeo di Giovanni, another Florentine painter of the late fifteenth century. The engraving consists of two separate parts that form a whole.

---

[1] Chap. 36, Bk. VIII of *Istorie Fiorentine*, in *Tutte le Opere*, a cura di F. Flora e C. Cordiè (Verona, 1949–50), II, 432.

[2] Reproduced in *Botticelli / Des Meisters Werke in 155 Abbildungen*, hrgs. v. Wilhelm von Bode ("Klassiker der Kunst in Gesamtausgaben," Bd. 30; Stuttgart, 1926), p. 39, and in Arthur M. Hind, *Early Italian Engraving, A critical Catalogue . . .* , Part I: *Florentine Engravings and Anonymous Prints of Other Schools* (4 vols.; London, 1938), Vol. II, Pl. 112–13.

[3] *Alessandro Filipepi commonly called Sandro Botticelli* (London, 1908), p. 83.

Fig. 1a and 1b.—Ascribed to Sandro Botticelli or to Bartolommeo di Giovanni, Bacchanal. British Museum Print Collection. (Photograph by the Museum.)

According to Horne, it has no direct connection with the carnival procession of Lorenzo de' Medici, even though the subject is the same. Rather is the design similar to that found on several well-known ancient sarcophagi, of which a fine example is now in the British Museum.

It is possible that Piero di Cosimo (1462–1521?) was the artist responsible for planning the *trionfo* and decorating the *carri*. Vasari tells us that "Piero in his youth, because he was whimsical and full of extravagant ideas, was employed a great deal in the *mascherate* made for the carnival . . . . And they say he was one of the first who thought of sending them out in the form of *trionfi*, or at least he improved them greatly by embellishing the plot not only with music and poetry suitable to the subject, but also with incredible splendour of retinues on foot and on horseback, of costumes and finery adapted to the story."[4]

In the dedication of *Tutti i Trionfi*,[5] il Lasca (Antonfrancesco Grazzini) relates that men of Lorenzo de' Medici's epoch, "masking themselves, and using the carnival [as an excuse] to play the roles of ladies, were accustomed to walk about during the Calendimaggio: and thus disguised in the habits of women and young girls sang *canzoni a ballo*. Considering that manner of singing to be always the same, *il Magnifico* thought of varying it, and not only the song text, but also its ideas, and the manner of writing the words, contriving songs with various other meters; and then composing music for them with new and diverse melodies: and the first *canto*, or *mascherata* sung in this form was about men who sell *berriquocoli e confortini* [sweet cakes made with honey]. . . ." Certainly Lasca is correct in stating that Lorenzo was not the originator of the carnival song. One of the many in vogue before he began his reform was "Ferravecchi, rami vecchi,"[6] published in both issues of the first printed edition of these poems, *Canzone per andare in maschera* (see below), but already cited, in the 1485 edition of *Laude*,[7] as a *canto* to which various lauds by Feo Belcari could be sung.

Not only the aforementioned *trionfi* by Lorenzo but also *mascherate*

[4] *Le Vite de' più eccellenti Pittori, Scultori ed Architettori,* ed. G. Milanesi (Firenze, 1879), IV, 134ff.

[5] *Tutti i Trionfi, Carri, Mascheaate* [sic] *ò canti Carnascialeschi andati per Firenze, Dal tempo del Magnifico Lorenzo vecchio de Medici; quando egli hebbero prima cominciamento, per infino a questo anno presente 1559.* In Fiorenza MDLVIIII [Lorenzo Torrentino].

[6] Published by Charles S. Singleton, *Canti Carnascialeschi del Rinascimento* ("Scrittori d'Italia"; Bari, 1936), p. 5.

[7] *Iesus. Laude facte & composte da più persone spirituali....* Impresso nella Magnifica città di Firenze per Ser Francesco bonaccorsi a petitione di Iacopo di maestro luigi de morsi / Nellanno. MCCCCLXXXV. Adì primo di marzo. (Copies in the British Museum; the Bibl. Nationale, Paris; the Bibl. Nazionale, Florence; and in other Italian libraries). This "cantasi come" may also be found in the contemporary laud manuscripts, Florence, Bibl. Nazionale, Magl. XXXV, 119, fol. 281ᵛ, and Rimini, Bibl. Civica Gambalunga, 4.A.II.16, fol. 165.

of the most varied content were sung on the streets of Florence during the festive season. Judging primarily and correctly from the manuscript sources, A. Simioni considers eleven *canti carnascialeschi* to be authentic works of *il Magnifico*.[8] Only nine are ascribed to the poet in several of the key manuscripts, the list being prefaced by a virtually identical statement in two of them,[9] "Canzone andate in Maschera nel 1489 a tempo del Magnificho Lorenzo de Medicj / Et composte le infrascrippte Canzone dal Magnifico Lorenzo." At the end of the ninth poem the scribe has written: "Qui comincia altre canzone composte da altri Autorj Degni." In the orthography of Ms. Ant. 158, the titles and incipits of Lorenzo's contributions to the joyful carnival season are the following:

(1) Canzona de' septe pianetj: "Septe pianeti siam, che l'alte sede"
(2) Canzona della Bachaneria: "Quanto è bella Giovineza"
(3) Canzona di choloro che andorono colle Maschere drieto: "Le cose al contrario Vanno"
(4) Canzona de' Cialdonj: "Giovanj siam, maestri molti buoni"
(5) Canzona de' confortini: "Berriquocholi, Donne, et confortini"
(6) Canzona delle Forese: "Lasse in questo Carnasciale"
(7) Canzona degli Inestatori: "Donne, noi siam Maestri di nestare" (also dated 1489 in the Ms. Pal. 207, but 1488 in Pal. 206)
(8) Canzona del zibetto: "Donne, questo è uno Animale perfetto" (dated 1489 in Pal. 207)
(9) Canzona de' profumi: "Siam ghalanti di Valenza"

Both authorship and date of an additional *canto carnascialesco*,
(10) "Donne siamo, chome vedete, giovinette vaghe e liete," entitled the "Chanzona delle donne e delle cichale che andarono cho'l carro l'anno 1488 che la fe Lº de Medici," are established in the Florentine Ms., Bibl. Nazionale II.IX.42, fol. 84ᵛ. Concluding the list is a poem ascribed to "Mᶜᵒ L. D. M." in the Bibl. Mediceo-Laurenziana, Ms. Redi 129, fol. 84:

[8] Lorenzo de' Medici il Magnifico, *Opere*, a cura di Attilio Simioni ("Scrittori d'Italia," Bari, 1914; 1939), Vol. II. See also Simioni's article entitled "Intorno alle canzoni a ballo e ai canti carnascialeschi di Lorenzo il Magnifico," in *Raccolta di studi di storia e critica letteraria dedicati a Francesco Flamini da' suoi discepoli* (Pisa, 1918), pp. 497–536.

[9] Florence, Bibl. Mediceo-Laurenziana, Ms. Ant. 158 (apparently not known to Simioni or other contemporary scholars) and Bibl. Nazionale, Ms. Magl. VII, 735. The other manuscripts referred to are in the Bibl. Nazionale, Pal. 207, 208, 206 (omits No. 8 of my list), and Venice, Bibl. Marciana, Ms. Cl. It. 243 (omits No. 3). The last-named codex, beautifully illuminated on parchment, was written in Florence only four years after Lorenzo's death, hence particularly deserves to be weighed in the evidence. On the verso of the last folio is inscribed: "Scripsit Ioannes Ugolini Florentie die VIII. Ianuarij MCCCCLXXXXVI."

(11) Canzona de' fornai: "O donne, noi siam giovani fornai"[10]

An additional nine *canti* assigned to Lorenzo de' Medici by Lasca in 1559[11] lack confirmation of his authorship in earlier sources, hence must be considered inauthentic: the *canto de' romiti, delle filatrici d'oro, di mogli giovani e di mariti vecchi, de' calzolai, di mulattieri, de' votacessi, delle rivenditore, di poveri che accattano per carità*, and *di faciatori d'olio*. The *canto delle rivenditore*, for example, beginning "Buona roba abbiam, brigata," is ascribed to Poliziano in an authoritative manuscript of the Bibl. Mediceo-Laurenziana, Pl. XLI, 33, fol. 81. Michele Messina tentatively assigns various other carnival songs to Lorenzo, but his evidence is not convincing.[12] The lugubrious "Dolor, pianto e penitenza" is labeled "Canzona de' morti fatta far Lorenzo per il Trionfo della Morte" in one of the Florentine manuscripts analyzed by Messina,[13] but this statement must refer to Lorenzo, Duke of Urbino, who "had the *canzona* made." In his life of Piero di Cosimo, Vasari describes in detail the occasion on which the Triumph of Death was enacted and comes to the conclusion that it forecast the imminent return to power of the Medici, headed by the Duke of Urbino, which took place in September of 1512, "because at the time this trionfo was made they were exiles, like the dead, so to speak, who were soon to be resuscitated."[14] The phrase "fatta far Lorenzo" in Magl. VII, 1041 would seem to prove that Vasari was correct, in which case the *trionfo* probably formed part of the preceding carnival.[15] On the other hand, Francesco Zeffi (d. 1546), who wrote the life of Lorenzo Strozzi prefaced to the latter's *Le Vite degli Uomini Illustri della Casa Strozzi*,[16] claims that *his* Lorenzo "was the inventor and leader of the *carro della morte*" in 1506. In either case the carnival song in question, ascribed to Antonio Alamanni by Singleton,[17] could not possibly have been produced by the author of "Quant'è bella giovinezza."

[10] Three manuscripts contain all eleven of the *canti* considered to be authentic: Bibl. Med. Laur., Acq. e Doni 264; *ibid.*, Pl. XLI, cod, XXV; and Bibl. Apostolica Vaticana, Ms. Vat. Lat. 3219. In respect to the dates indicated above, Emilio Bigi, editor of *Scritti Scelti di Lorenzo de' Medici* (Torino, 1955), p. 231, states that Lorenzo's carnival songs were written *before* 1486, but offers no proof.

[11] *Tutti i Trionfi*, sign. a v[v].

[12] "Rime inedite di Lorenzo il Magnifico e del Poliziano?" in *La Bibliofilia*, LIII (1951), 23–51.

[13] Bibl. Nazionale, Magl. VII, 1041, fol. 71[v], in "Alcuni Manoscritti Sconosciuti delle Rime di Lorenzo de' Medici il Magnifico," *Studi di Filologia Italiana*, XVI (1958), 298. Messina had misread the phrase as "fatta per Lorenzo."

[14] *Vite*, v. IV, 135–37.

[15] For a similar description of the *trionfo* by Serafino Razzi, see B. Becherini, *Catalogo dei Manoscritti Musicali della Biblioteca Nazionale di Firenze* (Kassel, 1959), p. 103.

[16] (Firenze, 1892), p. xi.

[17] *Canti Carnascialeschi*, pp. 238 and 478–79. Also Lasca, *Tutti i Trionfi*, sign. a vii.

The aforementioned, earliest printed edition of carnival song texts, entitled *Canzone per andare in maschera per Carnasciale facte da più persone*,[18] appeared during the last decade of the fifteenth century without indication of date, place, or publisher. Although most of its contents must be assigned to other authors, the edition contains three *canzone* by Lorenzo the Magnificent: "Donne sian, come vedete," "Sian galanti di Valenza," and "Quanto è bella giovineza," but only the first is identified as "facta dal M[agnifico] L[orenzo]." The frontispiece (see Fig. 2) depicts a well-dressed man with a full purse (Lorenzo, in Ferrari's opinion) who is listening attentively to a group of masked men and boys as they sing a carnival song. Several young women leaning out of open windows are obviously enjoying the performance, whose text may possibly have been "Berriquocoli, donne, e confortini," because two of the singers are holding *ciambelle*, a kind of sweet bread or honey cake made in circular form, empty in the center. These are closely related to *berriquocoli* and *confortini*, sweet cakes made in rhomboid and various other shapes. All three are occasionally named in the same context, as in the *Rime piacevoli e Lettere* of Alessandro Allegri:

Il qual di berriquocoli e ciambelle,
Di melarance dolci e confortini,
Farò gremito . . .[19]

The woodcut gives us valuable information about performance practice, for the masquers are singing their parts without benefit of instrumental accompaniment. In general, carnival songs of the period are vocally conceived throughout the parts, even though a full text may appear under only the upper voice of the musical source. Although most of the earlier, three-voiced examples are purely homophonic, several of them, including the anonymous *canzona* "per scriptores," beginning "Orsù, [or]sù,"[20] and

---

[18] Copies of one issue may be found in the Bibl. Riccardiana, Florence, and the British Museum; of the other, identical in contents but differing in minor details, in the Bibl. Nazionale, Florence. Severino Ferrari published the Riccardiana version, with the textual variants of the Bibl. Nazionale copy in footnotes, in his *Biblioteca della Letteratura Popolare Italiana*, Anno I, Vol. I (Firenze, 1881–82). More recently G. Biagi edited the Riccardiana copy in *I Canti Carnascialeschi di Lorenzo il Magnifico secondo l'unico* [sic] *exemplare Riccardiano* (Milano, 1925 [Bibl. Umanistica, Ser. I, I–II]). Dietrich Reichling, in *Appendices ad Hainii-Copingeri Repertorium Bibliographicum* (Monachii, 1908), fasc. IV, p. 14, Nos. 1157 and 1158, dates the Bibl. Nazionale copy at about 1485, the other a year later. Both dates are obviously too early, given the presence of Lorenzo's *canti* in each volume. (See above.)

[19] (4 vols.; Verona & Firenze, 1605–13), Vol. III, sign. H 2 (fol. 30).

[20] Perugia, Bibl. Comunale, Ms. G.20., fol. 57'–58. Modern transcriptions appear in Joh. Wolf, *Music of Earlier Times* (New York, 1946), pp. 49–51; *Das Chorwerk*, Heft 43, pp. 6–7; and the *Historical Anthology of Music*, ed. A. Davison and W. Apel, (Cambridge, Mass., 1946), I, No. 96. Charles Singleton published the text in his *Nuovi Canti Carnascialeschi del Rinascimento* (Modena, 1940), p. 119.

Fig. 2.—Anonymous, Frontispiece of Canzone per andare in maschera per
Carnasciale facte da più persone. *Biblioteca Nazionale, Florence.*
(Photograph by Pineider.)

170

"Siamo, donne, tre i romeri,"[21] contain short imitative passages that contrast effectively with the prevalent homorhythmic texture. A shift from duple meter to a danceable triple and back again characterizes both of these compositions, as well as the setting of Lorenzo's "Siam galanti di Valenza."[22]

The group of carnival singers must occasionally have been accompanied by the lute, however, judging from the inscription at the head of "Quant'è bella giovinezza" in a manuscript of the Bibl. Nazionale in Florence, Magl. VII, 1225, fol. 45: "Chançona chonposta dal magnificho lorenzo de medici che questo charnascale fece fare el trionfo de bacho dove chantavano l'infrascritta chancone[.] chomposte da leuto fu choxe bellissime." Not only does this statement reveal an important aspect of contemporary performance practice, since it refers to *several* singers (*chantavano*) and an accompanying instrument, but it also proves that a musical setting of the *Trionfo di Bacco ed Arianna* existed and was highly regarded by Florentines of the late fifteenth century. It has been seen above that two manuscripts identify Lorenzo's *trionfo* with the carnival of 1489. The conflicting date of 1490 attached to a copy in the Bibl. Riccardiana, Ms. 2723, fol. 78[v] seems less likely because this manuscript is a haphazard compilation, containing only four anonymous carnival songs in a mass of strambotti.

It is almost an understatement to say that the music for Lorenzo's *canti carnascialeschi* has been preserved only in fragmentary form. Anonymous, incomplete versions of four songs can be found at the end of a manuscript in the Bibl. Nazionale, Florence, B.R. 230 (olim Magl. XIX, 141):[23] the discants only of "Le cose al contrario vanno" (fol. 151[v]), "Donne, noi siamo maestri di nestare" (fol. 149[v]), "Siam galanti di Valenza" (fol. 144[v]), and the two lower voices only of "Giovani siam maestri molti buoni" (fol. 145). The lost music for "Berriquocoli, donne, e confortini" must have been popular in the early sixteenth century, for it is cited ("cantasi come") in *Laude vechie & nuove*[24] as the composition to which an anonymous laud, "Chi vuol, Gesù, fruir con tutto'l core," can be sung. Alfred Einstein's statement that "only an inconsequential scrap of the tune has been transmitted in a quodlibet ("Donna tu pur invecchi"[25]), found in Codex Magl.

---

[21] Florence, Bibl. Nazionale, Ms. Panc. 27, fol. 110'–11.

[22] Published by Federico Ghisi, *Feste musicali della Firenze Medicea (1480–1589)*, (Centro Nazionale di Studi sul Rinascimento; Firenze, 1939), pp. 9–13, where the text is mistakenly attributed to Jacopo da Bientina.

[23] See Becherini, *Catalogo dei Manoscritti Musicali*, pp. 60–68. Unfortunately the author has followed older sources in attributing three of Lorenzo's carnival song texts to Jacopo del Bientina, and several poems by other authors to *il Magnifico*.

[24] A petitione di Ser Piero Pacini (Florence, 1510?), sign. l vi. (Copies in the Bibl. Marciana and the Bibl. Nazionale, Florence).

[25] Published by B. Becherini, "Tre incatenature del cod. fiorentino Magl. XIX, 164–167," in *Collectanea historiae musicae*, I (Firenze, 1953), pp. 79–96.

XIX, 164–167, No. 41"[26] is wishful thinking, for only the first word of
Lorenzo's poem, "Berricuocoli," appears here in another context, set to
what is rather far removed from a tune: repeated notes in the tenor. Otherwise, musical settings of the *canzona delle forese, delle cicale,* and *de' fornai*
must have been current around 1490, for they are named in the earliest
edition known to contain lauds by *il Magnifico,* an undated and unidenti-
fied publication entitled *Laude facte & composte da più persone spirituali.*
(See below.)

This list, including nine of the eleven *canti* by Lorenzo considered to
be authentic, provides evidence that part-song versions of most of them
existed at one time and probably were contained in the chief remaining
codex, B.R. 230, before it was mutilated, many leaves having been lost or
stolen, and the remainder rebound, to some extent incorrectly. Complete,
anonymous settings of two *canti* whose texts have been wrongly ascribed
to Lorenzo also have been preserved in this manuscript: "In questa vesta
obscura" (fol. 105^v–6) and "Buona roba abbiam, brigata" (fol. 147^v–48).
Both have been published under Lorenzo's name by P. M. Masson.[27] The
discant only of a third composition in this category, "Donne no' siam
dell'olio facitori," is attributed to Alexander Agricola on fol. 142^v of the
manuscript.

Apparently most of the musical sources of carnival songs perished in
the pyre of secular, "immoral" books, manuscripts, art objects, and musical instruments collected and set afire as the result of preaching by Savonarola's aide, Fra Domenico da Pescia, in February of 1497. Manuscript B.R.
230 is of later origin, having been compiled after the Medicis had been
returned to power. Charles Singleton shows that it must have been copied
out after 1513,[28] for it contains two *canti* intended for the carnival cele-
brating the election of Giovanni de' Medici (Leo X) to the papacy in that
year, "Volon gli anni, i mesi, e l'ore" by Antonio Alamanni,[29] and "Colui
che da le legge alla natura" by Jacopo Nardi. Two villotistic compositions
contained in the manuscript also help to determine its date, "Quando lo
pomo vien dallo pomaro," of disputed authorship, and "So ben che la non
sa" by Don Michael V.[Pesenti]. The former, ascribed to Tromboncino,
was first printed by Petrucci in *Frottole Lib. XI* (1514), whereas the latter
made its initial appearance in Andrea Antico's *Canzoni, Sonetti, Strambotti
et Frottole Lib. III°* (Roma, 1513). The scribe of B.R. 230 might have obtained these compositions from other sources, but judging from all available

---

[26] *The Italian Madrigal* (Princeton, 1949), I, 33.
[27] *Chants de Carnaval Florentins* (Paris, 1913), pp. 1–11.
[28] *Canti Carnascialeschi del Rinascimento,* p. 479.
[29] Text and music published by Ghisi, *Feste Musicali,* pp. 34–38.

evidence he compiled the manuscript in Florence during the early years of Leo X's reign.

In the aforementioned *Laude facte & composte da più persone spirituali*[30] eight lauds (all except my No. 9, below) are listed under Lorenzo's name. Highly significant from the musical viewpoint is the fact that the editor has indicated the titles of carnival songs and other compositions to which these lauds were to be sung. Since most of the *canti* in question are by *il Magnifico* himself, and these are dated in the manuscripts of his poetry, the lauds must have been written later than 1489 and the edition published no earlier than the following year, if not after Lorenzo's death.[31] Based upon the evidence provided by this publication and by such contemporary manuscripts as Pal. 208 of the Bibl. Nazionale, Florence, nine lauds can be ascribed with certainty to the poet. The rubric "cantasi come," attached to each incipit except the last, has been taken from the edition of about 1490–95:

(1) "Poi che io gustai, Iesu, la tua dolcezza." Cantasi come "Tanta pietà mi tira e tanto amore" (a well-known laud by Feo Belcari).
(2) "Io sono quel misero ingrato." Cantasi come *la canzona delle cicale*.
(3) "O maligno & duro core." Cantasi come *la canzona de' Valentiani*.
(4) "O Dio [o] sommo bene." Cantasi come *el fagiano* (the *canzona del fagiano*[32]).

---

[30] *Et oltre aquelle che gia perlo tempo passato furono impresse se facta hora in questa nuova impressione una aggiunta di piu daltrettante.* (Copies in the British Museum, the Bibl. Nazionale, Florence, etc.) The "aggiunta" refers to the anthology collected and edited by J. de' Morsi in 1485 (see above), which contains none of Lorenzo's lauds. Except for a few insignificant omissions and differing ascriptions the undated "nuova impressione" duplicates de' Morsi's edition until sign. g 4ᵛ, when the newly-added lauds by Lorenzo and others begin. G. C. Galletti (ed.), in *Laude Spirituali di Feo Belcari, di Lorenzo de' Medici, di Francesco d'Albizzo.... comprese nelle quattro più antiche raccolte* (Firenze, 1863), prints the additions of the "nuova impressione" on pp. 111–208 with the closing comment: "Finiscono le Laude composte da diversi, stampate già nel secolo XV.... si crede in Firenze per Antonio Miscomini nel 1489 per cura e a spese del Magnifico Lorenzo de' Medici." For the reasons mentioned above this tentative date is too early, but Eugenia Levi, in *Lirica italiana antica* (Firenze, 1905), p. 67, and Simioni, *Opere*, II, 343, cite Galletti's remark without the "si crede" and leave us with the impression that the lauds were composed before 1489.

[31] Modern bibliographers believe that the edition was published about 1495 in Florence by Bartolommeo de' Libri, a date that would correspond more closely to the contents. See Ministero della Pubblica Istruzione, *Indici e Cataloghi*, Nuova Serie I, *Indice Generale degli Inconaboli delle Biblioteche d'Italia* (Roma, 1954), Vol. III, and *Catalogue of Books printed in the XVth Century Now in the British Museum* (London, 1930), Part VI, p. 656.

[32] Printed by Singleton, *Canti Carnascialeschi*, p. 130.

(5) "Quanto è grande la bellezza." Cantasi come *la canzona delle forese.*

(6) "O peccatore, io sono Dio eterno." Cantasi come *la canzona de' fornai.*[33]

(7) "Vieni a me, peccatore." Cantasi come "Tu m'hai legato, amore."

(8) "Peccatori su tutti quanti." Cantasi come *la canzona de' visi adrieto.*

(9) "Bene arà duro core."

It is evident from this list that five of Lorenzo's lauds were marked to be sung to his own carnival songs, and another (my No. 4) to one of anonymous authorship. The directives must not be accepted without a test of their validity, however. In the *Laude* editions of about 1510[34] and 1512,[35] for example, "O peccatore, io sono Dio eterno" appears with the mistaken rubric "cantasi come *la canzona delle forese,*"[36] an impossible feat since the laud is a ballata in iambic, endecasyllabic lines with a two-line ripresa and four-line stanzas, whereas the song of the *forese* is a barzelletta in trochaic, octosyllabic lines with a ripresa of four lines and stanzas of eight. Obviously the editors misread the "cantasi come *la canzona de' fornai*" attached to the same laud in the edition of about 1490–95. In a similar instance, Simioni prefaces "Vieni a me, peccatore" with "Cantasi come *Amore io vo fuggendo,*"[37] having copied the rubric from Francesco Cionacci[38] without checking the poetical patterns involved. The laud is a ballata grande, whereas "Amore io vo fuggendo nocte e dia" is a strambotto ascribed to Serafino Aquilano in Giunta's edition of 1516.[39] The only grain of truth in Cionacci's reference lies in the fact (not known to him or to Simioni) that Serafino's strambotto was actually set to music at the time and may be found in Florence, Bibl. Nazionale, Ms. Magl. XIX, 121, fol. 22–23.

The earliest sacred source for the music of Lorenzo's most famous laud, "Quanto è grande la bellezza," is a manuscript in the Bibl. Nazionale, Florence, Rossi-Cassigoli 395, fol. 5ᵛ–6, which contains only the discant and tenor parts of the composition, each in a different hand. It is impossible to

---

[33] This is also entitled "Lalda del Mᶜᵒ Lorenzo de' Medici sopra *la canzona de' fornai*" in the Bibl. Mediceo-Laurenziana, Ms. 25 of Pl. XLI, fol. 273ᵛ.

[34] *Laude vechie & nuove.* A petitione di Ser Piero Pacini da Pescia, sign. fol. viiᵛ.

[35] *Opera nova de Laude facte & composte da piu persone spirituali....* Stampata in Vinegia per Giorgio de Rusconi a instantia de Nicolo dicto Zopino. MDXII a di. iiii. Marzo, fol. xlviiᵛ (Copy in the British Museum).

[36] Simioni reprints the error in *Opere,* II, 147.

[37] *Ibid.,* II, 142.

[38] *Rime Sacre del Magnifico Lorenzo de' Medici il Vecchio....* Firenze, 1680 (Ediz. Seconda; Bergamo: Pietro Lancellotti, 1760), p. 59.

[39] *Opere Dello elegantissimo Poeta Seraphino Aquilano* (Firenze: Philippo di Giunta, 1516), fol. 175.

determine exactly when the music was copied, but the notation (diamond-shaped notes in manuscript) points to the first half of the sixteenth century. The codex itself was compiled by a monk from Pistoia who wrote on the title page: "Questo libro è ad uso di Fra leone forteguerri di Pistoia dell'ordine de' Predicatorj.... Il Qual prese l'habito nel M.D.XXII A di XII di Septembre a hora una di Notte." Some years later, in 1563, Fra Serafino Razzi compiled an anthology[40] of laud texts and their music which contains Lorenzo's poem on fol. 10^v–11, set for three voices. A bassus has been added to the discant and tenor, which are virtually identical to those in the Ms. Rossi-Cassigoli, except for small variants, important because they show that neither of the monks copied from the other.

Razzi contributed uniquely to the literature by collecting and publishing "the proper music and manner of singing each laud, as was in use by those of earlier times (*gli antichi*), and is the custom in Florence," according to the title page. Several centuries later, Eugenia Levi printed both text and music of "Quanto è grande la bellezza" in her *Lirica Italiana antica*,[41] immediately following the text of Lorenzo's *Trionfo di Bacco ed Arianna*, having had the composition transcribed from Razzi's publication by G. Gasperini. Referring to the juxtaposition, she makes a highly significant observation (p. xxxi): "The laud which accompanies this melody extends itself for a number of strophes equal to those of the preceding poem by the Magnificent Lorenzo and in the same meter. Given the practice of the century . . . to set sacred and profane rhymes to the same melody, would it not be possible that we have here what served also to accompany the very beautiful *trionfo di Bacco e Arianna*? In all of the initial editions of Lauds of the 15th and 16th centuries [however] 'Quant' è grande la bellezza' carries the indication, 'Cantasi come La canzona delle forese.' " Referring to Lorenzo's laud in *I Canti Carnascialeschi nelle fonti musicali del XV e XVI secolo*,[42] Federico Ghisi dashes cold water on Signora Levi's hypothesis: "It does not seem probable that to the music of this laud would be sung also the *trionfo di Bacco ed Arianna*, given the similarity between the textual incipits. If that were the case there would be an inscription at the bottom: "Cantasi come il trionfo di Arianna ecc.' "

Of course, neither Levi nor Ghisi possessed any documentary proof that the music printed by Razzi for "Quant'è grande la bellezza" was of

---

[40] *Libro Primo delle Laudi Spirituali da diversi excell. e divoti autori, antichi e moderni composte.... Con la propria Musica e modo di cantare ciascuna Laude, come si è usato da gli antichi, et si usa in Firenze*. Nuovamente Stampate. In Venetia, ad instantia de' Giunti di Firenze. M.D.LXIII.
[41] P. 235.
[42] (Firenze, 1936), p. 94.

secular origin, nor were they aware of the earlier religious source, Fra Leone Forteguerri's codex. Since they had no way of knowing whether Razzi might not have composed the music himself, their hypotheses, on one side or the other, were pure speculation. In order to demonstrate that the music for Lorenzo's laud is the contrafactum of a carnival song, its identity as a secular composition must first be established. In other words, someone must find a *canto carnascialesco* set to the same music. A careful search through Mr. B.R. 230, which contains no religious compositions whatsoever, has yielded the required evidence. Close to the end of the manuscript, on fol. 151ʳ, near the fragmentary remains of Lorenzo's other carnival songs and immediately adjacent to the discant of his *Canzona degli Innestatori*, I have discovered two lower voice parts, without a text — that correspond to the tenor and bass of "Quanto è grande la bellezza" in Rossi-Cassigoli and Razzi. Since B.R. 230 is entirely secular, this must be the carnival song that served as a musical model for Lorenzo's laud.

But is it the *Trionfo di Bacco ed Arianna* or the *Canzona delle Forese*? Each of the lauds paired with a secular or religious composition in the earliest edition of Lorenzo's religious poems (see above) matches its model perfectly as to poetical pattern, length of lines, and rhyming scheme, *except* "Quanto è grande la bellezza" and its presumed mate, the *Canzona delle Forese*. The latter is a barzelletta with stanzas of *eight* trochaic, octosyllabic lines and the following rhymes in the ripresa: -ale, -iti, -iti, -ale. Since the volta of the stanza also ends in -ale, the refrain (not indicated in the poetry) must be the entire ripresa of *four* lines. The laud, on the other hand, not only has a similar incipit to that of "Quant'è bella giovinezza," but is also cast in exactly the same poetical form, a barzelletta consisting of a ripresa whose end-rhymes are identical in both poems: -ezza, -ia, -ia, -ezza, and a series of *six*-line stanzas, each of which ends in -ia. This connects with the second half of the ripresa, used as a refrain of *two* lines. It is obvious that the *Canzona delle Forese* differs so markedly from this pattern that it could not be set appropriately to the music of "Quant'è grande la bellezza." Even the complex laud in the form of a *ballata mezzana*, "Poi ch'io gustai, Iesu, la tua dolcezza," exactly matches its model, Feo Belcari's laud, "Tanta pietà mi tira e tanto amore,"[43] except for a minor change of rhyme in the ripresa, a,a,b and a,b,b. Both poems contain a ripresa of three iambic lines, of eleven, seven, and eleven syllables, respectively, and stanzas of nine lines, all of eleven syllables except the penultimate, septisyllabic one, rhyming c,d,c,d,c,d,e,e,b. Lorenzo'e laud, "O Dio, o sommo bene, or come fai," and the *canzona del fagiano* are both examples of the *ballata minore* with a

---

[43] Published by Galletti, *Laude Spirituali*, p. 1.

ripresa of two iambic, endecasyllabic lines rhyming a,a, and stanzas of six similar lines, rhyming b,c,b,c,c,a.[44]

Such an exact correspondence, even to the smallest detail, cannot have been accidental. Apparently Lorenzo de' Medici based each laud on a specific model, intending that the sacred poem be sung to pre-existent music which would have fit only its original pattern of verses. This fact should help to determine the identity of the carnival song that served as a point of departure for "Quanto è grande la bellezza." Since the laud is so close a parody, in the Renaissance sense of the term, of the *Trionfo di Bacco ed Arianna*, *il Magnifico* must have intended it to be sung to the same music, perhaps with the pious hope (very late in life) that it would replace the secular work entirely. The linking of Lorenzo's laud to the song of the *forese* in the early editions must have been a mistake, perhaps a purposeful one to avoid profaning the collection with a reference to Bacchus and Ariadne at a time when the influence of Savonarola was becoming more and more powerful. The naming of heathen gods and mythological personages in a compilation of Christian lauds would have been anathema to the editor of *Laude facte & composte da più persone spirituali*, as it must have been to the good Fra Leone Forteguerri and to Fra Serafino Razzi, provided that the monks knew of the laud's secular origin. It is quite probable that such a hedonistic, un-Christian poem as Lorenzo's *trionfo* had been forced into virtual oblivion during the hegemony of the Dominican monk because it ran dangerously counter to the latter's philosophy.

[44] Apparently the unique source of "Tu m'hai legato, amore," the textual model for Lorenzo's laud, "Vieni a me, peccatore," is the Cod. 33 of Pl. XLI, fol. 79 of the Bibl. Mediceo-Laurenziana, where it is ascribed to Lorenzo de' Medici in a later hand. It is difficult to prove that these ballate are absolutely identical in form, since only the telescoped ripresa and a single stanza of the secular poem have been preserved, to my knowledge. The laud has a ripresa of four iambic lines, rhyming a,b,b,a (7,7,7,11 syllables) and stanzas of six lines, c,d,c,d,d,a (7,7,7,7,7,11). The secular poem closely resembles it, even to the rhyme at the end of the incipit, but seems to lack at least one verse, hence does not lend itself to accurate comparison:

Tu m'hai legato, amore,
(.... ento?)
et io ne son contento:
tanta dolcezza sento dentro al core.

La più gentile et bella
che sia sotto la luna
(.... ella?)
sempre amerò quest'una,
perchè m' ama forte
fin doppo morte: et sarà mie Signore.

It was Federico Ghisi who first realized that the missing voice-parts of certain carnival songs might well be found in Razzi's publication because of the correspondence between lauds and secular compositions indicated in the early editions of *Laude*. Although the text of Lorenzo's "O maligno e duro core" was clearly marked "Cantasi come la canzona de' Valenziani" in the aforementioned edition of about 1490–95, neither Razzi, who printed the laud music in his anthology, nor Levi, who reproduced it in *Lirica Italiana antica*, nor G. Gasperini, who used the ripresa only in an article,[45] was aware of the music's secular origin. Ghisi was able to publish the complete version of "Siam galanti di Valenza" in his *Feste Musicali*,[46] having obtained the lower voices from Razzi's book. In the introduction to *Feste* (p. ix) he writes: "In the *Primo Libro di Laudi Spirituali* of Razzi (1563): 'Quanto è grande la bellezza' by Lorenzo the Magnificent is sung to the *canzona delle forese di Narcetri*, and another laud: 'O maligno e duro cuore,' also by il Magnifico, has the same tonal setting as the carnival song *de' valenziani*. I have done nothing more than to restore its primary carnival text to the original melody conserved as a laud. As proof of this musical identity the soprano of the *canto de' valenziani*, contained in the codex Magliabecchiano cited above [B.R. 230], shows itself to be identical with that of the laud." The last part of this statement is perfectly valid, and finds further confirmation when one compares the poetical patterns of "Siam galanti" and its contrafactum, but Ghisi moves onto uncertain ground when he attempts to draw similar conclusions from the pairing of "Quant'è grande la bellezza" and the *Canzona delle Forese* by the early laud editors. Without possessing the kind of indisputable evidence provided by "Siam galanti," he proceeded to publish the song of the *forese* set to the music of "Quant'è grande" as found in Razzi's publication,[47] which cannot be matched with the original music of the *Canzona delle Forese* because not a single voice part of the latter exists in B.R. 230 or in any other source.

Razzi does *not* indicate, as Ghisi definitely states, that Lorenzo's lauds should be sung to the music of carnival songs. He simply prints the music, followed by the text, entitled "Laude di Lorenzo de' Medici," with no mention whatsoever of the *Canzona delle Forese* or *de' Valenziani*. For that matter, only once does he point to the secular origin of a laud, that beginning "Tre virtù siamo," whose text is marked "cantasi come *Tre ciechi*

---

[45] "La Musique Italienne au XVe Siècle," in the *Encyclopédie de la Musique et Dictionnaire du Conservatoire*, ed. A. Lavignac, Vol. I²: *Histoire de la Musique. Italie-Allemagne* (Paris, 1914), pp. 622–23.

[46] Pp. 9–13.

[47] *Ibid.*, pp. 5–8.

siamo."[48] In the dedication to Caterina de' Ricci printed at the beginning, the publisher, Filippo Giunti, actually criticizes the old method of indicating a laud's music, as follows: "It has only been during the past few months that the Rev. Father Fra Serafino Razzi da Marradi has collected for me, almost as a pastime, a book of the most beautiful old and modern [lauds], and has added the music with which to sing them, abandoning that silly way of saying: *Cantasi come* this, and *come* that."

In the other important source of the music for "Quant'è grande la bellezza," the Ms. Rossi-Cassigoli 395, many of the laud texts are marked "cantasi come" a secular composition, such as "Oimè, lass' oimè!" But this is not true of *il Magnifico's* laud, whose text appears on fol. 80 with the rubric: "Composta da Lorenzo Vecchio de' Medici.... Cantasi a modo proprio." ("Sing it to its own music.") No mention is made here of the *Canzona delle Forese*, which would have been an innocuous enough reference, unlike the *Trionfo di Bacco*, had it been traditionally appropriate.

Ghisi's setting of "Lasse in questo carnovale" (the *Canzona delle Forese*) to the music of "Quanto è grande la bellezza" calls for an artificial adjustment that proves its lack of authenticity. In order to fit an eight-line stanza to music intended for one of six lines, he uses the repeated section at the close of the composition for the end of the volta (lines seven and eight of the stanza). Such a repetition is indicated in all three of the musical sources but is clearly intended for the refrain of the poem, not the volta. In the secular manuscript, B.R. 230, the scribe writes out the repeated music once, without any text, since this must have appeared in the missing cantus part; in the cantus of Ms. Rossi-Cassigoli 395 the initial word of the laud refrain, "Ciascun," is placed at the end of the music with a *custos* pointing to the proper pitch. The spot at which the repetition begins is marked by a *signum congruentiae* and the word "replica." For the sake of further clarity both text and music of the refrain's initium are written out in the tenor part, as well as in Razzi's setting. Had the secular music been intended for the *Canzona delle Forese*, the composer would normally have extended the setting of the stanza by two more phrases, in order to take care of lines seven and eight — but he did not.

As I have mentioned earlier, the volta of "Lasse in questo carnovale"

---

[48] Fol. 120^v–21. The laud music is a simplified arrangement, with a slightly different incipit, of Ioan Domenico da Nola's tongue-in-check *mascherata* about three blind men who claim to be "povr' inamorati." See the latter's *Canzone Villanesche a Tre Voci Novamente Ristampate Libro Secundo* (Venetijs apud Antonium Gardane, 1545). A modern transcription of the *mascherata* has been published by Alfred Einstein in *A Short History of Music* (London, 1948), and in the Italian edition, called *Breve Storia della Musica* (Firenze, 1960).

ends in "male," showing that the *entire* ripresa is to reoccur at that point. To repeat only half of it would make no sense:

Lasse, in questo carnasciale
noi abbiam, donne, smarriti
tutt'a sei nostri mariti;
e sanz'essi stiam pur male.

If the composition in B.R. 230 is the *Canzona delle Forese*, would not the scribe have called for a second repetition of the entire ripresa music? The fact remains that none of the sources indicates such a second repetition, since two of them are settings of the laud, which has stanzas of six lines and a refrain of only two, and the composition in B.R. 230 obviously fits the secular poem that is exactly equivalent to the laud, "Quant'è bella giovinezza."

How were barzellette with stanzas of eight lines actually set to music in Lorenzo's time? Evidence is provided by the very next composition in Ghisi's edition, "Siam galanti di Valenza."[49] Here the musical phrase for the *mutazioni* (2+2 lines) is repeated, while that for the volta (lines 5–8) is through-composed. Most certainly the anonymous composer does not use the second part of the ripresa music for the end of the volta, as does Ghisi in the other instance. The marked difference in mood between ripresa and stanza of "Quant'è bella giovinezza," reflected so well in the tonal setting (see below), is not apparent in the *Canzona delle Forese*, in whose ripresa some peasant women explain that they have lost their husbands during the carnival, then (in the stanza) identify themselves as "Di Narcetri noi siam tutte." For these various reasons one must conclude that the secular composition in B.R. 230 and its contrafactum in the other two sources do not fit the song of the *forese*, but pertain unequivocally to the *Trionfo di Bacco ed Arianna*. (See Ex. 1.)

The original music for "Quant'è bella giovinezza" exactly matches that in Razzi except for a lengthened and an extra note in both voices of the ripresa (m. 9 and 14, respectively) that make the secular version more rhythmic and regularly accentuated. The earlier religious source, Ms. Rossi-Cassigoli 395, also contains the lengthened notes of the third phrase, but corresponds to Razzi in the fourth. I have inserted another C into m. 14 of the discant and its repetition at the end to conform with the lower voices of B.R. 230. The F in m. 25 of the discant is misprinted C in Razzi, although the custos points to F there also. Otherwise the differences are of minor importance.

Because of the regular verse pattern in trochaic, octosyllabic lines, the

---

[49] *Feste Musicali*, pp. 9–13.

Ex. 1. Quant'è bella giovinezza   Transcribed and edited by Walter H. Rubsamen

Questi lieti satiretti,
delle ninfe innamorati,
per caverne e per boschetti
han lor posto cento agguati;
or da Bacco riscaldati,
ballon, salton tuttavia.
　　Chi vuol esser lieto...

Queste ninfe anche hanno caro
da lor essere ingannate;
non può fare a Amor riparo,
se non gente rozze e ingrate;
ora insieme mescolate
suonon, canton tuttavia.
　　Chi vuol esser lieto...

Questa soma, che vien drieto
sopra l'asino, è Sileno;
così vecchio è ebbro e lieto,
già di carne e d'anni pieno;
se non può star ritto, almeno
ride e gode tuttavia.
　　Chi vuol esser lieto...

Mida vien drieto a costoro;
ciò che tocca, oro diventa.
E se giova aver tesoro,
s'altri poi non si contenta?
Che dolcezza vuoi che senta
chi ha sete tuttavia?
　　Chi vuol esser lieto...

Ciascuno apra ben gli orecchi,
di doman nessun si paschi;
oggi sian, giovani e vecchi,
lieti ognun, femmine e maschi;
ogni tristo pensier caschi;
facciam festa tuttavia.
　　Chi vuol esser lieto...

Donne e giovinetti amanti,
viva Bacco e viva Amore!
Ciascun suoni, balli e canti,
arda di dolcezza il core!
Non fatica, non dolore!
Ciò che ha a esser, convien sia.
　　Chi vuol esser lieto...

anonymous composer has set the text in four-bar phrases, except for the last part of the ripresa and its repetition in the refrain, which has been extended to six measures by means of a melisma on the penultimate syllable. In setting Lorenzo's *trionfo* he contrasts the Aeolian mode of ripresa and refrain with a stanza that modulates from Ionian to Mixolydian, Phrygian, Dorian, and back to Mixolydian, thus demonstrating a mastery of musical form that is highly advanced for his time. As in most barzellette the form is tripartite because the ripresa music recurs at the end, but here the feeling of balance and proportion inherent in the ternary form has been enhanced by the scheme of modulation. Although one should beware of interpreting Renaissance changes of mode from the standpoint of the affections associated with major and minor in more recent music, the shift from what is essentially A minor for the melancholy "Quant'è bella giovinezza" to C major for the carefree "Quest'è Bacco e Arianna," coupled with a danceable dotted rhythm in the bass, seems fully expressive of the textual content.

Such mastery of technique points to a professional composer, possibly to Henricus Isaac, who had been called to Florence late in 1484 by Lorenzo himself and who was employed as "chantore" or "chonponitore" by the baptistry of San Giovanni and the cathedral from July of 1485 until March of 1493, according to documents discovered by Frank A. D'Accone.[50] Isaac, called Arrigo in Italy, undoubtedly was the most prominent musician in *il Magnifico's* service during the years associated with the latter's carnival activities. Lasca, in the dedication of *Tutti i Trionfi*, identifies him as the composer of Lorenzo's first *canto carnascialesco*, that of the sellers of *berriquocoli e confortini*: "composed for three voices by a certain Arrigo Tedesco, director at that time of the choir of San Giovanni, and a musician of the greatest reputation in those days. But not long thereafter he made some for four voices: and then, little by little, the number increased of those

[50] "Heinrich Isaac in Florence: New and Unpublished Documents," *Musical Quarterly*, XLIX (1963), 464–83; and "The Singers of San Giovanni in Florence during the 15th Century," *Journal of the American Musicological Society*, XIV, No. 3 (Fall, 1961), 338ff.

who composed notes, as well as words [of carnival songs]." Apparently his fame had spread far and wide, and Lorenzo held him in the highest esteem, judging from the latter's dispatch to the Florentine ambassador in Rome, Piero Alamanni, dated June 25, 1491: "Thank the magnificent Venetian ambassador for having requested these songs [by Isaac] of me. . . . And I shall have the aforementioned songs put in order and sent to you soon, I believe by the first cavalcade. Had I known what manner of song most pleases him, I should better have been able to serve him, because Arrigo Isach, their composer, has made some of various kinds, grave and sweet, skilful and full of ingenuity. . . ." And on July 8, 1491, to the same person: "I send to you by this cavalcade a book of musical compositions by Isac according to the request of the magnificent Venetian ambassador, to whom you should give the book, inquiring whether I can do anything else for him and presenting my respects to him. . . ."[51]

In view of the foregoing, it seems logical to assume that Isaac was called upon to provide music for many of his patron's carnival songs, but whether this included "Quant'è bella giovinezza" is doubtful. None of the known settings of Italian texts by Isaac are barzellette in trochaic, octosyllabic lines. The composer's habit was to separate homophonic phrases by rests and usually to include a section in triple time, neither of which holds true for the *Trionfo di Bacco*, which contains not a single rest in any voice part nor any portion in contrasting meter. From the stylistic viewpoint, therefore, Isaac's authorship is open to serious question. Another possibility is Alexander Coppinus, some of whose four-voiced *canti carnascialeschi* have been preserved in Ms. B.R. 230. According to D'Accone,[52] Coppinus was a choirboy in Florence as early as 1477, hence, he must have been born around 1465 and could well have been the composer of Lorenzo's carnival songs some twenty-five years later.

Anonymous as the composer of "Quant'è bella giovinezza" must remain, nothing can detract from the beauty and nostalgic quality of his music—a truly appropriate and qualitatively equivalent setting of the poem that is to many the epitome of the Italian Renaissance. One can only hope that future generations of Italians who memorize these verses will be given the opportunity to sing them also.

[51] Florence, Archivio di Stato, Medici-Tornaquinci, Filza 3, Nos. 123 and 126. I am grateful to Professor Frank D'Accone, who quickly responded to my request for a transcription of pertinent portions of these letters, which were first mentioned in the *Catalogue of the Medici Archives .... which will be sold by Auction by Messrs. Christie, Manson & Woods.* (London, 1918), p. 90, No. 329, and p. 92, No. 340. Their present location has only recently been revealed by Pier G. Ricci and Nicolai Rubenstein, *Censimento delle Lettere di Lorenzo di Piero de' Medici* (Ist Naz. di Studi sul Rinascimento; Firenze, 1964), pp. 177–78.
[52] Howard M. Brown, Walter H. Rubsamen, and Daniel Heartz, *Chanson and Madrigal 1480–1530*, ed. by James Haar (Cambridge, Mass., 1964), p. 79.

PART II: SCIENCE

"And God said, Let the waters under the heaven be gathered together unto one place, and let the dry land appear; and it was so.

"And God called the dry land Earth, and the gathering together of the waters called he Seas: and God saw that it was good."

Genesis 1: 9, 10

"Be it known that on this globe here present is laid out the whole world according to its length and breadth in accordance with the art geometry . . . wherefore let none doubt . . . that every part may be reached in ships, as is here seen."

Martin Behaim, inscription on his globe of 1492, made in
Nuremberg—the oldest surviving terrestrial globe known

"The sea yields the world to the world by this art of arts, navigation."

Samuel Purchas, *Purchas his Pilgrims* (London, 1625)

SCIENCE AND THE TECHNIQUES OF NAVIGATION IN THE RENAISSANCE ⁊❧ D. W. WATERS

t may be asked, and, indeed, I have asked myself, what profit is there in studying the history of ships and the sea? What, in addition to the training of the mind by the discipline of historical research—which can be done just as well with any other subject—makes the sea, seamen, and ships a fit subject for serious study? I would answer, the history of civilization and of science, and I would answer so because civilization and the products of science make the world physically tolerable; because, while science is itself the product of civilization, civilization is, basically, the product of the activities of seamen sailing in ships across the seas. Here, I suspect, some will protest, "But man is a land animal; civilization is the product of agriculture." Let me put my case this way. The seas of the world are, as the Bible says, one—one great sheet of water enveloping four-fifths of the globe's surface—yet, as Martin Behaim declared on the first known terrestrial globe, everywhere accessible to men in ships. The lands of the world are legion—there are the five great continents and islands innumerable as the sands on the seashore. Historical civilization is a riverine phenomenon, and its growth, until the development of mechanical transportation in the nineteenth century, depended upon accessibility to water and, in particular, to the sea. The sea was the great, and by far and away the greatest medium for the exchange of merchandise, the seminal fluid and the lifeblood of civilization. In short, up to a century ago, civilization was centered virtually exclusively upon the sea; civilization was a peripheral phenomenon of man's activities upon the sea; rather like driftwood and jettisoned cargo littering the shores of the seven seas, civilization littered—or tidily adorned—the seaboards and river-banks of the world.

The sea has been the great medium for the dissemination of civilization: its unpalatable salt waters have nourished the cultures of societies into precocious growth; indeed, they have cross-fertilized cultures otherwise doomed to sterility and thereby conceived and given birth to new ones.[1] I conclude, therefore, that civilization and its offspring, science, are the products of man's activities upon the sea and that, in consequence, seamen and ships and the use of the seas for communication merit academic study in the contexts of the humanities and of the history of science.

[1] W. H. McNeil, *The Rise of the West* (New York, 1965), repeatedly illustrates this fact.

The unity of the sea divides the lands of the earth; ships unite them; only ships carry the cargoes in the quantities and in the variety necessary for the sustenance and growth of the many societies scattered about the world's land surface. Clearly, how ships are conducted—are navigated—from one place to another is a matter of fundamental importance to the maintenance, let alone growth, of civilization and, hence, of science. As that rather underrated writer of the late Renaissance in England, Samuel Purchas, put it, "The sea yields the world to the world by this art of arts, navigation." In short, the history of navigation is a key to unlock the history of civilization and of science.

What then is navigation? Navigation is the art and science of conducting a ship across the sea between assigned places in a safe and timely manner.

The art and science of navigation! Wherein lies the art? The art lies in the navigator himself, in his professional aptitude and skill as a seaman in using the navigational aids at his disposal and in ship-handling. And the science, when or where or how is that manifested? The science lies in the tools of his trade, in their purpose; in their design, compilation, and manufacture; and in the manner in which they are intended to be used; moreover, like a well-laid rope, nautical science is made up of several well-tested strands—observation, organization, speculation, and experimentation—which, if successfully manipulated, if truly laid, permit repetition, form, as it were, a cable of confidence, in a sea of uncertainty. I say a sea of uncertainty because once a man is out of sight of land he is lost without knowledge of the signs of the sea, and except he be offshore but a little distance, aids to enable him to orientate himself and establish at least his approximate position over the seabed beneath him are essential. He needs aids at least to measure depth of water, the passage of time, distance sailed, and direction steered.

I propose to consider the role of science and, when possible, of particular scientists and organizations in Portugal and Spain in improving existing and in devising new techniques and instruments of navigation in the Renaissance, that is, from about the beginning of the fifteenth to about the middle of the sixteenth centuries, and to explain these.[2] One of my aims is to invite a reassessment of the contributions of Lusitanian administrators during the Renaissance to the adoption by practical men—seamen—of scientifically devised instruments and scientifically prepared data and instructions for day-to-day use; and another is to invite a reassessment of the influence of Lusitanian scientists of the Renaissance concerned with nautical science upon scientific inquiry in Europe in the sixteenth century, of their con-

[2] By *science* I mean organized positive knowledge whose advancement depends upon refinement of measurement.

tribution, in short, to the scientific basis of modern civilization, to what is called the "Scientific Revolution."

The most important thing a pilot must know has always been whether he has enough water under his keel for his vessel to float in safety. From time immemorial he has used the sounding lead of 7, 14, or 28 pounds and the sounding line marked at fathom—6 feet—intervals. In the Mediterranean the tidal effect is barely perceptible, the depth of water is virtually constant, and except in the region of the Nile Delta, the water is limpid clear so that the seabed can be seen at considerable depths; off northwest Europe, however, the action of the tides is pronounced: it causes a marked rise and fall of the sea level, in most places twice a day, while the continual scouring of the shallow seabed by the tidal streams renders the waters opaque. To the northern seaman the lead and line was, therefore, always vital for finding depth of water. (See Fig. 1.) In the often poor visibility of northern seas and their relatively low coastlines the lead was also an indispensable position-finding aid in the role of a seabed sampler. Armed, as the term is, with tallow in a depression in the bottom of the lead, it brought up samples of sand, mud, shells, or evidence of rocks, which by their texture, color, and smell would tell the experienced pilot, in conjunction with the depth of water measured, the locality in which he probably was. Bird flight and flocks, fish, the color and scend of the sea—all evidence stored in his memory over the years— also aided him, as a tracker in forests or plains is aided by sights and scents of places familiar to him but indistinguishable to the stranger. Because of the ever-present problem of depth of water and because the flux and reflux of the tides cause powerful tidal streams of ever-varying speed and direction whenever they flow, tidal lore loomed large in the Atlantic seaman's life. By the fifteenth century the set of the tidal streams and the rise and fall of the tides, which in places exceeded 5 fathoms—30 feet—had been established and had been codified to form a logical system of well-organized positive facts. Set down in writing, it was in constant and widespread use. It was based upon the observed fact that everywhere the highest and lowest tides—spring tides—occurred at or a few days after full and new (or change) moon, that the tides in most places occurred twice a day and approximately forty-eight minutes later each day so that a regular cycle of daily times of high and low water extended over a period of fifteen days or half a lunation. To reduce the problem of tidal prediction to manageable proportions, the time of high (sometimes low) water at various ports and havens of Europe on days of full and change (new) moon were recorded (as a result of generations of confirmatory observations), and simple rules were evolved for calculating the state of the tide at any given place at any time based upon the age of the moon at the time in question. It was in the sixteenth century that instru-

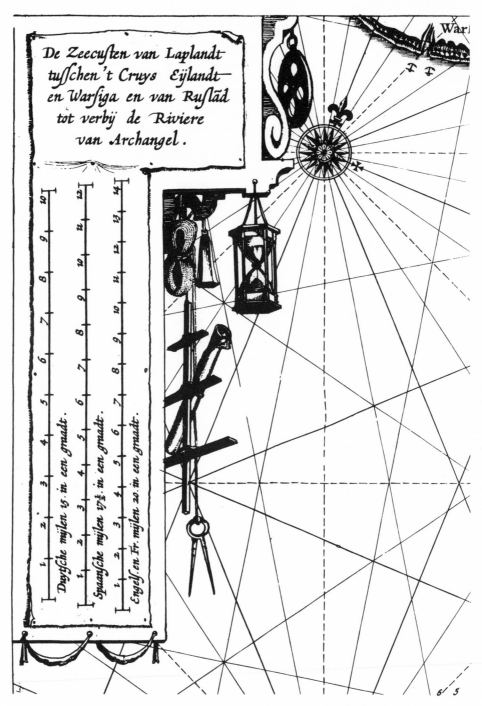

Fig. 1.—*Sandglass, lead and line, and chart. From W. J. Blaeu,*
The Sea Mirrour *(1625).*

ments to do the computations mechanically were invented for seamen—perhaps the earliest calculating machines.

Thus the pilot was directly concerned with the science of astronomy for the purpose of tidal prediction. In practice he was almost equally concerned with it for measuring time, for telling time by day and by night and for determining when there would be moonlight, the practical implication of which was highlighted, if I may use the term, as late as the close of the eighteenth century, by the nomenclature of the famous Lunar Society of Birmingham, whose members assembled for scientific discussion on moonlit nights in order to facilitate travel.

It follows that an almanac was almost indispensable to the Renaissance pilot. I say almost because, once again, he could learn by rote the rules for calculating the moon's age to determine the tides and moonlight, and how to do his sum by digital calculation, that is, with fingers and thumbs, in short, by rule of thumb. But how, without a mechanical clock on board ship, did the pilot tell the *time* of high or low tide? He did so by knowing the *bearing* of the moon (instead of the time) at the occurrence of high water and so expressed this in terms of the winds, rhumbs, or points of the magnetic compass. Thus the tidal establishment of a place—the time of high water on days of full and change moon—he expressed as, for example, "North-south moon," if it occurred at noon and midnight; as, "South-east, North-west moon," if it was at 9:00 A.M. and 9:00 P.M., and so on. For calculating the state of the tide when only the sun was visible, the pilot had another set of rules with which, knowing the age of moon, he could calculate its angular separation from the sun and, hence, its bearing, a calculation known as "to shift the sun and moon."

You will, of course, recall that, when the moon is full, it is in line opposite—in opposition—to the sun and that each day it slips back, as it were, about 12° away from the sun in an easterly direction; at the end of fifteen days it will again be in line but in conjunction with the sun and will be new. It is a consequence of this easterly retardation of the moon, which is equivalent to about forty-eight minutes of time each day, that the time of high water at a place occurs about forty-eight minutes later each day. The moon's retardation is a consequence of the earth's annual revolution about the sun in a west-to-east direction.

Although the moon was associated with tidal phenomena, the cause of the tides was not known, and it would be the end of the seventeenth century before it would be explained as the combined effect of the attractive forces of the sun and moon on the earth, in which that of the moon is dominant. Nevertheless I have devoted some time to the tides because the tide tables and associated almanacs compiled for the seamen of the Renaissance were scientific publications of crucial importance to the day-to-day business of transporting the bulk of Europe's commerce—the lifeblood of its communities.

On shore, public mechanical clocks had become widespread from the fourteenth century, indicating time from midnight by two periods of twelve equal hours each day, the time being based, of course, upon the apparent daily rotation of the sun around the earth. On shore the northern seaman kept this time and, at sea, he measured it approximately by taking the bearing of the sun. Thus noon was to him "South sun," 6:00 A.M. was "East sun." By this means he also correlated the tidal "bearings" with shore-side "hours," and in the sixteenth century, as portable and private as well as public mechanical tower timekeepers proliferated, high and low water were expressed in hours and minutes instead of bearings, with the exception of tidal establishments.

Another mechanical device introduced at this time which assisted this change was the nocturnal. This instrument consisted essentially of two volvelles, the larger having a handle, the lesser rotating around a hollow tube with a radial arm. Engraved peripherally with the months and days of the year on the larger volvelle and with the hours on the lesser, the nocturnal enabled the pilot to tell the time by the stars, the relative position of the pointers of the Little Bear, Kochab, or of the Great Bear, Deneb and Merak, to the polestar. By converting sidereal time into solar time mechanically, it was, in effect, an analogue computer of which the earliest dated example, one of 1517, is in the National Maritime Museum. Previously mariners had learned by rote the midnight position of selected stars at the beginning and in the middle of each month and had interpolated by memory to find the hour at other times. Thus, for example, when Kochab was on the meridian, vertically below Polaris, on the twentieth of October, he knew it was midnight.

For purposes of watch-keeping and for estimating the ship's speed and, hence, distance sailed, the pilot used a sand—or running—glass to measure the passage of time, usually in half-hour, hour, and four-hour intervals (the duration of a watch). Sand or filings partially filling an upper glass bulb emptied into a lower, similar one attached to it. The turning of the half-hour glass when it emptied was marked by the striking of the ship's bell as many times as the glass had been turned from the start of the watch.

Before I leave the matter of time, I should like to make two assertions which seem supportable by available evidence. First, the development of portable sundials capable of indicating local time by the sun in any latitude received great stimulus from the needs of the cargo-laden Renaissance navigator regularly plying the Atlantic, from the early sixteenth century, in all latitudes from the soundings off the English Channel to the Cape of Good Hope. Of these dials, the universal equinoctial ring dial perfected by the

English mathematician William Oughtred around 1600, out of the astronomical ring described by Gemma Frisius in the 1530's, was the most practical and popular.

My second assertion is that oceanic navigation demanded for its perfection the invention of an accurate seagoing mechanical timekeeper. From 1530, when Gemma Frisius propounded the method of determining longitude at sea with the aid of a mechanical timekeeper, all the great practical improvements in timekeeping, such as the gilding of watch parts against corrosion, introduced in the mid-sixteenth century, and the use of the pendulum and balance spring in the mid-seventeenth century, were inspired by attempts to perfect the means to find the longitude, primarily at sea.

In the thirteenth century the magnetic sea compass had been evolved from the iron-needle and lodestone into an efficient, self-registering, direction-indicating instrument fit for shipboard use in all weathers, as Professor Lane has so brilliantly demonstrated by his studies of the movements of shipping recorded in the medieval Italian port and other mercantile books. In the Renaissance the steering compass, mounted in gimbals and housed for protection in a binnacle, underwent no significant change, but the northern system of dividing the card into thirty-two points based upon four cardinal points—North, East, South and West—began to supplant entirely the Mediterranean eight-rayed card based upon the names of the eight prevalent Mediterranean winds. (See Fig. 2.) The great practical compass

Fig. 2.—Sea compass typical of the Renaissance period. (Courtesy of the Trustees of the National Maritime Museum.)

developments of the period were the detection by observation of magnetic variation—the deflection of the compass needle from the true meridian—in the middle of the fifteenth century and the development of instruments and techniques to measure magnetic variation at sea for oceanic navigation in the sixteenth century—made urgent by the needs of the Magellan voyage of circumnavigation of 1519–22. This practical work culminated at the end of the century in the development of amplitude and azimuth compasses and in the preparation of amplitude tables for the use of seamen in order to simplify the process of finding the variation. The measurement and tabulation of magnetic variation in the oceans were practiced by Portuguese seamen from the 1530's, and the data were used as aids not only to correct direction-keeping but also to determine progress in an east-west direction across the oceans. Many of these observations were published in English, French, and Dutch in 1599 and stimulated wide interest among scientists.[3]

All this measurement of magnetic phenomena was essentially scientific work; moreover, it was work the practical results of which were fed back into the art of navigation specifically to reduce the penumbra of ignorance surrounding it, which made its practice so hazardous. Thus, in the Renaissance, the light of scientific knowledge began to shine—fitfully, perhaps, nevertheless to shine—over the dark surface of the seaman's ignorance and aid him to make safer and speedier passages to the haven where he would be.

Before examining some of the other problems which the discovery of variation created, I should like at this stage to turn from the science of terrestrial magnetism and revert to the science of astronomy, but to do so indirectly by way of sailing directions and tide tables. I suppose that, next to astronomical tables and planispheres, nautical sailing directions and

---

[3] *The Haven Finding Art*, trans. E. Wright (London, 1599). Meanwhile, the practical problem of making reliable sea compasses had led Robert Norman to the scientific discovery of magnetic dip. "I have," he wrote in 1581, "set downe whatsoever I could finde by exact triall, and perfect experiments . . . foundyng my arguments onely upon experience, reason and demonstration, which are," he truly added, "the groundes of Artes," that is, science. It was Norman's pioneer work which inspired Dr. William Gilbert to undertake his great experiments in magnetism, published in *De Magnete* in 1600, while William Borough's published magnetic observations of the 1580's—he was a patron of Norman's and wrote a pamphlet on the compass for his book *The Newe Attractive*—followed up by William Gellibrand fifty years later, led to his discovery, published in 1635, of the secular–periodic–change of magnetic variation.

For a full discussion of this and the scientific development of navigation in England down to 1640 see D. W. Waters, *The Art of Navigation in England in Elizabethan and Early Stuart Times* (Hartford, 1958). It also contains a very extensive bibliography of contemporary works and of secondary sources many of which are relevant to the broader theme of navigational developments in the western hemisphere in the late fifteenth, the sixteenth, and the early seventeenth centuries.

charts—of which more anon—and tide tables constitute the earliest examples of organized, objective statements of observed scientific fact.

The exciting thing about the Renaissance is and was—as many men who participated in it were acutely aware—that, besides rediscovering much knowledge common to the ancients, men also found and learned to handle usefully knowledge unknown to or imperfectly assembled or processed by the scholars of the ancient world. The discovery of variation by measurement is a good example.

It is, I suppose, debatable when the Renaissance started, but as a maritime historian, I see a start in the thirteenth century with the revolutionary developments in navigation which, for the first time, enabled the ports and cities of northwest Europe to be linked with those of the Mediterranean directly by sea and which, in addition, made all-the-year-round sea-trading practicable for the first time in human history. Certainly, by the fifteenth century the art of navigation as developed by Italian scholars and seamen in the Mediterranean in the thirteenth century had been assimilated by Portuguese seamen whose shores, of course, flanked the Atlantic Ocean. By the 1420's, if not earlier, they navigated using, in addition to the lead and line, the tide table, the sea compass, the *portolano* chart, the *portolano* manuscript sailing directions giving magnetic compass bearings between places and the distances in miles, a sandglass to measure speed, a *toleta di marteloio* or traverse table to calculate the courses and distances to be sailed on various winds in order to make a desired landfall, dividers to prick off on the chart the ship's progress, and, of course, Arabic or Hindu numerals—algorisms— and geometry and trigonometry to calculate courses and distances to be sailed and actually made good. Unfortunately no Portuguese—or Spanish— and only a few Italian pilot books of the fifteenth century survive in addition to one Low German book and an English one. All these contain tidal information—the establishments of ports, direction of tidal streams—in addition to sailing directions.

The first pilot book to be printed was published in 1490 and was, I might add, in Italian, as might be expected: *Questa e vna opera necessaria a tutte li naviganti chi vano in diuerse parte del mondo*, Venezia, Bernardino Rizo, 6 Novembre, 1490; of this several copies survive. It is attributed to the Italian navigator Alvise Ca' da Mosta.

Meanwhile, the Portuguese had developed an entirely new form of navigation involving the use of nautical astronomy to assist in position-finding in the oceans. In consequence, they had invented new instruments; printed new tables, new rules, new books; developed new techniques and taught their pilots to use them in order to find their way with assurance across the trackless wastes of the unfathomable oceans whose bounds no man had yet discovered.

When, after the capture of Ceuta in North Africa by the Portuguese in 1415, Prince Henry the Navigator resolved to seek to circumvent the Saharan caravans which brought gold, spices, and slaves from the south to the northern littoral by finding a seaway to their sources of supply and to advance, at the same time, the cause of Christianity against Islam, he unwittingly also started a scientific revolution. Did I say "a"; am I too bold if I say "the"? In 1420, or thereabouts, he obtained the services of the celebrated Catalan cosmographer, cartographer, and maker of nautical instruments, "Mestre Jacome de Maiorca," probably the converted Jew formerly known as "Jafuda Cresques" and son of Abraham Cresques, author of the famous Catalan atlas of 1375, on which sources of the Saharan caravans' cargoes were indicated.[4]

When Prince Henry's seamen went south down the northwest coast of Africa into uncharted seas, the prevailing northeasterly wind prevented, as a general rule, return by the same inhospitable route: for hundreds of miles it was flanked with reef-ridden and waterless shores; thus, the nascent science of meteorology was given of necessity a new and urgent impetus toward growth. Cape Bojador was rounded in 1434. By then the Portuguese seamen must have discovered the key to Atlantic exploration, to Atlantic navigation and trade until the development in the nineteenth cenury of steam ship-propulsion: the pattern of the prevailing winds. They found by observation and experience—another word for *experiment*—first, that there were prevailing winds in the eastern Atlantic and close inshore off the coast of northwest Africa and, second, that they were systematic, those out in the ocean

---

[4] The Catalan atlas of 1375 contains on the first two leaves cosmographical information about the sun, the moon, the tides (with a circular diagram for determining the establishment of a number of ports); rules and diagrams to find the golden number and dominical letter, to determine the dates of the new moon and of the movable feasts, to tell the time at night by the revolution of the guards of the Little Bear; an exposition of the Ptolemaic system of the universe; and various astrological notations. It has been well called "the quintessence of the geographical ideas of the Middle Ages." See A. Cortesão, *Cartografia Portuguesa Antiga* (Lisboa, 1960), pp. 104–8. Professor Cortesão's studies in early Portuguese hydrography are of the highest order of scholarship and have brought a wealth of previously obscure and scattered material into the light of knowledge through a series of lavishly illustrated and lucidly organized publications, notably: A. Cortesão, *Cartografia e cartografos portugueses dos séculos XV e XVI* (2 vols.; Lisboa, 1935) which is basic; A Cortesão, "Cartographic indications of otherwise unknown early Portuguese voyages," *Actas II of Actas of the Congresso Internacional de Historia dos descobrimentos, Comissão Executiva das Comemorações do V centenário da morte do infante D. Henrique* (Lisboa, 1961), 111–33; A. Cortesão (ed.), *The Suma Oriental of Tome Pires . . . 1512–1515* and *The Book of Francisco Rodriques . . . written and drawn in the East before 1515* (2 vols.; Hakluyt Society, 1944); and A. Cortesão, and A. Teixeira da Mota, *Portugalia Monumenta Cartographica* (6 vols.; Lisboa, 1960). It contains facsimile reproductions of all known surviving Portuguese charts with elucidating text and a first-class index.

circulating in a clockwise direction. This was a scientific discovery of the first order and of immense, immediately practical importance—as the Portuguese authorities recognized by keeping it secret for as long as possible. It meant that their seamen could sail south with the assurance that, by sailing west into the ocean, they would find favorable winds to bring them north again and finally east to a Portuguese port. But how were they to find the port infallibly after many days of sailing immense distances on diverse courses out of soundings and out of sight of land? That was the scientific problem by the mid-fifteenth century.

A few years before the capture of Ceuta, an Italian scholar had brought into Italy in translation the long-lost geographical treatise by the second-century Greek scholar Ptolemy, describing and illustrating the mathematical construction of maps with a lattice of degrees of latitude and of longitude. Provided the requisite celestial observations had been taken correctly, this enabled the position of any place on the earth's surface to be defined scientifically in unambiguous numerical terms. Methods of determining latitude —the angular distance of a place north or south of the equator measured from the earth's center—had long been familiar in the Iberian Peninsula, where the thirteenth-century Alfonsine astronomical tables had been prepared as part of the magisterial *Libros del Saber de Astronomia del Rey Alfonso* of 1252–56, and others were in widespread use among astronomers.[5] Seamen were to be taught the use of this aid to position-finding. Therefore, in the mid-fifteenth century, astronomers, perhaps "Mestre Jacome" among them, began to teach Portuguese seamen the use of heavenly bodies for position-finding. This is easy to say, but it was hard to do. The astronomer was a scholar, mathematician, and theorist, and the pilot was a practical man of action; nevertheless, they had two things in common: both were trained observers—the one of the skies, the other of the seas and shores— and each was skilled in mathematical calculation using Arabic numerals (algorism) and also in plotting positions on charts or planispheres. Moreover, the motions of the heavenly bodies, although complex, were known empirically by pilots and learned by rote, as we have seen, for time-and tide-finding. Furthermore, seamen had used, time out of mind, the sun, stars, and moon for approximate direction-finding. But instruments and techniques for position-finding under the rigorous conditions of fifteenth-century shipboard life had to be evolved.

Hitherto the pilot had worked by "dead reckoning," plotting his estimated courses and distances made good on a portolan chart drawn on a system of magnetic compass bearings and estimated distances. This had been

---

[5] Isaac ibn Sid, of Toledo, prepared the tables. See Alfonso X, *Libros del Saber de Astronomia,* ed. M. Rico y Sindbas (3 vols.; Madrid, 1863–64).

introduced in the mid-thirteenth century. To compensate for the lack of parallel rulers, which were not invented until the 1580's, on it were drawn the characteristic network of directional rhumb lines; it had distance scales of leagues and miles for plotting. The pilot was now to be taught "altitude sailing." At first, probably in the 1450's, he was taught to use the height of the polestar above the horizon as a rough guide to change of position north or south. Thus, Ca' da Mosta, the Venetian pilot already mentioned, when on a trading expedition in Portuguese service, recounts how, in 1454, after leaving the Canaries, the ships sailed south for 200 miles, then, standing in toward the land, coasted, sounding constantly, until, off the River Gambia, "The polestar was about the third of a lance above the horizon." But the pilot's main problem once the *volta do mar largo*, as the sweep into the Atlantic to pick up returning winds came to be called, was discovered was how to find Lisbon. The simplest solution was to tell him the altitude of the polestar at Lisbon and teach him how to measure it—using, as was done, the simplest instrument, a quadrant with two vanes or sights and a plumb line measuring up to 90°. (See Fig. 3.) But the polestar was then about $3\frac{1}{2}°$ distant from the true Pole, so that in the course of twenty-four hours, its altitude at Lisbon changed by 7°. The problem was therefore not simple, for it was necessary to be able to determine the Pole's altitude at any time of night. Once again the familiar guard of the Little Bear, Kochab, was called in. The pilot was told that the polestar was not always in the same altitude at Lisbon and that, because the difference was small, the star's circumpolar movement could not be determined at sea owing to the unsteadiness of the ship but that it could be by observation of the larger sweep made by the "7 stars"—the Little Bear—in whose tail the polestar was to be found. (See Fig. 4.)

He was given the rules in the form of an imaginary figure of a man in the sky, the positions of the guards being related to his head, feet, right and left arms, and the intermediate spaces 45° "above" or "below" the arms. Simple diagrams were prepared on which the height of the polestar at Lisbon in relation to the positions of the guards was shown in Arabic figures; for example, "41" indicated that Polaris was 3° above the Pole when the guards were "in the feet," that is, on the meridian below the polestar, because the altitude of Lisbon is 38°. But confusion was sometimes caused by uncertainty as to the direction in which the figure was facing, and the rule was later refined into terms of the "west" or the "east arm" to avoid ambiguity.

Thus, sailing north, once being in the height, "altura," of Lisbon, the pilot had but to steer east, remember the rule of the North Star, and keep a good lookout in order to reach the Tagus. Here, at least, is a logical explanation of the rule of the polestar expressed in terms of its *altitude at*

Ebairo das tauoas precedentes, acharas q̃ se põe a letra Dominical, z o Bisserto, o qual a algũs não he muy claro, z por tãto breuemẽte diremos q̃ cousa he Bisserto z como serue em aq̃lle anno as letras Dominicaes, z quãdo se jejũa a vigilia, z se celebra a festa de sam Mathias. Pera o qual nota, q̃ o sol cũpre seu curso ou mouimẽto em .ccclxvj. dias z .vj. horas, ao qual chamamos hũ anno. E desta demasia destas .vj. horas, em quatro annos faz hũ dia, z daqui vẽ que se causa o Bisserto. E em bo quarto anno dia de sam Mathias, cõtamos sobre letra dous dias, z o primeiro dia se jejũa a vigilia, z em o .ij. dia se celebra a festa, saluo se cae em sabado, q̃ não se pode rezar, z passase aa segũda feyra, z jejũase ao sab.. do: z em aq̃lle anno teremos duas letras Dominicaes, z a primeira serue des do dia da Circuncisam ate o dia de sam Mathias, a segunda depois ate o cabo do anno.

¶ Seguese o regimento pera se poder reger pelo Quadrante ou Astrolabio pela estrella do Norte

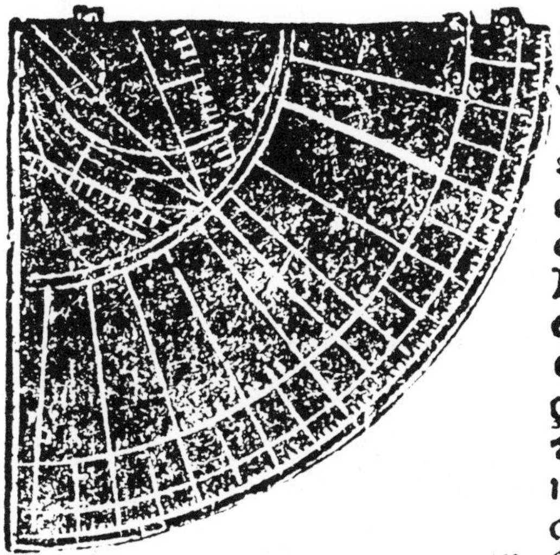

Saberas q̃ quando tomares a altura do Norte q̃ tanto mõte estar muyto ao este como pouco, z polo cõseguinte em leste, porq̃ tudo he bũ. ¶ Saberas q̃ em Lirboa todalas horas não estaa o Norte é hũa altura. ¶ Jté, as guardas na cabeça tẽ ho Norte .xxxvj. graos z bũ terço, z este nã se ba de meter em cõta, por causa da quantidade q̃ he pequena, z ho Mar nam segura a tomar esta conta, z assi has de fazer as outras alturas q̃ tomares. As guardas estando na linea esquerda, acharas o Norte em .xxxvj. graos. Se forem as guardas na luica do pee esquerdo, acharas o Norte em .xxxix.

Fig. 3.—Quadrant. *The earliest illustration of the seaman's. From Valentim Fernandez,* Reportorio dos Tempos (1518). (Courtesy of the Trustees of the National Maritime Museum.)

Fig. 4.—Shooting the Little Bear with a cross-staff. From Pedro de Medina, Regimiento de Navegacion (1563). (Courtesy of the Trustees of the National Maritime Museum.)

Lisbon. But as the coast was explored and islands were discovered, other points of call became important and the rule of the polestar was soon reformulated and the diagrams redrawn to tell the pilot the amount the polestar was above or below the Pole in degrees and, later, fractions of a degree in the eight positions of the guards. With this improved rule he could apply the correction to any observation of the polestar so as to find the "height" of the place of observation, wherever that might be in the northern hemisphere.

Out of "altitude" sailing now came another important development—the integration of angular celestial measurements with linear terrestrial measurements. The pilot was now taught to mark on his quadrant on leaving his port the altitude of the polestar and "one, two, or more days later," to observe the polestar's altitude with the guards in the same position, to note the angular difference, and to convert this into leagues. One degree, he was told, was equal to 16⅔ leagues—16 leagues and 2 miles—on the distance scale on his chart so that, probably with an abacus, he could calculate from his astronomical observations how far "south or south-west," as it was put, if the elevation was lower, he had sailed in leagues and miles.

Certainly in the 1450's Ca' da Mosta was measuring the altitude of the polestar, and he wrote highly of Prince Henry's encouragement of the study of the skies:

".... sapere che il primo inventore di far navigari á tempi nostri questa parte del mare Oceano verso mezzodi delle terre dé Negri della bassa Etiopia, è stato lo illustre signor Infante don Enrico de Portogallo.... il quale ancorché degli studj suoi nelle scienze delli corsi dé cieli, e di astrologia grandemente...."[6]

[6] J. Bensaúde, *Regimento* (1914), p. 11, n. i. Bensaúde made fundamental contributions to the study of nautical science in the Renaissance with *L'Astronomie Nautique au Portugal à l'Époque des Grandes Découvertes* (Berne, 1912) and with his wonderful series of facsimile editions of rare early nautical books: *Histoire de la Science Nautique Portugaise à l'Époque des grandes Découvertes*, comprising Vol. I: *Regimento do Estrolabio e do Quadrante tractado da Spera do Mundo* (Lisboa, about 1509); Vol. II: *Tractado da Spera do Mundo. Regimento da Declinaçam do Sol* (Lisboa, about 1517; Manual de Évora, Geneva, n.d.); Vol. III: *Almanach Perpetuum Celestium Motuum (Radix 1473). Tabulae Astronomicae Raby Abraham Zacuti, Astronomi Johannis II. et Emanuelis serenissimorum Regum Portugaliae* (Augsburg copy, edition of 1496, Leiria; Munich, 1915); Vol. IV: *Tratado del Esphera y del Arte del Marear, compuesto por Francisco Faleiro, natural del Reyno de Portugal* (Sevilla, 1535; Berne and Munich, 1915); Vol. V: *Tratado da Sphera com a Theorica do Sol da Lua e ho primeiro livro da Geographia de Claudio Ptolomeo. Tirados novamente de Latim em Lingoagem pello Doutor Pero Nuñez. Tratado que ho Doutor Pero Nuñez fez em Defensoam da Carta de Marear. Tratado que ho Doutor Pero Nuñez fez sobre certas Duvidas da Navegação* (Lisboa, 1537; Berne and Munich, 1915); Vol. VI: *Almanach Perpetuum Celestium Motuum (Radix 1473). Tabulae Astronomicae Raby Abraham Zacuti* . . . *Canons en Espagnol, traduction de Joseph Vizinho* (edition of 1496, Leiria; Geneva, n.d.); Vol. VII: *Reportorio dos tempos: tresladado de Castelhano em Portuguez per Valentim Fernandez. Com o regimento da declinação do sol* . . . *Tirada pontualmente do Zacuto pelo honrado Gaspar Nicholas* . . . (Lisboa, 1563; Geneva, 1919).

The oldest rule of all of the new navigation is found in the *Reportorio dos Tempos*, first published in Lisbon by the German scholar-printer Valentim Fernandez, in 1518. The bulk of the work was an astrological almanac which he had translated, for lack of time, he explained, in which to do more, from an almanac in Castilian—the *Reportorio de los tempos*, by André de Li, published in Seville in 1495. But to it Fernandez had added various rules of nautical astronomy and some tables. In fact, possibly because he was a keen historian, he included the oldest navigational rules while giving the latest nautical tables—as became a scientifically minded publisher with an eye to sales.

"Seguse o regimento pera se poder reger pelo Quadrante ou Astrolabio pela estrella do Norte" ["Here follows the rule to be able to guide yourself with the Quadrant or Astrolabe by the polestar"] is accompanied by the oldest surviving illustration of an instrument of the new navigation, a quadrant. It is, in fact, a simplified astronomer's quadrant, and one rule explains: "Saberas q em Lixboa toda las horas não estaa o Norte ē hūa altura" ["Know that at Lisbon the polestar is not always at the same altitude"]. Here too are given the length of a degree of 16⅔ leagues. But it also includes the next great leap forward in navigational development—the use of the sun "de saber quanto contra o nauio pola altura do Sol."[7] This became necessary as the equator was approached in the 1460's. At first the pilot was taught to measure its altitude at midday when it was south by the compass and at its maximum altitude for the day, to mark this altitude on the quadrant, and to remeasure it at noon "one, two, or more days" later "wherever he was." The difference, converted into leagues, it was explained, told him how far he had sailed "south or south-west" if the sun's altitude was higher, or how far north if it was lower. This crude rule suggests, incidentally, that the voyages must have been made about midsummer, when the sun's change in declination is slow, and that the use of the sun in this way was developed before the Gulf of Guinea—where the coast runs to the east—was reached; that is to say, in the 1460's. This is the more probable because the polestar was, in practice, lost to sight navigationally several degrees before the equator was reached—it was crossed in 1471—so that the need for some celestial body other than the polestar for altitude-finding had been felt urgently in the sixties.

Once Cape Verde is passed, the coast begins to trend to the southeast, finally running east for hundreds of miles to form the coast of Guinea. From here, therefore, navigation by "altura" began to break down; moreover, experience had already probably begun to show that the length of a degree

---

[7] Bensaúde, *Reportorio,* p. 142.

had been underestimated. A length of 17½ leagues of 4 miles to a league was in use among most pilots by the sixteenth century and had probably been introduced by the 1480's. Now navigation by "altura" was to be superseded by the universally applicable and more accurate navigation of "running down the latitude." "Latitude sailing" was, in fact, with but few later refinements, to become and remain for 300 years the chief method of oceanic navigation.[8]

The movement of the sun across the sky in the course of a year is not simple to apply to navigation. The combined effects of the tilt of the earth's axis and of its annual revolution around the sun cause the sun to appear to change its daily noon altitude. This change—solar declination—is greatest at the equinoxes—March and September, when it amounts to almost ½° a day—and least at the solstices—December and June, when, indeed, for several days, there is virtually no apparent change. The detailed alterations are impossible to memorize. Moreover, it is not self-evident to an observer whether declination has to be added to or subtracted from his observed altude of the sun in order to obtain his latitude. Consequently, detailed astronomical tables, suitable for seamen to use, *together with rules* on how to use them in order to correct altitude observations into latitude, would have to be prepared and pilots would have to be taught their use.[9]

At the University of Salamanca in Spain the great Jewish scholar Abraham Zacuto was working from 1473 to 1478 on his epoch-making *Almanach Perpetuum*, and his pupil, José Vizinho, was in the service of King Alphonso V of Portugal. On the King's death in 1481, the prince royal, hitherto in charge of the affairs of Guinea, succeeded him as King John II. Four years later Vizinho was in Guinea establishing latitudes by solar observations and using solar declination tables that he must have derived from Zacuto's *Almanach*. These, of course, were still in manuscript in Hebrew. Vizinho, like a good scientist, was experimenting in person with the new solar navigation tables and rules evolved, probably by him in the late 1470's and early 1480's, for the determination of position at sea by latitude observation.

In this same year, 1485, a very important meeting of scientists, including

---

[8] See Appendix II.

[9] By 1443 the Portuguese had reached Arguin, on the southern edge of the desert, 1400 miles from Portugal. The first trading house was built there on a barren island two years later. By 1462 the Portuguese had charted the mouths of the Senegal and Gambia Rivers and that year passed the highlands of Sierra Leone, advancing into tropical country new to Europeans. The Guinea—now Ghana—coast was explored by the Portuguese in 1470 and named from the gold obtained there, *A Mina*, "The Mine," later corrupted into *Elmina*. The site of a fort, "the Castle of St. George of the Mine," was selected by Diogo da Azambuja in 1482, and the fort was constructed that year with the labor of artisans brought out from Portugal. It provided a secure anchorage, with watering and careening facilities for ships. See A. W. Lawrence, *Trade Castles and Forts of West Africa* (London, 1963), pp. 30–31, 103–5.

Vizinho, was held in Lisbon. Their task was evidently to codify the new, scientific navigation in a form easily understood and practiced by seamen. From all the evidence available it seems probable that this *junta* decided that Portuguese pilots must be grounded in the theory of the celestial sphere and that they should learn this from Sacrobosco's (John Halifax of Holywood, fl. 1220–56) *De Sphaera*, popular among the educated since it had first been compiled in the thirteenth century. It was probably further decided that they should have a simple table of the sun's daily declination at noon, expressed in degrees and minutes north or south of the celestial equator in the course of a leap year; that this table should be prepared from Abraham Zacuto's *Almanach Perpetuum*, the latest and the best source; and that the pilots, having been taught to observe the sun's altitude at noon, should be given a set of simple rules based on the direction of the observer's shadow on the deck to enable him to know whether to add or to subtract the solar declination to an observed altitude in order to deduce the latitude. Without these latter rules the declination tables were virtually useless; thus, their codification and the simple tabulation of solar declination were two of the greatest contributions of Portuguese astronomical science to navigation. However, since the success of this nautical science depended upon a pilot's knowing what latitude to run down in order to get to his destination and if he were above or below his latitude, two further tables were necessary if he were to be able to find out easily how to reach or return to it. In consequence, as part of his navigational equipment, the pilot was provided with a table of latitudes of places (at first of places between the equator and Cape Finisterre (lat. "43"° N.), and with a table or rule to raise or lay 1° of latitude, often known as the regiment of rhumbs, and which may have been inherited from the technique of altitude sailing. The table of rhumbs gave him the distance in leagues which had to be sailed along the various thirty-two points of the compass in order to increase or decrease his latitude by one degree. The preparation of this table was, of course, a mathematical problem and was accomplished, in fact, by geometrical construction, for the requisite trigonometrical tables for its calculation had not yet been compiled. But it also involved splicing the new nautical science thoroughly into the old in the shape of establishing and expressing as accurately as possible the relationship between linear and angular length on the surface of the earth. It was now, I suggest, that the Portuguese chose a length of 17½ leagues of 4 miles—70 miles in all—as equal to a degree of latitude.[10]

[10] Their experience of altitude sailing must have shown them that 16⅔ leagues was too little to allow on the meridian for a change of 1° of latitude, and there is cartographical evidence to suggest that from about 1485 the African coast began to be more correctly de-

It is also probable that it was at this time that The Regiment of the North Star was reformulated in terms suitable for the determination of latitude, that is as an angular correction to be applied to observations of the polestar according to the orbital position of Kochab at the time of the observation. For example: "Itē quãdo as guardas estã na lynha abayxo da loeste: esta a estrella do norte açima do pollo tres graãos & meeo" ["and when the guards are in line below the arm in the west, the North Star is $3\frac{1}{2}°$ above the Pole"] (*Manual de Munich*). (By the old regiment the latitude of Lisbon—actually $38\frac{3}{4}°$ N.—was given as $42\frac{1}{2}°$.)

The *junta* of 1485 which—if it did not prepare—initiated the new navigation was probably called in a spirit of some urgency after the return of Vizinho in March from taking astronomical observations in Guinea. Diogo Cão had already reached latitude $13\frac{1}{2}°$ S.; a certain Christopher Columbus, who had put forward proposals for a westerly voyage across the ocean sea to Cathay, having had them rejected by a council including Vizinho, had departed a year before, leaving, perhaps, a legacy of doubt. Ten years before that the theory had been discussed between the Florentine scientist Paolo Toscanelli and the Portuguese scholar Fernan Martinez de Roriz. De Roriz had later been King Alfonso's confessor in Lisbon. Toscanelli had supported his views with a chart having a grid of latitude and longitude. The pressure for advances in navigational techniques must have been building up toward the bursting point in Lisbon in the early 1480's.

Among the experts advising King João II was the German Martin Behaim. His chief role appears to have been to explain to the Nuremberg financiers the Portuguese achievements and plans in order to persuade them to provide enough money to pay for the Portuguese voyages and to assure them that they would get a return for their expenditure. This, of course, was the purpose for which his terrestial globe of 1492 was made in Nuremberg. Another expert was Diego Ortiz. Ortiz was a Spaniard who in 1469 became professor of astrology in the University of Salamanca. He fled to Portugal in 1475 as a political refugee. Here his talent as a cosmographer found favor at the court, and he eventually held a succession of bishoprics. He was on the *junta* which rejected Columbus' scheme of 1483, and he planned the great reconnaissance (1487–90) of the East by Pero de Covilhan. He was also in the "junta dos mathematicos" of 1485.

---

lineated. However, not all pilots adopted $17\frac{1}{2}$ leagues to the degree: some preferred 18 leagues, an estimate only 4 per cent as compared to $7\frac{1}{2}$ per cent in error. Both were underestimates, but the prudent pilot preferred to underestimate his distance made good. This reduced the risk of making a landfall unexpectedly. A. Fontoura da Costa, *A Marinharia dos Descobrimentos* (Lisboa, 1933, 1939, and 1960), deals at length with the problem of units of length used at sea in the Renaissance and has a wealth of illustrative material relating to it and the Renaissance nautical problems and their solutions.

Ortiz had a great friendship with the Spanish astronomer Juan da Salaya, who in 1481 made the first Spanish translation of Zacuto's *Almanach*. As Beaujouan has put it: "Les relations étaient sans doute beaucoup plus étroites qu'on ne l'imagine entre Zacut, Salaya, Ortiz et Vizinho."[11] He also points out that the source of the Portuguese version of Sacrobosco's *Sphere* was probably a Castillian version associated with the works of Zacuto and that, therefore, "On en reviendrait une fois de plus au rôle joué par les savants en Salamanque, lors de la preparation d'une première ébauche de manuel nautique vers 1483."[12]

The elucidation of the stages of navigational development in Portugal in the Renaissance—which, I hope I have made clear, for the Portuguese was not so much a Renaissance as a new learning—is the result primarily of study of the oldest surviving Portuguese pilot books and almanacs of which the oldest printed pilot book, of about 1509, is known as the *Manual de Munich* and the second oldest as the *Manual de Évora*. Of the former a unique copy is preserved in Munich and entitled *Regimento do Estrolabio e do Quadrante pera saber ha declinaçam e ho logar do soll.... asy pera saber ha estrella do norte....* That of Évora, of about 1517, is *Seguese ho regimmento da declinaçam do sol pera per ella saber ho mareãte em qual parte esta....*

The Munich manual is a later printing of a book first published after 1493 and in about 1495 from manuscripts used in the 1480's to discover the sea route around Africa to the Indies, e.g., the solar declination table is for the year 1483.

Thus it was no accident that Bartholomew Diaz, in 1488, after making his westerly sweep into the South Atlantic, was able to round the Cape of Good Hope and fix his position astronomically by astrolabe observations on the southeast coast of Africa—first recorded as being made by Diogo da Azambuja in 1481—to enable Vasco da Gama to follow ten years later and reach the Indies.

Indeed, if we examine the sources of success of these classic expeditions of unprecedented length of time and distance (Diaz sailed over 5,000 miles), we find that besides brilliance of leadership the navigational equipment and training of the pilots were for those days superlative. Vizinho the scientist had been behind the navigational preparations for Diaz's voyage, using the astronomical work of Zacuto. Ten years later, when Zacuto was a religious fugitive from Spain in Portugal, to ensure the navigational success of Vasco da Gama's expedition, for which special high-charged ocean-going ships had been designed and built (see Fig. 5), not only was Diaz one of the experts

[11] G. Beaujouan, *Science Livresque et Art Nautique au XVe Siècle* (Paris–Lisbonne, 1960), pp. 16–17.
[12] *Ibid.*, p. 20.

*Fig. 5.—Cornelis Anthoniszoon, Portuguese carracks, about 1520–30.
(Courtesy of the Trustees of the National Maritime Museum.)*

employed in the preparations but both of the leading astronomers in the Iberian Peninsula were called in, Abraham Zacuto and José Vizinho. Gaspar Correa, in his *Lendas da India* (unpublished until the mid-nineteenth century), details very precisely how King Manuel consulted Abraham Zacuto about the navigational problems of the ocean route to India, the advice he received, and the practical steps Zacuto took, on his instructions, to reduce the risks to within the scientific bounds then practicable. These included not only the preparation of more accurate—4-year cycle—solar tables, for the years 1497–1500, for pilots, but also the design and manufacture for them of special instruments of observation in the form of small simplified brass as well as larger wooden sea astrolabes.[13] (See Fig. 6.)

In his "Armada dos Alboquerques que passarão á India, o ano de 1503," Correa emphasized:

.... todas estas armadas.... navegando polo regimento que dera o judeu Çacuto, que já os pilotos tinhão experimentado, navegando para outras partes a que El Rey a isso os mandara.

[In all these fleets the navigation was done with the astronomical tables and rules prepared by the Jew Zacuto, in the use of which the pilots were very experienced, and with them they navigated to all the other places to which the King commanded them to go.]

This, indeed, is science being used efficiently at sea and, be it noted, in the Renaissance.

[13] After this voyage the Portuguese evolved the cross-staff for celestial observations, probably from the Arabian pilots' *kamal*, which they met for the first time and which was used, in effect, by Arabian pilots to find their "altitude" from star observations. The cross-staff was used by the Portuguese for star observations to supplement or, if the sun was obscured at noon, to take the place of sun observations. See L. Albuquerque, "A determinação de latitudes por alturas de estrelas no náutica dos descobrimentos," *Actas* II, 429–50; T. Chumovski, "Uma enciclopédia maritime arabe do século XV," *Actas* III, 43–55; J. Custodio de Morais, "Determinação das coordenadas geográficas Oceano Índico pelos pilotos portugueses e árabes no principio do século XVI," *Actas* II, 475–521; J. Denucé, *Les Origines de la Cartographie Portuguaise et les Cartes des Reinels* (Amsterdam, 1963); A. Fontoura da Costa, *La Science Nautique des Portugais à l'époque des Découvertes* (Lisboa, 1941); S. Franco Garcia, *Historia de Arte y Ciencia de Navegar* (2 vols.; Madrid, 1947); S. Franco Garcia, "La Legua nautica en la Edad Media," *Instt*oo. *Historico de Marina* (Madrid, 1957); J. Guillén, "Las cartas de dos graduaciones en España," *Actas* II, 163–69; B. Penrose, *Travel and Discovery in the Renaissance, 1420–1620* (Cambridge, Mass., 1952), which is invaluable; L. Pereira da Silva, *Obras completas* (3 vols.; Lisboa, 1943–46), which contains fundamentally important articles on the history of Portuguese nautical science in the Renaissance which did so much to elucidate its origins and growth; A. Teixeira da Mota, *A evolucão da ciencia náutica durante os séculos XV–XVI na cartografia portuguesa da epoca* (Lisboa, 1961); and A. Teixeira da Mota, *Methodes de Navigation et Cartographie nautique dans l'ocean indien avant le XVI*e *siècle* (Lisboa, 1963).

ORIZONTE

Fig. 6.—*Meridian altitude observation of the sun by sea astrolabe.* *From
Pedro de Medina,* Regimiento de Navegacion *(1563). (Courtesy of the
Trustees of the National Maritime Museum.)*

By about 1500, probably as a result of learning from Arabian pilots in the Indian Ocean the value of star observations, the Portuguese had evolved the "Regimento da altura do polo pelo Cruzeiro do Sul" for finding latitude and also the "Regimento das horas da noite pelo Cruzeiro do Sul" for finding time by observation of the southern stars. These observations were valuable not only when the sun was obscured at noon but also when, owing to its proximity to the zenith, it was difficult to determine from the observer's shadow whether the observed latitude was north or south of the equator.[14]

The achievements of Spain in the development as distinct from the practice and use of the new science of navigation have, I think, in general been underrated. The great navigational contributions of Spain in the sixteenth century were, I suggest, to impose, to use a phrase coined by Quiller-Couch, "the charm of order" upon the acquisition of navigational knowledge and then to make that knowledge available to seamen of all nations.

Ten years after Columbus' return from his first voyage, the Spanish monarchs Ferdinand and Isabella established the Casa de Contratación at Seville to control trade and navigation between Spain and the Indies. Through the various scientific and nautical officers within the organization —Piloto-Mayor (see below), pilots, cosmographers, and cartographers—the Spaniards consciously systematized instruction in the difficult art of oceanic navigation specifically to ensure that their Atlantic pilots would be qualified in navigation, astronomy, and hydrography by experts and that their charts and instruments should be of the best. The government keenly recognized that the safety, regularity, and benefits of the Indies navigation depended absolutely upon the skill, experience and knowledge of the pilots of its ships.

---

[14] The Regiment of the Southern Star, "O Regimento do Cruzeiro do Sul" or "O Regimento da Estrela do Sul," was probably evolved by Pèro Anes about 1500. The "Estrela do Sul" was the star γ^2 Octantis, whose polar distance in 1500 was 5°, in the constellation of Octans. It was used by pilots in the southern hemisphere as they used Polaris in the northern hemisphere for finding latitude. The pointers or guards of the Southern Cross were α and γ Crucis at the head and foot of the constellation Crux. Their large polar distances of 29°.7 and 35°.7 made them quite unsuitable for latitude observations but, because their Right Ascensions were almost 180° (α Crucis, R.A. = 179°.97; γ Crucis, R.A. = 181°.08) and that of γ^2 Octantis was all but 360°, the guards of the Southern Cross were diametrically opposite the "Southern Star." As a consequence they made admirable pointers for the Southern Star, and their large polar distance was ideal for framing the rules to correct the Southern Star's observed altitude. See L. M. de Albuquerque (ed.), *O Livro de Marinharia de Andre Pires (1500–20)* (Lisboa, 1963). The Introduction, by a professional astronomer, is indispensable to the student of Renaissance nautical science.

The key official was the Piloto-Mayor, an office filled for the first fifty years by a succession of experienced navigators of high renown. The first, in effect, was the Castilian seaman and hydrographer Juan de la Cosa, who served from the foundation of the Casa in 1503 to the actual creation of the post in 1508. It was then filled, until his death in 1512, by Amerigo Vespucci, an Italian skilled in the Portuguese navigations. Juan Díaz de Solís, a Portuguese, followed until, in 1516, Sebastian Cabot, a Venetian, succeeded him to remain in office for nearly forty years, when, indeed, he absconded to England in 1548, primed with all the secrets of the Casa, and was bribed to reveal them to the English merchants and seamen. The office, vacant for four years, was finally filled by the scholar Alonso de Chaves (1552–87), and from thenceforward a succession of scholars, not seamen, held the post of Piloto-Mayor.[15]

The main tasks of the Piloto-Mayor were the instruction, examination, and certification of pilots; the maintenance of a master chart of all known seas, the *Padrón Real*; supervision of the compilation of navigational charts and certification as correct of those issued for use in the Indies trade; and examination and certification of the instruments made and used by the pilots.

In 1519 the office of Maestro de Hacer Cartas was created to cope with the growing volume of work, and four years later (1523) still another office— that of Cosmógrafo, Maestro de Hacer Cartas, astrolabios y otros ingenios de navegación—came into being. The first Maestro de Hacer Cartas, it is worth noting, was Nuño Garcia de Toreno, who had worked in the Casa as an illuminator since 1512, while the first Cosmógrafo was none other than the outstanding Portuguese hydrographer Diogo Ribeiro.[16]

Clearly the competence of these early officials was of the highest order. Let me instance one of the lesser officials also, Andres de Morales, appointed

---

[15] J. Pulido Rubio, *El Piloto Mayor de la Casa de Contratación de Sevilla, Pilotos Mayores, Catedráticos de Cosmografía y Cosmógrafos* (Sevilla, 1950), a quarry of important information of over 900 pages but without an index. It reprints many original relevant documents.

[16] Diogo Ribeiro made "all the charts" for Magellan's expedition, and "he makes," it was reported in 1519, "the compasses, quadrants and spheres, but he does not go with the fleet. . . ." See Cortesão and Teixeira da Mota, *Portugalia Monumenta Cartographica*, I, 87.

He was appointed first Cosmographer of the Casa by a *cédula* of the Emperor Charles V on July 10, 1523. In 1525 he was at Corunna fitting out a fleet, making "the nautical charts, spheres, planispheres, astrolabes and other things for India . . ."; see *ibid*.

A little later Fernando Columbus, head of the Casa in Seville, was ordered to use Ribeiro and other cartographers to make a nautical chart ". . . on which are located all the islands and continents discovered up to now and that will be discovered from now on . . . [as] standards (*padrones*) for all charts and world maps that will be made, and the said pilots must be ordered and it will be their duty to use them for their navigations." Ribeiro died in 1533. See *ibid*., pp. 87–89.

a royal pilot to the Casa in 1515. He had sailed with Columbus, probably on his third voyage; he had been pilot to Rodrigo de Bastidas, 1500–2; he had accompanied la Cosa on his voyage of 1504–6; and he had lived for some years in San Domingo when employed by the Governor, Ovando, to explore and chart the Antilles. While Díaz de Solís was Piloto-Mayor, his charts were considered the best procurable. Morales first formulated the theory of the clockwise circulation of the ocean currents in the North Atlantic, which so greatly assisted the navigation between Spain and the Indies, particularly the return voyage with the Gulf Stream.[17]

After nearly fifty years the office of Catedrático de Arte de la Navegación y Cosmografía was created in 1552, held first by Gerónimo de Chaves, son of the Piloto-Mayor. His function was to ease the latter's duties by taking over from him the instruction in navigation of the pilots of "la carrera de las Indias."[18]

It was only a few years later, when Philip of Spain was the spouse of Mary Tudor, Queen of England, that the English pilot, Stephen Borough, was officially conducted around the Casa de Contratación and returned home so filled with enthusiasm for the excellence of its organization for teaching the techniques of navigation that he was to recommend the setting up of a similar establishment in England. It is, therefore, interesting to see what the official curriculum comprised. Gerónimo de Chaves was instructed in December, 1552, that he had to teach pilots the sphere, or at least the two books of the first and second parts; to teach

"The Regiment of the Sun" and how it was applied;
"The Regiment of the Polestar" similarly;
"The Use of the Chart" and how to plot the ship's track on it and to determine the position;
The use of navigational instruments and their manner of making, so as to be

[17] C. H. Haring, *Trade and Navigation between Spain and the Indies in the Time of the Hapsburgs* (Cambridge, Mass., 1918 [and 1966]), p. 300. This work contains an excellent study of the navigational functions of the Casa de Contratación at Seville.

A fundamental work recounting in detail and analyzing the shipping movements between Spain and America is H. et P. Chaunu, *Séville et L'Atlantique* (1504–1650), comprised of, in part, Vol. I: *Introduction Méthodologique* (Paris, 1955); Vol. II: *Le traffic de 1504 à 1560* (Paris, 1955); Vol VI (Vols. I and 2): *Tables statistiques*, (1504–1650), *with* Guy Arbellot and Jacques Bertin (Paris, 1956); Vol. VII: *Construction graphique* (1504–1650), (Paris, 1957). The publications of The Hakluyt Society, c/o The British Museum, London, include many reprints in English of journals of voyages and of navigational books of the Renaissance with scholarly introductions and annotations and provide a ready source of study material relating to nautical science otherwise often inaccessible.
[18] His *Chronographia & Reportorio de los Tempos* (Sevilla, 1548), was popular into the 1580's.

able to determine whether an instrument is without error: namely sea compass, astrolabe, quadrant, cross-staff;

The theory and practice of each; their manufacture and use;

The points of the compass and how much northeasting or northwesting there is in each place, because this is one of the most important things to know, as are the equations and rules of navigation;

The use of the sundial and nocturnal, and why these are most important in navigation;

From memory or writing the age of the moon on every day of the year, in order to know when and at what hour it will be high water, to enter rivers or to cross bars; and other things of this sort pertaining to the practice and use of navigation.[19]

Of course regulation of scientific knowledge of immediate practical value ran the risk of stultifying research and experiment through overregulation and the sheer inertia inherent in all administrative organizations with centralized control; of course, too, the inevitable—under the circumstances—happened: from about the 1560's Spanish innovation in nautical science ended. It is perhaps significant that from about this time the chief scientific offices in the Casa were all held by scientists without sea experience.

The Casa de Contratación at Seville, there is much indirect evidence to show, was organized along lines similar to those already laid down by the Portuguese authorities in the fifteenth century to control and exploit economically their rapidly growing overseas commercial empire and to ensure the preparation of the latest navigational aids and dissemination of these aids and of information among the pilots of the ships conducting the trade. Unfortunately, through a variety of causes—fire, earthquake, and official secrecy at the time—Portuguese records of the fifteenth and early sixteenth centuries relating to nautical science are tantalizingly few, and more has to be inferred from the evidence of achievement than can be proved by reference to written records. It is, however, certain that by the end of the fifteenth century the Portuguese were conducting their navigational enterprises through an official organization most conveniently referred to as the Casa Da Mina. Here was a flourishing hydrographic office with an *armazem* or repository for confidential sea charts and navigational instruments used by the pilots. From 1531, when Pedro Nuñes was cosmographer-major, he was responsible for its hydrographic service; for maintaining an up-to-date master chart; for issuing charts, navigational instructions, and instruments to pilots; and for the examination and licensing of official chart and instrument makers.[20]

---

[19] Haring, *Trade and Navigation between Spain and the Indies*, p. 303.

[20] A. Cortesão, *Cartografia Portuguesa Antiga* (Lisboa, 1960), pp. 166ff., but see J. I. de

Although by the 1560's innovation was dying within the field of Portuguese, as of Spanish, navigational technology, consider what dividends the
highly organized investment in nautical science represented by the Casa da
India of Lisbon and the Casa de Contratación of Seville had paid: the former,
the discovery of the direct sea route to India, the East Indies, and their
fabulous riches; the latter, the systematic exploration of a New World and,
within a score of years of its foundation, the circumnavigation of the earth
(1519–21), a feat unparalleled in the history of man and comparable in its
audacity and technological excellence with the first manned circuits of the
earth in space. Like Vasco da Gama's epic voyage those of Magellan and
Del Cano were accomplished successfully as a result of meticulous and far-
sighted preparations by the Casa de Contratación at Seville, involving application of the latest scientific innovations by the best scientific brains ob-
tainable.[21]

By the 1560's the effects of what was perhaps the second great Spanish
contribution to navigation in the sixteenth century had begun to be felt
outside Spain. This was the publication of a series of textbooks on, principally, nautical astronomy and the art of navigation, textbooks of such superlative quality that some among them were translated into Dutch, English,
French, and Italian and remained in widespread use for a century. This
achievement justifies the claim of that great scholar of Spanish navigation
Admiral Guillén: "los principales y tradicionales paises marítimos *estudiaron
la Náutica en obras españolas*" ["The principal and traditional maritime
countries of Europe learned their navigation from Spanish works"].[22]

The first Spanish manual published was written by a lawyer who had
spent many years in the West Indies and recognized and felt acutely the
importance of navigation to the growth and prosperity of the Spanish empire.
The latter part of the title of Fernández de Enciso's work betrays the Portuguese origin of its navigational material: *Suma de geographia que trata de
todas las partidas y provincias del mundo: en especial de las indias, y trata*

Brito Rebello (ed.), *Livro de Marinharia, Tratado de Agulha de Marear de João de
Lisboa, [c.1514]* (Lisboa, 1903), pp. xxix–xxxi, who declares roundly that from the time
of Prince Henry to the latter half of the sixteenth century, "o inicio das navegações,"
"a escola do piloto era o navio" for the Portuguese.

[21] Among the Portuguese in Spanish service before and during the preparation for Magellan's expedition and in the following years, there were, besides Magellan himself and
numerous pilots, the cartographers João Dias de Solis, Estevão Gomes, Diogo Ribeiro,
and João Rodrigues, the cosmographers Simão de Alçacova Sotomaior, Francisco and Rui
Faleiro, and many others either temporarily or permanently; see Cortesão and Teixeira da
Mota, *Portugalia Monumenta Cartographica*, I, 20, n. 6.

[22] J. F. G. T., "Los Libros de Náutica en los años del Emperador," *Revista General de
Marina* (Madrid, Tomo 155, Oct., 1958), p. 508.

largamente del arte del marear: juntamente con la esphera en romance: con el regimiento del sol y del norte: nueuamente hecha. It was published by Juan Cromberger at Seville in 1519, with solar declination tables prepared originally for Vasco da Gama for the four-year cycle 1497–1500; two editions, with tables for the years 1529–32, followed in 1530 and a fourth in 1546. It was novel in that it included detailed sailing directions of the coasts and islands of the Spanish Indies from Cape St. Augustine in the North to the River Plate in the South. It thus contained the first overseas rutter (derrota) ever printed and, except for an English translation by John Frampton, of 1578, published in London as *A briefe description of the portes, creekes, bayes, and havens of the Weast India*, remained the only printed rutter of the New World, indeed of the New Discoveries, until Linschoten's *Itinerario* was published in Holland in 1596, the English translation of it in 1598, and the third volume of Hakluyt's *Voyages* two years later. The Portuguese published no *roteiros* until the seventeenth century.

Enciso's colophon expresses so perfectly his intention and his philosophy of navigation that I cannot resist quoting it in full:

Fenece la Suma de geografia con la esfera en romance y el regimiento del Sol y del Norte por donde los mareantes se pueden regir y gobernar en el marear. Asi mesmo va puesta la Cosmografia por derrotas y alturas, por donde los pilotos sabran de hoy en adelante muy mejor que fasta aqui ir a descobrir las tierras que hubieren de descobrir. Fué sacada esta suma de muchos y auténticos autores, conviene a saber: ... Tolomeos, a Erastótenes ... y otros muchos; y la experiencia de nuestros tiempos que es madre de todas las cosas ...

[Here ends the Suma de geografia with the treatise of the sphere in the Castilian tongue and the Regiments of the Sun and the North Star by means of which mariners can position and direct themselves at sea. Also included is cosmography for sailing directions and latitudes, by which the pilots can know whence and whither they are going far better than ever before in order to discover lands yet to be found. This suma has been compiled from many and reputable authors, such as: . . . Ptolemy, Eratosthenes . . . and many others; and with the experience of our own times, the mother of all knowledge. . . .]

Enciso's *Suma* is of particular importance, for it alone preserves the declination tables prepared by Zacuto for Vasco da Gama's voyage of 1497–99.[23]

The next manual—the second Spanish one—was written by a Portuguese

---

[23] The Manual de Evora, *Tractado da Spera* . . . . , of about 1517, has solar declination tables for the bissextile cycle 1517–20 based on Zacuto's *equationis, solis e declinationis* and places of the sun tables; those of the Manual de Munich, *Regimento do estrolabio* . . . , of about 1509, had a single table for the year March, 1483/84. See Fontoura da Costa, *A Marinharia dos Descobrimentos*, pp. 92–93, 98, 104.

scholar in Spanish service who was one of the scientists who many years before had contributed to the success of the first voyage of circumnavigation, Francisco Faleiro. His *Tratado del Esphera y del arte del marear: con el regimiēto de las alturas: cō algūas reglas nueuamēte escritas muy necessarias* was published in Seville in Castilian in 1535. It reprinted the solar declination tables for 1529–32, first found in the 1530 edition of Enciso's *Suma*; these were probably prepared by Ruy, the scientist-brother of Francisco Faleiro, who had been no less concerned with the scientific preparations for Magellan's voyage. The second part of Faleiro's *Tratado* is, I think, a most notable contribution to the art of navigation for the following reasons: for the first time he printed definitions of terms used in navigation, for example, his twelfth, "quando se dize derrota, se entiende el camino que por la mar se hazer o deue hazer," ["by the term *route* is to be understood the track or course steered across the sea or that has to be steered"]; because, for the first time, he set down clearly in print what the navigator has to know and use and do to find and keep track of his position; because, for the first time, he printed—with, moreover, a very clear diagram—the navigational consequences of magnetic variation upon the steering compass; because, for the first time, he discussed in print, and at length, the problem of magnetic variation and illustrated and described an instrument designed to determine it at sea and told how to use it; because, for the first time, he printed in simple diagrammatic form the rule to raise or lay 1° of latitude. (In fact, he provided two, one for 17½ leagues and one for 16⅔ leagues to the degree because, perhaps, some pilots still preferred the latter.) Here, at last, and for the first time, the means to acquire skill in the science of oceanic navigation were being provided to all men through the medium of the printed word and diagram without restriction, provided only that they were capable of reading Castilian. It was the first navigational *text*book published, the prototype "manual of navigation" indispensable ever since then to the would-be navigator—and instructor. Ten years later appeared one of the two most famous and influential textbooks of navigation ever written, Pedro de Medina's *Arte de Navegar*, published in Valladolid in 1545, in Castilian. In translation it was the navigation manual *par excellence* of virtually all European seamen— Italian, French, Flemish, Dutch, and, for a time, English—well into the seventeenth century. But the work that was most influential among Englishmen was Martin Cortes' *Breue compendio de la sphera y de la arte de navegar*, Seville, 1551 and 1556. Translated by Richard Eden as a result of Stephen Borough's visit to the Casa de Contratación, it was published in London in 1561, twenty years before John Frampton's translation of Medina; it was the work which navigated the English seamen to their meteoric rise to fame as, to use the words of a Venetian ambassador at the time of the Armada, "great sea dogs." More popular in Spain was Medina's *Regimiento* *de Navegación,* Seville, 1552 and 1563, probably because it was written expressly to enable pilots of the *carrera de las Indias* to qualify under the "Cátedra de Cosmografía y Arte de Navegar para la enseñanza de los pilotos" of the Casa created that very year, 1552. The simplicity, lucidity, and excellence of illustration of the *Regimiento* in both editions certainly place it in the first rank of good textbooks, and the rarity of surviving copies testifies to its popularity as a working navigational manual.[24]

I want now to turn to the role of science in hydrography in the Renaissance. There is plenty of evidence from Zurara and other sources that the Portuguese pioneer pilots were equipped with the best charts available before Cape Bojador was reached in 1434 and that hydrography was practiced in Portugal by then. All but three of the Portuguese charts of the fifteenth century have been lost, so that the cartographical evidence of Portuguese discoveries and hydrographical skill is largely found in Italian charts based on Portuguese originals. But one chart of about 1471 survives to show the navigational problem of determining position encountered then because of the easterly trend of the coast and the winds which necessitated a sweep out into the Atlantic here, the *volta da Guinea.* Fortunately another only recently discovered chart of about fourteen years later survives, also. This is of the time when the problem of navigation by *altura,* Diogo Cão's explorations to the south, and other pressures probably caused the scientific *junta* of 1485 to initiate the new navigation of running down the latitude. The delineation of the African coastline has, apparently, been improved as a result of astronomical observations, and a more realistic outline in place of the characteristic conventionalized series of curves of the portolan chart is noticeable. This chart is signed "Pedro Reinel . . ." and so first gives us the name of a Portuguese hydrographer.

The technique of running down the latitude called for charts with a scale of latitude, the oldest known of which is in fact Portuguese and is ascribed to Reinel. It is preserved in the Bayerische Staatsbibliothek at Munich and can be dated about 1500 or a little earlier. It is particularly interesting, for it exemplifies well the technical hydrographic difficulty of integrating the scale of latitude, based on accurate astronomical instrumental measurement, into a chart drawn originally on the basis of estimated linear distance and magnetic bearings uncorrected for variation. The chart is of the northeast Atlantic and Mediterranean. The scale of latitude is on the

---

[24] See M. Cortés, *Breve Compendio de la Sphera y de la Arte de Navegar,* (Sevilla, 1551; edición facsímil, Zaragoza, 1945); P. de Medina, *Suma de Cosmografía.* Edición facsímil del Manuscrito en la Biblioteca de la Catedral de Sevilla, 1947); P. de Medina, *Arte de Navegar* (Valladolid, 1545), edición facsímil; P. de Medina, *Regimiento de Navegacion* (Sevilla, 1563; edición facsímil, 1965).

left-hand edge. Based on the latitude of Lisbon (38° 44′ N.), it extends from 16° N. to 61° N. and, while accurate for the Atlantic coasts of Portugal and Spain, it is 4° too long overall, so that at the extremities the latitudes are about 2° out—Land's End is 52° N. instead of 50° N.

The Mediterranean is orientated in typical portolan chart style, the east-west axis being twisted through about 10° N.E.–S.W. as a result of the effect of variation on the compass bearings upon which it was based. The latitude of places at the east end is therefore 5° too high. However, the longitudinal distance Gibraltar–Alexandretta amounts to 43½° (instead of 41½°, the difference being explained by the reduced degrees of latitude). As Heinrich Winter has pointed out, what is also new—in addition to the latitude scale—is that the Mediterranean is for the first time given a length of approximately as many degrees of equatorial latitude as the number of its degrees of longitude.[25] The Portuguese knew the length of the equatorial degree from recently observed latitudes along the meridian, and the hydrographer has, in effect, produced an approximate equidistant rectangular or equidistant-cylindrical projection akin to that of Marinus of Tyre, applying for the first time the length of the equatorial degree to the longitudinal extension of the Mediterranean. In other words he has, to all intents and purposes, converted the bearing and distance portolan chart into a chart constructed as a mathematical projection of a part of the earth's sphere, one drawn on a plane surface in accordance with Ptolemaic principles.

We turn now to another important step in hydrography resulting from latitude sailing and first found in four planispheres by the Portuguese hydrographer Diogo Ribeiro, dated 1525, 1527, and 1529, that is, when he was *Cosmógrafo* and chief instrument-maker of the Casa de Contratación in Seville. They are all examples though copies of the *Padrón Real*;[26] all contain cosmographical information, drawings of a sea quadrant, an astrolabe (the earliest known of a mariner's) a *Circulis Solaris* or solar-declination diagram, and, for the first time, as a result of the demands of latitude sailing, the Mediterranean delineated with its longitudinal axis correctly orientated. (See Fig. 7.) This is, indeed, proof of Ribeiro's scientific knowledge and probity. In a large framed inscription he explains that "the *Levante* which we usually call what is contained inside the Strait of Gibraltar, is situated and laid down by its height [that is, latitude], according to people who have been in some of its parts and taken [the height of] the Sun: and in the rest I follow the cosmographers who have specially spoken of the latitude of some places. . . ."[27] Thus we can say that, from the time that

[25] Cortesão and Teixeira da Mota, *Portugalia Monumenta Cartographica*, I, Pl. 7.
[26] *Ibid.*, I, 92.
[27] *Ibid.*, I, 93, n. 18.

Fig. 7.—Diogo Ribeiro (Portuguese hydrographer), Planisphere, detail, the Mediterranean correctly orientated, 1529. (Courtesy of the Trustees of the National Maritime Museum.)

Diogo Ribeiro was *Cosmógrafo*, the *Padrón Real* of the Casa de Contratación at Seville began to be drawn on the basis of latitude observations and, as a consequence thereof, of true as distinct from magnetic bearings.[28]

But it was a long time before other hydrographers adopted Ribeiro's innovation. In 1559 Diogo Homem drew a chart of the Mediterranean in this manner, and from the 1560's the practice began to become general, but in the 1530's the failure of Diego Gutiérrez, one of the cosmographers in the Casa de Contratación of Seville, to draw his charts in conformity with the *Padrón Real*, which was then drawn on the basis of latitude and true bearings, led to a bitter dispute in the Casa de Contratación and the rejection of his charts. In the event, the hydrographic principles involved were thrashed out, notably by Sebastian Cabot, the Piloto-Mayor, Pedro de Medina, then one of the cartographers, and Diogo Ribeiro. It is a fascinating early example of a scientific approach to and solution of a problem of great practical importance with far-reaching results.

While the hydrographic effect of magnetic variation was to incline the east-west axis of the Mediterranean about 10° N.E.–S.W. off the coasts of what are now Nova Scotia, Newfoundland, and Labrador, the very large magnetic variation experienced at the time of their discovery by the Cabots, Venetians sailing from Bristol from the 1490's, and Portuguese pilots sailing from 1501 from the Azores, notably the brothers Gaspar and Miguel Corte Real, was to make the trend of the coast northerly instead of northwesterly. Furthermore, when charted on the basis of compass course and estimated distance, their plotted positions could not be reconciled with that of their observed latitude. If their latitude was measured from a scale drawn on a meridian in the eastern or central Atlantic, they were shown as lying too far south. This, of course, was the result of pilots steering courses with compasses affected by the prevailing westerly variation so that instead of, for instance, making good west when they steered west by compass, they steered, depending on the amount of variation experienced, W.S.W. or even S.W. It was Pedro Reinel who first hit upon an interim solution now known as "the oblique meridian," first found in a chart of his about 1504.[29] (See Figs. 8 and 9.) On this chart he drew, to seaward of the coasts, a meridian, with a scale of latitude on it, at such an angle (22½°) that when the pilot ran down his latitude to any place on these coasts he reached it. This was endorsing the scientific accuracy of latitude sailing indeed. Another device sometimes found on sixteenth-century charts is the double meridian or latitude scale. Two

---

[28] For a succinct account of other features see G. R. Crone, *Maps and Their Makers* (1953), pp. 93–94.

[29] Cortesão and Teixeira da Mota, *Portugalia Monumenta Cartographica*, I, Pl. 8.

Fig. 8.—Pedro Reinel (Portuguese hydrographer), Planisphere with oblique meridian, about 1504. (Courtesy of the Trustees of the National Maritime Museum.)

223

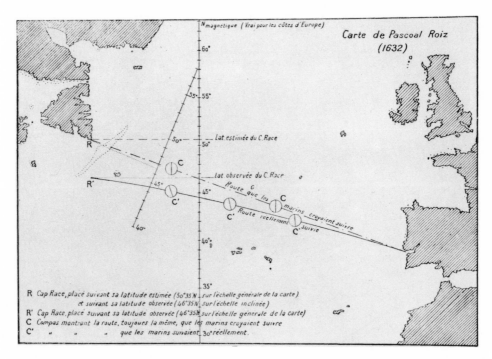

Fig. 9.—Oblique meridian, explanation. (Courtesy of the Trustees of the National Maritime Museum.)

latitude scales, one on either side of the chart, were drawn with the scale of one shifted several degrees above that of the other. A mid-century chart of the Mediterranean by the French hydrographer Jean Rotz exemplifies this.[30]

It was Diego Gutiérrez's continued use of the oblique meridian which led to the great hydrographic dispute in the Casa of the 1530's.

Why was the device of the oblique meridian necessary? The answer lies partly in the fact that few pilots knew methods of measuring magnetic variation at sea and, in consequence, had to steer magnetic courses and take, report, and plot magnetic bearings of places, and partly because variation was not understood.

The Portuguese pilot João de Lisboa wrote at length in his manuscript *Tratado da algulha de marear* of 1514 on magnetic variation, its consequences, and the need to correct it.[31] The oblique meridian was a hydrographical emendation but not a solution. A solution depended upon providing the pilot with the means to determine magnetic variation at sea by

[30] Reproduced in Waters, *The Art of Navigation*, Pl. XXIII.
[31] De Brito Rebello, *Livro de Marinharia*, pp. 20–24.

measurement, that is, scientifically. This statement is again a simplification of the problem as then seen. Not only was the cause of variation in dispute but its actuality was also, many seamen and some scholars maintaining that magnetic variation was a matter of compass manufacture and not a terrestrial or, as some thought, a celestial phenomenon.

João de Lisboa early in the century recommended measuring it by observation of the polestar or Southern Cross (knowing by the guards when they were on the meridian), and he used a primitive variation compass for measuring the angles between the meridian and the compass needle.[32] In 1519 the celebrated botanist Felipe Guillén invented an instrument which, by measuring the shadow cast by the sun, enabled the variation to be measured, while for Magellan's voyage, one of the Faleiros invented an "instrumento de marcar" or instrument for measuring the sun's azimuth (bearing) or that of the polestar or Southern Cross. Probably the most successful of these early variation instruments was designed by the Portuguese scholar *Cosmografo-mór* Pedro Nuñes, in 1537, for use by the scholar-navigator D. João de Castro during his voyage to India. This was mounted in gimbals and was used to measure equal altitudes of the sun before and after midday. Half the difference (if any) between the azimuths gave the deflection of the compass needle from the meridian.

Thus on June 18, 1538, D. João found the variation as follows:

Forenoon observation: alt. 20°, shadow = 24½° S.W.
Afternoon observation: alt. 20°, shadow = 56° S.E.
                                 Difference = $\overline{31½°}$
                                 ½ Difference = 15¾° N.E.

Later he found it by observing the sun's amplitude (bearing at sunrise and sunset); thus

August 24, 1538
At sunrise: on the horizon, shadow = 90° W.
At sunset: on the horizon, shadow = 74½° W.
                         Difference = $\overline{15½°}$
                         ½ Difference = 7¾° W.[33]

I have so far avoided the problem of longitude determination at sea, for the very good reason that the Iberian seamen avoided it also. The problem is and was known by scientists to be one of measuring the difference of time between places. In the fifteenth—and sixteenth—centuries the only possible way was by observation of conjunctions of celestial bodies and eclipses, but none of sufficient accuracy to produce reliable results could be

---

[32] *Ibid.*, pp. 38–39.
[33] Fontoura da Costa, *A Marinharia dos Descobrimentos*, pp. 189–90.

made with the instruments available. Mechanical timekeepers were found quite unreliable at sea. Practical solutions depended upon the invention by Galileo in the seventeenth century of the astronomical telescope, exploitation of his earlier discovery of the isochronism of the pendulum and of Hooke's (or Huyghens') balance spring, to mention only some of the prerequisite technological developments. Nevertheless, from the time of the preparations for Magellan's voyage, some charts did incorporate a longitude scale; the earliest known is on a planisphere of about 1519 by Jorge Reinel (son of Pedro) who, although Portuguese, was involved in the preparation of charts for the voyage though under rather obscure circumstances.[34] Each of Diogo Ribeiro's four planispheres had a longitude scale, but they were not generally used at sea. From the same voyage stems the first circumpolar chart.

Navigation by plane chart was never easy. Because the charts were drawn on a plane surface, the larger the sea area covered and the higher the latitude into which the chart extended, the greater were the errors inherent in its method of construction, simply because it is impossible to represent the curved surface of a sphere on a flat or plane surface without distortion. When a mathematical projection is made, it is possible to correct the effect of the positional distortions by the aid of suitable scales, but because the plane chart was not a projection but a geometrical construction which ignored the curvature of the earth, correction was impossible. The scale of leagues served for the measurement of distance over the whole chart whose meridians, had any been drawn in, would have been shown as parallel, equidistant straight lines and not converging lines, and the parallels of latitude, if drawn, would likewise have appeared as parallel, equidistant but horizontal lines. One solution was for a pilot to carry a globe with him, with parallels of latitude and meridians of longitude on it on which to keep track, and certainly John Cabot used a globe in the 1490's. But to be of much use, a globe had to have rhumb lines on it to give direction, which at first no one knew how to draw, and it needed to be too large for convenient stowage on board ship; nevertheless, small globes from 12 to 18 inches in diameter were carried. A geometrical diagram, such as João de Lisboa used early in the 1500's, showing the length of a degree of longitude in different latitudes was useful to convert departure (distance sailed east or west) into difference of longitude—for the convergence of the meridians on to the poles causes their linear distance apart to diminish with increase in latitude.[35] Pedro Nuñes first printed the equivalent diagram in 1537 and Martin Cortes the table—and a

[34] Cortesão and Teixeira da Mota, *Portugalia Monumento Cartographica*, I, Pl. 12.
[35] The length of a degree of longitude varies as the cosine of the latitude; thus in latitude 60° one degree of longitude is only 30 miles long (Cos. 60° = ½).

diagram with an explanation—in 1551.[36] But the labors of mathematical calculation without tables confined such *expertise* to the few. The rhumb or directional lines drawn on plane charts did not lead a pilot where he expected. The reason for this was not understood by seamen until after the Portuguese navigator Martin Afonso de Sousa returned from his voyage of 1530–33 along the coast of Brazil and complained to the *Cosmografo-mór*, Pedro Nuñes, about the difficulty of navigating with the plane chart. The scientist replied, probably in 1534 and 1536, in two treatises, the first, *Tratado sobre certas Duuidas da navegação*, the second, *Tratado em defensam da carta de marear;* and these were both published for the general benefit of seamen in 1537 by Germão Galharde of Lisbon, together with a translation of Sacrobosco's *Sphaera Mundi* as *Tratado da Sphera....*[37]

Besides explaining the problems of distance and of the convergence of the meridians, Nuñes elucidated for the first time the characteristic of rhumb-directional-lines: they cut all meridians at a constant angle. As the meridians converge, unless rhumb lines are themselves meridians—N.–S. lines—or are parallels of latitude—E.–W. lines—they are spiral, not straight, lines. Nuñes illustrated this characteristic with a woodcut diagram which, although not precise in detail, illustrates it vividly. The manner of constructing plane sea charts made it impossible to delineate the spiral nature of rhumb lines, but as a result of Nuñes' researches into them (see Fig. 10), a young Netherlands scientist, Gerhard Mercator, was able to draw them for the first time on the gores of a terrestrial globe, of 16½ inches diameter, which he published for the use of seamen in 1541 and which was popular for half a century (see Fig. 11), but the bulk and fragility of globes made them unhandy instruments of navigation at the best of times.

Nuñes also pointed out that, because of their spiral nature, rhumb lines do not provide the shortest course between places but that this is provided by great circles—circles whose planes pass through the center of the earth. Again science was coming to the help of the navigator. But a practical difficulty about great circle sailing—apart from the problem of favorable winds and currents—is that to follow a great circle track necessitates frequent alterations of course for the very reason that, unless the great circle being followed is a meridian or the equator, it cuts all meridians at different angles. On Mercator's globe these characteristics were easily demonstrated and the difficulties could be solved with the use of a flexible half meridian or *quarta altitudina*, but not so on the plane chart.[38]

[36] Fontoura da Costa, *A Marinharia dos Descobrimentos*, pp. 217–18; Waters, *The Art of Navigation*, pp. 73, 76.
[37] Bensaúde, *Histoire de la Science Nautique Portuguaise*, Vol. V: *Tratado da Sphera*.
[38] This device is clearly shown on the celestial globe in Holbein's painting "The Ambassadors," in the National Gallery, London. For the first known terrestrial globe see E. G.

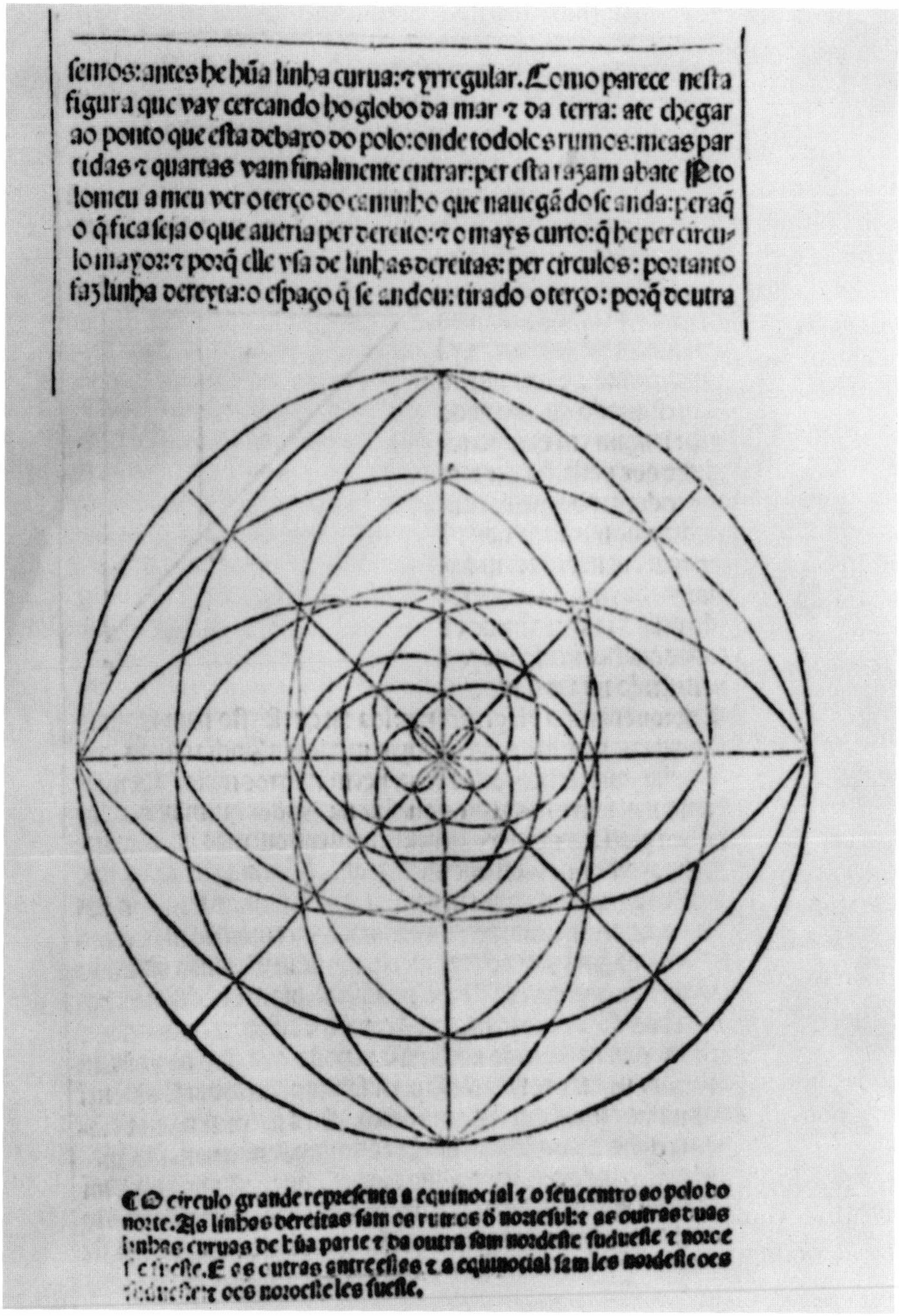
femos: antes he hũa linha curua: ⁊ yrregular. Como parece nesta figura que vay cercando ho globo da mar ⁊ da terra: ate chegar ao ponto que esta debaxo do polo: onde todolos rumos: meas partidas ⁊ quartas vam finalmente entrar: per esta razam abate Ptolomeu a meu ver o terço do caminho que nauegãdo se anda: peraq o q fica seja o que aueria per dereito: ⁊ e mays curto: q he per circulo mayor: ⁊ porq elle vsa de linhas dereitas: per circulos: portanto faz linha dereyta: o espaço q se andou: tirado o terço: porq de outra

O circulo grande representa a equinocial ⁊ o seu centro ao polo do norte. As linhas dereitas sam os rumos d nordesul: ⁊ as outras duas linhas curuas de hũa parte ⁊ da outra sam nordeste sudueste ⁊ noroeste sueste. E as outras entre estas ⁊ a equinocial sam les nordeste oes sudueste ⁊ oes noroeste les sueste.

Fig. 10.—Pedro Nuñes, Rhumb lines, 1537. (Courtesy of the Trustees of the National Maritime Museum.)

Fig. 11.—Gerhard Mercator, Rhumb lines on a globe, 1541. (Courtesy of the Trustees of the National Maritime Museum.)

Since a chart was the navigator's great stand-by on which he could actually see where he thought he was in relation to the land whether it was in sight or not and since a chart could be rolled up out of harm's way when not in use, some hydrographical solution to the flat chart problem was what the educated navigator most earnestly desired in the mid-sixteenth century. Gerhard Mercator produced it. In 1569 he published a great engraved chart of the world.[39] The claim of this immense wall planisphere, 78 by 53 inches in size, to imperishable fame lies in the fact that it was the first "chart on which *directions* were true, a chart on which a ship's course could be plotted accurately with a ruler on a straight line" and, I should add, distances and

Ravenstein, *Martin Behaim; his life and his globe* (London, 1908), which is basic. The standard work on globes is still L. Stevenson, *Terrestrial and Celestial Globes, their History and Construction* (2 vols.; New Haven, 1921). The largest collection of globes in the world is in the National Maritime Museum, Greenwich, England.

[39] The first portolan engraved on copper plate was Diogo Homem's, printed in Venice by Paolo Forlani (*La discrittione dell' Europa, et parte dell' Africa, et dell' Asia, secondo l'uso de naviga(n)ti del S. Giacomo Homen Portughese* [Oct. 1, 1569]).

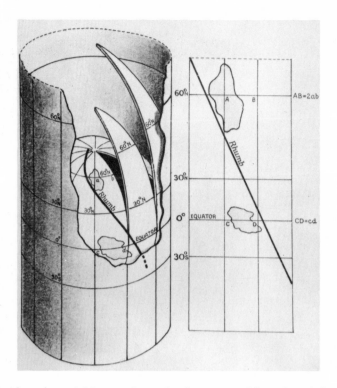

Fig. 12.—*Explanation of Mercator's projection (after Edward Wright, 1599).*
(Courtesy of Henry C. Taylor, of New York.)

latitude could be plotted accurately also.[40] But Mercator gave no explanation of the mathematical construction of his chart or of the distortion of its appearance of the land masses and seas in high latitudes, nor did he supply instructions as to how the navigator had to use its scale of "augmented" degrees of latitude for plotting distances. (See Fig. 12.) I doubt whether any seamen ever used it at sea.[41]

[40] See *The World Encompassed* (Walters Art Gallery: No. 132; Baltimore, 1952), Pl. XLI. It is a valuable annotated, illustrated catalogue of an exhibition of early maps and charts.
[41] As a matter of historical accuracy, Mercator was not the originator of augmented or increasing degrees of latitude. Two compass sundials with a map of Europe and North Africa engraved on the back with latitude scales of increasing degrees are preserved in Nuremberg. They were constructed in 1511 and 1514 by, it is believed, Erhard Etzlaub, an instrument, dial, and compass maker of Nuremberg; that is, he devised Mercator's projection nearly sixty years before Mercator. Even so, scientists and seamen had to wait nearly thirty years more before charts on "Mercator's projection" became practical tools for navigation. In 1599 Edward Wright, an English astronomer, scientist, and mathematician with practical sea experience, published in London *Certaine Errors in Navigation,*

In conclusion, I must tidy up the hydrographical story by drawing in the ends of some of the points which I have left untied. An undesirable feature of manuscript charts was their proneness to error introduced by careless or ignorant copyists. The printed chart eliminated this source of error and therefore helped to make hydrography more scientific. The first chart printed for the use of seamen on board ship was woodcut; it was a true portolano chart of the eastern Mediterranean, from the Strait of Messina and head of the Adriatic to the coast of Syria and the Sea of Marmora. Published in Venice in 1539 by G. A. Vavassore, a second edition appeared in 1542, and in 1558 a new edition was published by Matheo Pagano.[42] Of these charts unique copies are preserved of the 1542 and 1558 editions, none of that of 1539.[43] The first portolano chart to be engraved on copper plate appears to have been published in Venice by Paolo Forlani in 1569 and to have been taken from a manuscript chart compiled in 1560 by the Portuguese hydrographer Diogo Homem.[44] Nevertheless, sea charts engraved on copper plates would seem to have been introduced in the 1540's, for the only chart surviving of those produced by Sebastian Cabot is from a copper plate engraving printed in 1544.[45] Also the first French chart printed was from Nicolas de Nicolay's copper-plate engraving of his four-sheet "Nova et exquisita descriptio navigationum ad praecipais mundi partes," restricted to Europe and part of the coast of North Africa. It too was published in 1544.[46]

Printed sailing directions began to be published in northern Europe in the early sixteenth century. Between 1502 and 1510, in what may be termed the traditional make-up but confined to the waters of northern and western Europe, appeared the small French *Le routier de la mer* and, in 1520, the much more extensive *Le grand routier de la mer*—both having been compiled originally by a French pilot, Pierre Garcie, in 1485. An English translation of *Le routier* by Robert Copland, *The Rutter of the Sea*, was pub-

---

in which he explained the mathematical principle of what is now known as "Mercator's" but for long was called "Mr. Wright's projection," together with tables of meridional parts to enable charts to be constructed easily on it and an explanation of the manner of using the latitude scale—which increased in proportion to the amount that the meridians were distorted by being drawn as parallel straight lines—for plotting distance and position. A notable "Mercator" world chart was published at the same time. It is probably Shakespeare's "new map" referred to in *Twelfth Night*, Act III, Scene 2: "He does smile his face into more lines than is in the new map with the augmentation of the Indies."

[42] See L. Bagrow, *History of Cartography*, ed. R. A. Skelton (London, 1964), pp. 115–16. It is a valuable text which reprints a wealth of early charts and maps.

[43] The unique copy of 1542 is in the National Maritime Museum, Greenwich.

[44] Bagrow, *History of Cartography*, p. 116.

[45] *Ibid.*, p. 114. It is in the Bibliothèque Nationale, Paris.

[46] *Ibid.*, p. 117.

lished in London in 1528 and, from the 1530's, Dutch *leeskaarten* began to be printed. From the 1540's the latter, like Garcie's pioneer *Le grand routier* of the 1520's, was illustrated with woodcut coastal elevations to facilitate the identity of landfalls. Of these work-a-day pilot books a very few copies survive today.[47]

It was the succession of Charles of Ghent to the throne of Aragon in 1516 and, through the insanity of Queen Joan, in effect to that of Castille also that brought the Low Countries into the economic web spun by the Casa da Contratación at Seville, their seamen into direct and frequent contact with Spanish pilots practicing the new navigation, and their scientists—particularly after Charles became Emperor in 1520—such as Gemma Frisius and his pupil Gerhard Mercator, at Louvain University, into direct contact with the problems of oceanic navigation. Probably the most immediate result of the Spanish connection was the tremendous enriching of the Low Countries as a result of the port of Antwerp's becoming not only the staple town of the Portuguese African and Far Eastern trade but also the main European entrepôt for that of the Spanish empire. A navigational consequence was the demand for a printed navigational textbook in Dutch, reliable printed sailing directions in Dutch, and a printed chart for the North Sea and Baltic. So, probably in 1544, appeared Cornelis Anthoniszoon's *Onderwijsinghe vander zee om stuermanschap te leeren* . . . ["Instructions of the sea, to learn navigation (steersmanship) . . ."] with his *Heir beghint die Caerte van die Oostersee See* . . . ["Here begins the Chart (sailing directions) of the Baltic . . ."] and his woodcut *Caerte van Oostlant* published in Amsterdam a year before, which a note engraved on the sole surviving copy —of the second edition of 1652—makes clear was intended as a sea chart.[48] But the great era of Dutch cartography started in the 1580's, with what may be termed the large scale production of copper-plate engraved sea charts, and of these I shall say no more than that the first of the great series of "waggoners" was Lucas Janszoon Waghenaer's *Spieghel der Zeevaerdt* (1584) [*The Mariners Mirrour* (1588)] and that the production of cheap plentiful engraved charts marked a great technological advance in the practice of navigation.

[47] D. W. Waters, *The Rutters of the Sea* (New Haven, 1967). This reprints in facsimile these two French routiers, the English rutter translated from the earliest one and a transcription of an English MS. rutter of about 1460–80.

[48] Anthoniszoon was an artist—the first marine artist as well as a hydrographer. Of the first edition of his *leeskaart* no copy survives. The earliest now preserved is an edition of 1558 in the Harvard University Library. The first printed Dutch *leeskaart* is *Die Kaert van der zee* of 1532, containing sailing directions between Spain and the Baltic; a second edition appeared in 1540/41. See also J. Keuning, "Cornelis Anthisz.," *Imago Mundi*, VII (1954), and C. Koeman, *The History of Lucas Janszoon Waghenaer and his 'Spieghel der Zeevaerdt'* (Lausanne, 1964).

I began by claiming that civilization and science are the products basically of ship-borne trade and that this trade is made practicable by the art and science of navigation—by the use made by seamen of the means provided by scientists to enable them to conduct ships about the seas in a safe and timely manner. Professor Lane has exemplified this thesis by his brilliant studies of medieval Mediterranean ship-borne commerce.

I have attempted to show the scientific basis of some of the more important techniques of navigation developed in the Renaissance by, principally, Portuguese and Spanish scholars—scientists as we would say today. The consequences of these inventions were revolutionary, epoch-making; some might say cataclysmic. With their use, new worlds were discovered, new riches revealed, new foods cultivated, new medicines distilled, new arts assimilated, new commerce carried on, and the East was linked directly to the West by sea; from them the "Scientific Revolution" received tremendous impulse and momentum in the sixteenth century. I hope I have made the point that the techniques, instruments, almanacs, textbooks, charts, and globes evolved, invented, and compiled for the use of the seamen of Portugal and Spain and adopted or adapted later by other European seamen were the products of scientists, many of whom never set foot aboard a ship if—probably—they could avoid doing so, employed and paid to solve navigational problems. (See Fig. 13.)

But who paid them? Who were the administrative men of vision who saw that their employment would be gainful? As so often, we come back to the seldom romantic, often shadowy figures of the managers of affairs. Beaujouin, with the devastating clarity of the French scholar, has roundly stated, "The birth of nautical astronomy was much less a problem of science than of organization," and after detailing the specific preparations made using existing scientific knowledge, he concludes, "There, indeed, are the sources and first fruits of what, today, we would call 'scientific politics.'" King John the Second of Portugal (r. 1481–95) had, he points out, the immense merit of having known, *before any other head of state*, how to organize the technical exploitation of contemporary theoretical knowledge.[49] Teixeira da Mota has said much the same thing, namely, that the great merit of the Portuguese was to inject existing scientific knowledge, which was confined to a small and esoteric group, into the practice of navigation and to make that knowledge widespread for practical ends.[50] King Manuel (r. 1495–1521), who followed

[49] "La naissance de la navigation astronomique à été beaucoup moins un problème scientifique qu'une question d'organisation. ... Jean II de Portugal [a] l'immense mérite d'avoir su—avant aucun autre Etat—organiser l'exploitation technique des connaissances théoriques de son époque"; Beaujouan, *Science Livresque et Art Nautique au XVᵉ Siècle,* pp. 13–14.
[50] A. Teixeira da Mota, *A Evolução da ciência náutica* (Lisbon, 1961), p. 10.

LAPIS POLARIS, MAGNES.

Lapis reclusit iste Flauio abditum   Poli suum hunc amorem, at ipse nauitæ.

Fig. 13.—*I. Stradanus (sixteenth century), The scientist and nautical science.
(Courtesy of the Trustees of the National Maritime Museum.)*

John II, had this same ability. So, too, had those remarkable Spanish monarchs Ferdinand of Aragon and Isabella of Castille.

It has been claimed that the Casa da Contratación of Seville, founded in 1503, was "la primera institución científica cultural" ["the first scientific cultural institution"] in Europe.[51] I believe it was; I believe that it was because its organizations for navigational and hydrographic research and instruction were laid on sound scientific lines and lasted for two centuries—far into the epoch of the "Scientific Revolution" which it did so much, unwittingly and for mundane purposes, to initiate. I said "unwittingly to initiate." Am I right? When I read the dedications—the introductions—to the earliest manuals of navigation—Enciso's, Faleiro's, Medina's, Cortes'—I have my doubts; still more do I when I reflect upon the motives of the able administrators who set up the Casa da Mina of Lisbon and the Casa de Contratación at Seville. They were men who had a vision, but saw as through a glass, darkly, what the New—scientific—Learning might, in the fullness of time, enable men to achieve.

APPENDIX I

INVENTORY OF THE MAGELLAN EXPEDITION[52]

Cartas e instrumentos náuticos	*Maravedis*
Pago a Nuno Garcia para comprar pergaminho para as cartas	1.125
Uma dúzia de peles de pergaminho fornecidas a Nuno Garcia	900
Outra dúzia de peles de pergaminho fornecidas a Nuno Garcia	864
7 cartas facturadas por ordem de Rui Faleiro	13.125
11 cartas que fez Nuno Garcia, por ordem de Fernão de Magalhãis	11.250
6 cartas encomendadas por Rui Faleiro das quais uma destinuda a El Rei	13.500
6 quadrantes de madeira feitos por Rui Faleiro	1.121
1 astrolabio de madeira feito pelo referido Faleiro	750
1 planisfério que Magalhãis mandou executar expressamente para El Rei	4.500
Pago ao dito Magalhãis por 6 astrolábios metálicos com as respectivas réguas	4.500
Pago ao dito por 15 bússolas	4.080

[51] Pulido Rubio, *El Piloto Mayor*, p. 433.
[52] Visconde de Lagoa, *Fernão de Magalhais (A sua vida e a sua viagem)* (2 vols.; Lisboa, 1938), pp. 263–64.

Pago ao dito por 15 quadrantes de madeiro com aplicações de bronze	1.875
Estojo com compasso doirado enviado ao Rei conjuntamente com a otrás citada carta	476
Estojo em coiro para o planisfério	340
12 ampulhetas compradas pelo capitão	612
2 bússolas que o capitão possúi	750
6 pares de compassos	600
Pago a Nuno Garcia por duas bússolas	750
Pago por concertar uma bússola escangalhada	136
4 caixas grandes para 4 compassos, asquais fez Rui Faleiro	884
16 bússolas e 6 ampolhetas, enviadas de Cádiz por Bernaldino del Castillo	6.094

APPENDIX II

LATITUDE SAILING IN THE SIXTEENTH CENTURY

The narrative of Frobisher's return passage from the Northwest in 1577 gives an admirable illustration of the use of the new nautical science in conjunction with the old traditional methods of pilotage, that is, latitude sailing by celestial observation outside soundings, combined with lead and look-out when within soundings, the 100-fathom line:

Having spend foure or five dayes in traverse of the seas with contrarye winde, making oure souther way good as neare as we could, to raise our degrees to bring ourselves with the latitude of Sylley, we tooke the heighth the tenth of September, and founde ourselves in the latitude of [cypher] degrees and ten minutes. The eleaventh of September about six a clocke at night the wind came good south-west, we verde [veered] short and sette oure course southest. And upon Thursday, the twelfth day of September, taking the height, we were in the latitude of [cypher] and a halfe, and reckoned oureselves not paste one hundred and fiftie leagues short of Sylley, the weather faire, the winde large at west-south-west, we kepte our course southest. The thirteenth daye the height [latitude] being taken, we founde ourselves to be in the latitude of [cypher] degrees, the wind west-south-west, then being in the height of Sylley, and we kept our course east, to run in with the sleeve or channel so called, being our narrow seas, and reckoned as shorte of Sylley twelve leagues. Sonday, the fifteenth of September, about foure of the clocke, wee began to sounde with oure lead, and hadde grounde at sixty-one fadome depth, white small sandie grounde, and reckned us upon the backe of Sylley, and set our course easte and by north, easte north-easte, and north-east away. The sixteenth of September, about eight of the clocke in the morning sounding, we had sixty-five fadome osey sande, and thought ourselves athwart of Saint Georges Channel a little within the bankes. And bearing a small saile all nighte, we made many soundings, whiche were aboute fortie fadome, and so shallowe that we coulde not well tell where we were. The seaventeenth of September we sounded and had fortie fadome, and were not farre off the landes end, branded sande with small worms and cockle-shells, and were shotte betweene Sylley and the landes ende, and being within the baye, we were not able to double the pointe with a south and by east way, but were fayne to make another boorde [*tack*], the wynde beeyng at southweast, and by weast, and yet could not double the poynte, to come cleere of the landes ende, to beare along the Channell: and the weather cleered up when we were hard aboorde the shore, and we made the landes ende perfite, and so put up alongst Sainte Georges Channell: and the weather beeying very foule at sea, we coveted some harborough, bycause our steerage was broken, and so came to anker in Padstowe roade in Cornewall.[53]

In short, on his return Frobisher made his southing until he arrived in the latitude of Scilly, when he steered an easterly course along the parallel of latitude of his intended landfall until he sighted it.

[53] G. J. Marcus, *A Naval History of England,* Vol. I: *The Formative Centuries* (London, 1961), p. 62.

PHILOSOPHIES OF SCIENCE IN FLORENTINE PLATONISM ᷒ GEORGE BOAS

n considering any idea that is expressed in words, one can attend either to what it says or to the manner in which it is said. If anyone wishes to correlate these two aspects of ideas with content and form, no harm will be done if at the same time it is remembered that the two are reciprocally interdependent. That is, at least in most Indo-European languages, the form of a sentence modifies its meaning and the meaning of a sentence often determines its form. The two following sentences contain the same words but their meaning is very different and they are logically independent: "The cat sees the mouse," and, "The mouse sees the cat." I leave it to others to determine what is content and what is form in these two examples. But almost anyone would admit that a human being has to say them, that they have meaning only to human beings, and that only human beings who understand and can speak English could either apprehend or express their meaning. This has given rise to the philosophical question of how we know what form to give to the expression of our ideas, as well as the subsequent question of how we can tell whether they are true or false. Historically these questions go back to what historians have called the problem of Plato's *Meno*. But since I have no intention of beginning with the twin egg, let me say briefly that the problem is how we can know anything if we know nothing to start with.

As the Italian philosophies of the Renaissance developed, the two emphases became clearer. On the one hand you have the Neo-Platonists, who were more interested in those factors of knowledge which, they believed, could not have come from experience. On the other you have the Empiricists, who sought the origin or subject matter of ideas. I shall call Neo-Platonists men like Ficino, Pico della Mirandola, and Leone Ebreo. I shall call Empiricists Telesio, Campanella, and Pomponazzi. But such labels are misleading. No one in the whole history of philosophy was ever purely anything. In John Locke, for instance, who is often spoken of as the founder of British empiricism, we find the first two books of his *Essay* given over to the source of our ideas and an attempt to refute Herbert of Cherbury's epistemology. But when you come to the fourth book, you find that the mind is given powers of forming complex ideas through its own inherent processes, and the result is not very different, except in terminology, from what his Platonistic opponents had been preaching. But these opponents also, though they believed in something that they called innate ideas, were thinking with one exception of the form in which ideas were expressed rather than in their subject matter. The exception was clearly the idea of God. The other innate ideas were axioms of arithmetic, principles of comparison and contrast, and formulas believed to be basic to the formulation of sentences—in short, syntactical rules.

What a philosopher will call empirical varies. I should myself call a mystic a pure empiricist. And it would not be farfetched to say that a philosopher, if such there be, who thought that all knowledge came from sensory percepts was a mystic. The mystic needs no criteria to tell him that what he sees in the beatific vision is God. And the sensationalist needs none to tell him that what he perceives is a tree, a stone, a human being. The mystic asks no question about how he knows that he is face to face with God; the sensationalist asks none about how he knows how to classify his percepts. Such knowledge is supposed to carry its own verification with it. No color or sound or other percept has a temporal dimension that is perceptual; hence the knowledge of a past is always read into direct experience and empiricists of this type have always had to handle the problem of the past along with analogous problems.

I imagine that most people would agree that no single percept tells us anything beyond itself. As soon as it is translated into a sentence, principles of classification are involved whether one is aware of them or not. It is only in recent times that any philosopher has been willing to admit that human beings live in time, that they learn things, and that, on the basis of what they have learned, they anticipate the future. At the risk of putting too much emphasis upon the obvious, it should be always recalled that the child who is learning to speak is also learning how to organize his percepts. He is told that a certain object is a dog or a cat or a book. In short he is given by others the right name, always a common noun or adjective, for his perceptual experiences. He might of course develop a language of his own, but whether he could or not, in fact he inherits his language and absorbs it from other human beings. Consequently a habit of framing what I can call only by the old name of ideas is instilled into him from infancy and, like all habits, becomes compulsive. A dog, for instance, could be classified in a variety of ways, and they are not all zoological. He is a friend of man, we are told, a cat-chaser, a toy, a bow-wow, one of my possessions belonging only to me, a terror, a destroyer of goods, a household nuisance, and so on. Most of these properties are not mutually exclusive, and if we think of a dog as primarily a zoological specimen, that is probably because we are oriented from childhood toward zoological names. I doubt very much that a child would spontaneously think of his dog as a member of the genus Canidae and lump him together with wolves, foxes, and jackals. That would be of no interest to him. To a dairyman a cow is a milk-giver and not a member of the genus *Bos*. These various classes into which we situate our experiences depend, I suppose, on what sort of problem we happen to be dealing with. But I confess that the scientific taxonomist has the upper hand here, and it is often his label that is used rather than one that comes from folklore. One sees this clearly among amateur horticulturalists. They are not growing flowers either to sell or for botanical purposes; they grow them because they are pretty, smell good, and are at times difficult to bring to maturity. But nevertheless it is often a matter of pride with them to call their plants by their scientific names, especially when their interlocutors are ignorant of them but do know the common names.

To return now to the Renaissance, it may be of some interest to point out a few of the essential and characteristic details of the world of science as conceived at that time. First, and perhaps most important because seldom rejected, was the theory of the four elements: Earth, Water, Air, and Fire. This, as everyone knows, was an all-embracing theory, for with each element were correlated not only certain perceptual characteristics, but also dynamic properties on which a physics could be erected, and psychological traits extending throughout the animal kingdom, human beings, and even the planets. Since the details of the theory are expounded in dozens of texts, I shall merely say that as far as intellectual simplicity and elegance go, it was not far from perfection. Second, the main scientific interest was astronomy and after that alchemy. That these two interests were closely allied with magic is neither to be denied nor deprecated. For magic, insofar as it is of interest here, was a set of rules for gaining power over the world, and that was also Bacon's program and has remained the program of the applied scientists, engineers, physicians, and advertising men. Both astronomy and alchemy were riddled with superstition at that time, for it is impossible to distinguish between Renaissance astronomy and astrology, or between alchemy and chemistry. Third, human beings were still a special kind of animate object. They had not as yet been integrated into the animal world, and hence one could talk scientifically and not only theologically of human nature as *sui generis*. As the image and likeness of God, man had peculiar properties, and as the culmination of the six days of creation, he had a very special place in the universe. The traditional way of describing him was to say that he was midway between the angels and the beasts, and his main problem was the rise towards angelicity and flight from animality. Fourth, the universe as a whole was not a collecton of galaxies, but a closed system in which there were natural ranks, ranks which were not only proximity to God, but also levels of worth, *dignitas*. In the *Divine Comedy* these ranks, running from the lowest circle of Hell to the highest point of Paradise, were indicated in detail and, though there were disputes in the Middle Ages and indeed later about the precise location of the Earthly Paradise, no one of the disputants, as far as I know, denied its existence or the existence of the Heavenly Paradise. Finally, it was generally believed that the course of human history was determined by man's relation to God, especially by his obedience to the Commandments. Failure was attributable to his sinfulness and success to his virtue, though there were naturally different theories about what success consisted in. These five beliefs were, as I say, held in common by all who thought about such matters and contained within them certain principles in terms of which all science must be expressed. It will be noticed that though some of these beliefs had no direct logical connection with Christian theology, others did and that this connection had to be kept in mind. The doctrine of the four elements, humors, temperaments, had no obvious relation to the Bible, but nevertheless the stars were made of fire and the stars declared the glory of God and the firmament showed His handiwork. If the realm of fire was not really up there bounding the universe on this side of Paradise, then the closed cosmos was destroyed.

Among the principles which could not be rejected was one which both Professor Bober last year and Professor Singleton have spoken and which was repeated over and over again in the Middle Ages. It was contained in a verse from *Wisdom* (xi, 21):

Omnia in mensura, et numero et pondere disposuisti.

That verse determined in part the new science of Galileo and the mathematical school. For number, measure, weight, being all quantitative, provided a definition of all things, a definition which enabled scientists to skim off the perceptual world those qualities which were not measurable by number and weight. One of the philosophical problems of Renaissance scholars therefore became that of accounting for these non-quantitative beings. By Galileo's time they were relegated to a realm known as the mind, or later, the subjective. Here then was the beginning of that bifurcation of nature, to use Whitehead's term, which previously had designated one world that was open to everyone's inspection and the world of dreams, illusions, and hallucinations.

Among the three items mentioned was number. Consequently the interest of a man like Pico della Mirandola in the Cabala was understandable.

It was a futile interest from our point of view, for it led to nothing that would enable a man to control the world and refashion it to his desires. But there had always been and indeed always has continued to be a twofold interest in science, the interest in power over events and the interest in contemplating order. But order itself could be of several kinds, of which two were of especial importance: the order found in arithmetic and geometry and that found in teleology. Though the physicists of the sixteenth and seventeenth centuries refused to discuss the latter, sooner or later they would have to answer such questions as, "What is it all for?" I do not say that the question is any longer inevitable, but at that time men still believed that the final cause of the universe was the glory of God and that a hierarchy of beings led directly to Heaven. The minerals existed for the vegetables, the vegetables for the animals, and the animals together with everything below them existed for man. But men too had a purpose, the adoration of their Creator. The existence of God may not be a necessary hypothesis for an astronomer, but after all the men of the Renaissance were not so highly specialized as we are. They could not say, as we so frequently do, "We leave such things to the philosophers." For the scientists had their philosophic moments, and one is always uneasy about theories which seem to be in conflict with inherited beliefs. The reception of the Darwinian theory of the origin of species is a case in point. But this situation was not new. And if you have on your hands a body of revealed truth which you sincerely believe must never be contradicted, you are bound to seek some path that will lead to reconciliation.

That path, according to men like Ficino and Pico, was mapped by the author of the *Timaeus* and Plotinus, to say nothing of the minor Neo-Platonists, such as Iamblichus and Proclus. A world of weight and measure could give one no basis for a system of values, a system which would include virtues and vices as well as beauty and ugliness. But the third member of the triad, number, was a source of beauty to all Neo-Platonists, for it was supposed to reveal forms that were the hiding place for all values. We know very little of the teaching of primitive Pythagoreanism, but by the end of the Pagan period its evidence is clear. We find, for instance, in Philo Judaeus a long disquisition on the significance of numbers, just plain cardinal numbers, in which God's reasons for making the world as it is were concealed from vulgar eyes but revealed to those of the wise. To the wisdom of Philo it was clear that if there was to be any creation at all, it must come into being in some order. And the Bible tells us that it did come into being in six days. Why six? Because six is the first perfect number. Its perfection consists in its being equal to the product as well as the sum of the first three integers, 1, 2, and 3. Going back to a Pythagorean source that can be found in Aristotle's *Metaphysics*,[1] he knew that the odd was male and the even female. Hence 6 is both, since it is composed of both odd and even numbers. Why Philo thought that other integers differed from that, I do not know. But in any event he did, and he continues, "If the cosmos was to be the most perfect of all things that come into being, it must be composed in accordance with a perfect number, 6, and to have in itself the future genesis of things produced through the union of two, it must be imprinted with a mixed number, the first which is both odd and even, containing the idea of both the sowing male and the female which receives the sperm" (*Op. mundi* III, 14). In fact, it is not impossible that this part of Philo's treatise was inspired by the verse in *Wisdom* to which we have referred. But to correlate the Neo-Platonism of the Italian philosophers with this sort of numerical symbolism is only to indicate their folly, unless one appreciates to the full the importance of number as an indication of order and as a source of beauty. Renaissance science as it developed put more emphasis on weight and measure, and these too are calculated in numbers. One could indeed say, without more exaggeration than is essential if one is to group thinkers instead of expounding their views one by one, that the intellectual choice was made at the end of the fifteenth century between thinking of the cosmos as embodying beauty and the other values or thinking of it mainly as something that was to be controlled by man for his mundane satisfaction. The very word cosmos had the connotation of *beauty*, and there was sufficient Scriptural authority for maintaining that its beauty was such an inherent part of it as to induce philosophers to emphasize it. Emphasis on beauty would not lead to science as it was practiced in the seventeenth and eighteenth centuries, but it would lead to an appreciation of the world as a work of divine artistry.

Whether one uses the word *science* or *philosophy* to name the main interest of these men is unimportant, for the two terms at that time and in fact up to the end of the eighteenth century meant the same thing. They both denoted knowledge, and the difference between the Florentine Platonists and the Paduan physicists lay in the focus of their attention. Since the growth of laboratory technique the two foci have moved into opposition. The physiologist no longer harps on the beauty of the human body, the adjustment of means to end as exemplified in our organs, the significance of man's erect posture in religion, nor does he dwell upon our manual peculiarity as the source of our ability to use tools. Yet reasonable things have been said about all these items. As far as human welfare is concerned, we may have lost something by coming to think of ourselves as essentially physico-chemical conglomerations. It is, as far as I know, impossible to infer

---

[1] *Metaphysics* 986ᵃ, especially ll. 23ff.

from DNA or the position of our genes on the chromosomes the value of the arts or even of the sciences. Our elaborate social institutions sometimes seem presumptuous when one realizes that they are made by creatures whose organisms are largely controlled by the nucleic acids. On the other hand, one could never argue from our aesthetic and intellectual interests, assuming a fundamental difference between the two, that we carry within us these purely material agents. And if we insist on taking one element in our nature as basic, in the sense of furnishing evidence for everything we do, then we are a hopelessly divided set of animals. But on the empirical level we are both and neither one to the exclusion of the other. That there are causal relations between the two I am not denying, though no one to date has found out what they are. Yet the fact that the biochemist does not know why some men paint pictures or write poems or study biochemistry is no proof that these occupations are either illusory or inessential. To tack the adjective *essential* upon a man's main business in life is simply to play with words. The essence of things is determined by the purpose to which a student is going to put his information. The essence of a grain of wheat is one thing to the botanist, another to the dietician, a third to the farmer, a fourth to the speculator in the pit, a fifth to the baker, and a sixth to the man who is starving. No one of these essential characters is predominant, though some of them may be subordinate to others, in the sense that a regularly beating heart is essential to being anything.

There was, however, another side to Florentine Neo-Platonism which has often been disregarded. That is the side that insisted on the contribution of the human mind to knowledge. The language in which this was expressed is no longer fashionable, for we no longer believe in innate ideas or the transcendental unity of apperception. We are more inclined to accentuate those parts of knowledge which we believe to come into our minds from the external world, as if we were passive recipients of information. But in the Platonistic tradition it was seen that the sensory data themselves do not organize themselves without interference from the human being. It is we who put data together to systematize them. This was brought out as early as the publication of Plato's *Theaetetus*. The question that remains to be answered even today is whether we have a set of categories built into us, in terms of which we have to organize our data, or whether these categories change from time to time and from culture to culture. There seems to be a tendency today to grant that they may change, and what is now known as the sociology of knowledge seems to produce evidence of such change. Furthermore, whereas our forebears could think of Euclidean geometry as applicable to the cosmos as a whole, it is difficult for us to extend its theorems over so much space. But even if our basic principles of organization change, the question is still with us as to where they come from. To use an example that will not be emotionally traumatic, does such a principle as *ex nihilo nihil* come from observation, as colors and sounds do? Does one see, and by *see* I mean by using our eyes, nothing arising out of nothing? On the contrary, the principle was a corrective of what we see. For as far as the eyes are concerned, we see matter being destroyed every time we smoke a cigarette. In fact it requires some knowledge of elementary physics or chemistry to learn that our eyes have deceived us. Now it is true that Ficino and Pico would have located such a principle among the inevitable rules of thinking. They would have said that it can no more be set aside than the equality of identities can be set aside. In this they would have agreed with all the scientists of their time. But they would have explained such procedural rules as they did the axioms of mathematics: they were innate in the human mind and neither mutable nor acquired from experience. When Florentine Neo-Platonism entered England in the seventeenth century, this idea became part and parcel of the attack against empiricism. In Lord Herbert of Cherbury, for instance, the rules of thinking were the most certain of all epistemological data. They were listed right after the idea of God's existence. And that was the most certain idea possessed by man.

One of the cardinal principles of Renaissance Neo-Platonism was the subjectivity of relations. Pico states this as one of his *Conclusions*: *relatio nullam rem dicit extra animam.*[2] It will be noticed that he does not say *extra animum* but *extra animam*. I take it that he means that our relational judgments are not merely the assertions of an ego or subject or mind, but are an integral part of our souls. Now since relational assertions appear in mathematics as well as in all the sciences, one is forced to the conclusion that a large part of science is projected into the world of nature as a reflection of human psychic characteristics. I have found no explicit statement of this in Pico, no statement that says in so many words that terms like *equality, larger than, similar to*, and the like correspond to nothing in Nature, except this one *conclusio* or thesis which I have cited. It implies, however, that though we may see two colors, we do not see their similarity or difference but pronounce such relations because we find their equivalents in our souls. In English Neo-Platonism, on the other hand, the full implications of this point of view are clearly stated. The external world is simply a collection of atomic sensory data and, as we have said above, even John Locke, the strongest opponent of this, had to come round to something like it before the *Essay* was finished. Locke evidently did not fully comprehend what he was

---

[2] I quote from the *Opera Omnia* (Basle, 1557), p. 71. This is attributed to Mahumuth Tolletinus, but ultimately comes from *Theaetetus* 153ff.

saying. For if what he called the mind was making these relational judgments on its own and not because they were verified by any simple or complex idea passively received from the external world, then he was clearing
the way for Kant.

It is always dangerous to attribute to the dead intellectual motives which
are not overtly stated in their writings. But occasionally one is willing to
run the risk. If we take Pico as a leading and typical Italian Neo-Platonist,
we observe that his main interest was the discovery of how we think, what
our fundamental impulses consist in, and what are the universal motives of
mankind. Today such a study would be called social psychology, cultural
anthropology, or the sociology of knowledge. It might even end up in a
kind of cultural relativism, that bugaboo of most philosophers. But such
philosophers may console themselves, for none of the Florentine Neo-
Platonists went that way. Like most Occidental philosophers, they thought
that they were fair samples of the human race. Pico, however, in his attempt
to reconcile all the diversities of philosophy and religion and thus to un-
earth a set of universal traits, did assume that underneath or above or behind the diversity was unity. In another of his theses we find him saying,
*Singulare non intelligitur ab intellectu, nec secundum veritatem, nec se-
cundum etiam opinionem Aristotelis commentatoris Sc. Thomae,*[3] and this
is backed up later in his thesis, *In intellectu est hoc et illud, sed non
est hoc extra illud.*[4] Hence, as in Plotinus, everything was intimately
connected with everything else, and the separation between things was
attributable to our perceptions, not to Nature. That is, we see or touch,
for instance, individual things, but individuality disappears in the intellect.
It disappears because individuals are absorbed into classes for intellectual
purposes and the classes are absorbed into higher classes until we reach the
apex of the hierarchy, which in the terminology of these men was called the
One or Being. This One is utterly undifferentiated. Hence a philosopher or
scientist was to spend his time seeking the common unity of things and to
turn away from their diversities.

Unfortunately the unity in question existed on paper more than in
experience, and the philosophers in whom we are interested saw in this kind
of existence not simply fact but also value. In keeping with the Platonistic
tradition, there was held to be something inherently better in common
characters than in individuality. Hence the kind of problem that an empirical scientist would think of as critical seldom attracted the attention of
the Neo-Platonist. I mean the problems inherent in the observation that

---

[3] No. 5, p. 84, a *conclusio secundum propriam opinionem.*
[4] No. 17, p. 90. This is also *secundum propriam opinionem.*

things are not going according to rule. One can always obviate this difficulty by calling the deviation from the rule a trivial exception, or one can put the blame on the weakness of our senses and thus postpone answering the question. In the long run we all have to throw up the sponge and say, "These are the facts," or, if you prefer, "This is the will of God." But as Spinoza puts it, "The will of God is always the refuge of ignorance [*asylum ignorantiae*]," and one tries to stay out of that refuge as long as possible.[5] But there was another reason why Neo-Platonism was deaf to the appeal of experience.[6] Experience was attached to the material world, and matter was not only the lowest rung in the ladder of being but also an evil and an impediment to the uniformity of nature. In the world of pure ideas nothing could go wrong, though a human being might make an error in calculation or inference. The Neo-Platonists were skilled in the elaboration of ideal structures, like their contemporaries in architecture, painting, and sculpture. When we wonder, for instance, at their passion for perspective, we overlook the beauty that they were seeking and that they found in geometry. We should not underestimate this search, for it produced some of the most admirable works of art that the Occident has ever seen. But modern science was stimulated by the same motivation as magic or astrology, as we have suggested, and its practitioners wanted to control the world and not merely to look at it. Hence, though the Florentine Neo-Platonists perfected harmonious intellectual structures, they did very little toward perfecting man's technology.

When it was a question of understanding intricacies of the mutable world, they had recourse to allegory. The most impressive example of this is to be found in the *Dialoghi d'Amore* of Leone Ebreo. What later scientists would call attraction and repulsion, he called love and hate, using the old poetic terms of Empedocles.[7] Oddly enough, in spite of

---

[5] *Ethica*, Bk. I, Appendix; ed. Van Vloten and Land (The Hague, 1895), I, 69. See *Tractatus Theologico-Politicus,* in II, 27: *Ii igitur plane nugantur, qui, ubi rem ignorant, ad Dei voluntatem recurrunt.*

[6] It was standard practice to follow the example of the disciples of Plotinus and use the ancient myths and parables for metaphysical purposes. Ficino even went so far as to use the story of the Judgment of Paris as a parable of the Three Lives, the contemplative, the active, and the voluptuous: *sapientia, imperium, voluptas.* See Marcilio Ficino, *Supplementum Ficinianum,* ed. P. O. Kristeller, I, 80. An earlier Italian example of this technique is Coluccio Salutati's *De laboribus Herculis.*

[7] This requires modification. For Pico in his *In astrologia* iii. 19, points out that in practical matters, navigation, medicine, farming, we do better to rely on experience rather than on astrology. In such affairs we must deal with proximate causes. The Astrologer, he says (p. 502), *respicit . . . coelestem dispositionem, quae causa tantum universalis non efficit varietatem inferiorum, nisi pro materiae conditione causarumque efficientium inferiorum.* He continues, p. 503, to say *artifices autem quos memoravimus veris et naturalibus propriisque ac proximis causis indiciisque nituntur, et ideo raro excidunt a veritate, sed veris praesentionibus futurorum, et fidem suis methodis faciunt, et cum*

the mythological character of his writing, the two primordial forces of his universe were found in a strictly empirical manner. For no one could deny that love and hate were easily discovered, that everyone could easily recognize them and their effects, and that, by seeing in them two universal powers, he relocated human beings into the panorama of nature. No matter how objective an investigator tries to be, he will have to find some words to express his ideas. And if he tries to avoid this by going to mathematics, his symbols inevitably will reflect some basic empirical datum. For our purposes it makes little difference whether one speaks of force, weight, and motion, or of whatever algebraic symbols are chosen to represent them. The three things mentioned are felt by anyone who cares to pay attention to them, and their entrance into science came from daily life. One sees an analogue to this in the beginnings of modern chemistry. The early chemists are discredited nowadays to be sure; no professor of chemistry in a modern university speaks of spirits of wine or of the affinity of oxygen for metals. But go back to Robert Boyle and you will come upon fantastic names for various substances. One could with justice describe some of the nomenclature as echoes of alchemy, which in turn was an echo of Renaissance metaphysics. Some of these names are retained in modern non-scientific speech, as when we talk of *aqua vitae*, flowers of sulphur, or just plain mercury. We fortunately are not called upon to give the etymology of such words.[8] But if we were to indulge in this sport, we should find our steps turning toward astrology as well as to mythology.

The mathematical type of scientific inquiry which Professor E. A. Burtt has investigated so thoroughly and for the foundation of which Galileo is given so much credit was based upon a willingness to accept a large gap between science and experience. Every sentence that we pronounce, if it contains nouns and adjectives, will be bound to include such a gap. For unless we decide to express everything however incoherently related, we shall have to accept a genuine disparity between experience and science. The fact, for instance, that Euclid was able to speak of a circle without taking into account the material in which circularity was embedded, how far away from him it was, from what angle it was seen, what its color was, whether it was ugly or beautiful, meant that he was also willing to subtract from experience much that makes experience real. This ability was, to be sure, a great achievement, but it was also obtained at a heavy price, a price

---

proventu bonaque felicitate suas artes exercent. In short, he is not denying the hierarchical universe in which everything is intricately related to everything else, but insists that each level has its own causes and rules of explanation. See St. Augustine's attack on astrology in *Civitas Dei* iv. l.

[8] For more examples see Marie Boas, *Robert Boyle and Seventeenth Century Chemistry* (Cambridge, 1958), p. 157.

paid not by the intellect, but by our senses. On the other hand it was the one thing that made verbal communication possible. If one were to omit nothing from experience and insist on garnering the whole of it with every breath, one would have such a hodgepodge of perceptions on one's hands that no sense whatsoever could be made of it.

Ultimately the omissions or, if one prefers, the purifications of experience that are essential to science as we think of science are to be measured by what we intend to do with our knowledge. The engineer, the physician, the social reformer, to take but three applied scientists, cannot rise to such levels of abstraction as the physicist, the biochemist, or the sociologist. The likelihood is that all three will resort to a statistical generalization rather than to algebraic theorems. For they are confronted with material circumstances which modify the invariability of mathematical truths. I am not saying that the calculus of probability is not mathematical when I say this, and I trust that no one will think that I am. But even when you know what is the probability of a given type of event's occurring, you still do not know whether or not it actually will occur. All historical events are highly improbable; that is their differentia as contrasted with scientific events. Hence the physician takes the history of his prospective patient before he begins his diagnosis.

The Florentine Neo-Platonist was not so much interested in applying his knowledge to the control of the world as he was in finding an adequate picture of the world. And by "adequate" I mean a picture that would include as much as possible of the experienced qualities of the world. He took as his fundamental premise the incorporation in all experience of the moral and aesthetic values. The world had to be good and beautiful, and he must explain how it was that it had these qualities. This meant that he had also to utilize a technique of explanation which accepted the purposiveness of events (teleology) as well as a cosmic scheme which was hierarchical. The former he could derive from Plato, but the latter went back to Plotinus. In contrast with the kind of science which comes to us from Galileo and Newton and of which the philosophic apologists are Bacon and Locke, the men I have been discussing had a world which not only declared the glory of God but manifested His goodness. Teleology alone would not necessarily imply a single purpose in all events except in a very general sense. But when teleology was combined with a hierarchical cosmos, ranks could be identified with value, and the higher the purpose, the better it would be; the lower, the worse. Hence it was literally, not figuratively, lower to deal with minerals rather than with men. Science made progress by eliminating any considerations of value from its subject matter. The price it paid was the admission that the scientist had no responsibility for the effects of his occupation. It could be boldly said that what men did with the power that science gave them was no concern of the scientist. He had eliminated the values from his world, and to ask why things happened was a technical error. But it was only thus that he could purge the universe of anthropomorphism. In the fifteenth century, however, it was taken for granted that man was made in the image of God and that God's purposes, when not mysterious, were very much like man's, with the exception that they were never evil and were always fulfilled. It is worth recalling in this connection that the protagonists of the new science seldom saw the implications of their technique.

The purgation of teleology began in the Italian Renaissance and was not the contribution of Bacon alone. When Telesio began to explain all events as due to expansion and contraction, he also began to give non-teleological accounts of change. And when at the end of this trend Bruno was able to see each man as the center of the universe and an infinity of other worlds sprinkled about space, teleology was gone forever. It is too bad that there is not somewhere in Bruno a sentence to the effect that purpose is always a human purpose, that there are a variety of purposes in the world, each of which is the program of an individual human being, and that purpose, though real and as much a hard fact as the collision of material objects, is limited to the world of humanity. It is unfortunate that philosophers fail to foresee the needs of future historians, and all I can say is that the grounds for such an opinion existed in Bruno's books. But the most obvious source of the reaction against the theory of final causes is to be found in the various technological treatises which had begun to appear. For those were all concerned, as technology must be, with the aims of certain procedures. Yet their authors by the very nature of their program had to spend their time on means. And the means were either machines which would achieve the desired ends or processes which would eventuate in them.

The so-called universal man of the Renaissance, Michelangelo, Leonardo, Brunelleschi, Ghiberti, who has been celebrated by historians for his extraordinary knowledge of many arts, could not make an existential separation between knowledge and practice, though he could of course in speech. There were several discussions of the differences in both kind and quality between the liberal and manual arts, but each of these men practiced both. The various treatises on how to paint or write or design buildings, on the effectiveness of color in painting, and, before long, on the *ars combinatoria* demanded an exact knowledge of what we would call natural law, even when that law was, so to speak, unconstitutional. A man like Leonardo was not only interested in making careful drawings of clouds, leaves, and rocks, but also in designing all sorts of machines. For Michelangelo to design the dome of St. Peter's demanded a knowledge of engineering, and that knowledge could not be procured simply by inspiration.[9] Alberti's *Della Pictura*, which dates from the 1430's, has detailed suggestions which would serve the painter's ends as well as cautions to observe such natural phenomena as the green light that is reflected on the face of a person crossing a field of grass.[10] None of these men, with the possible exception of Leonardo, would appear in a history of philosophy, but at the same time they suggest vividly the intellectual atmosphere in which the philosophers lived. In short, what I am saying is that even if you are concerned with achieving a given purpose, you are forced to study the means which will accomplish that purpose. Those means turn out to be efficient or material, not final, causes. It is one thing to say that the planets move around the sun in perfect circles because the circle is nobler than any other figure; it is another to say that they so move because the sun attracts them as a function of masses and distance. But even if one says the latter, one can still maintain that their motion is *ad maiorem Dei gloriam*. It will be noticed in this connection that the main objection to magic was not that it would not work, but that it was wicked.

When one objects to the manufacture of cigarettes on the ground that smoking them will induce cancer, one is neither exclusively a teleologist nor a mechanist, but both. One is a teleologist in objecting to the end which the tobacco companies are serving by their product and also admitting the existence of one cause of cancer. Similarly the philosopher of science can combine in his views of natural events the concept of antecedent causation —or determinism, if one will—but also certain purposes which the causal series in question will further. Similarly a biologist who explains the action of the heart as serving to furnish blood to the rest of the body, or of the lungs as furnishing oxygen, or of the muscles as moving the bones, is not committed to denying the biochemical and mechanical elements in any of these processes. It is true that if we absorbed oxygen through our skin instead of through our lungs, we should have to conclude that one of the functions of the skin was something that we should call breathing. And we might even elaborate an analogy to the absorption of light through the leaves of plants. Purposes are of course limited and sometimes determined by the mechanisms that serve them, just as one cannot draw a picture with a dictionary or cook with pencil and paper. But that is irrelevant. The point is that to separate purpose from cause is useful in conversation, for sometimes we want to appraise ends as good or bad and not be bothered with means. Similarly we sometimes want to appraise means without considering ends. This often happens in both ethical and aesthetic discussions. But the fact

[9] See Paoli Rossi, *I Filosofi e le Macchine (1400–1700)* (Milan, 1962), pp. 24, 48, and 61.
[10] See John R. Spencer, *"Ut Rhetorica Pictura," Journal of the Warburg and Courtauld Institutes*, XX, Nos. 1–2 (1947), especially pp. 36ff.

that we can do this does not imply an existential separation between the two, as if an end could exist apart from the means of reaching it or a means without an end. Consequently when one is imbued with religious or metaphysical problems, there seems to be no reason for splitting the process in half and thinking only of one half and denying the existence of the other. To do so was the stumbling block of the men with whom we are concerned.

On the other hand there was a factor of realism in the teleological method of the Italian philosophers. For the only purpose discoverable in any empirical manner was human, just as the mechanical causes which they could discover were in the efficacy of the human will. Both terms are analogical when used in scientific explanations. That, I dare say, is why causative explanations were discarded later on for generalized descriptions. In such a system of explanation as that of Newton's *Principia*, the question of why things behave as they do is nonsensical. In the body of that work, as distinguished from the *General Scholium*, no problem is formulated about why gravitation should be measured by the product of the masses rather than by the difference between them or by the square of the distance rather than the cube or why the resulting fraction should not be inverted. Even to think of such a question is to think beside the point. One might just as well ask why we receive sensory impression through the sensory nerves rather than through the motor nerves. The *Principia* was concerned with finding the facts and not with integrating them into a larger conceptual scheme with religious relevance. This would, I think, be admitted by anyone. But to go beyond such a description is very tempting always and perhaps legitimate.

If I say "perhaps," it is because if we do venture in this direction, we land in the region of conjecture, indeed of myth, and we are all leery of such fantasies. We tend to forget that the philosophers of the Italian Reniassance believed in the Bible, not only as a source of ethical inspiration, but even as a document in geology, biology, and history. One of their basic problems was to make their new discoveries square with its teachings. It was, in other words, true. If one came upon a bit of information that did not meet its requirements, that information must be false, or allegorical. But few of us any longer think of Scripture in these terms, however devout we may be. It would be the height of hypocrisy to deny this. The intellectual struggle that went on in the sixteenth and seventeenth centuries was the struggle stimulated by this very situation. And though some of us think that the battle of free thought is won, we have only to read the daily papers to see how wrong we are. Hence when we find a man like Pico wondering why the world was created in six days rather than in five or eight, we can best understand his question as rooted in beliefs which he was unable to doubt. I might add that our contemporary scientists have their share of credulity too when they talk in awesome terms of the experimental method and similarly the students of the humanities when they talk about the Classics.

In fact the Florentine Neo-Platonists were the last philosophers to attempt the wedding of religion and science, or, if this is better, to attempt a generalized description of all cosmic facts. If they failed in their attempt, it may well have been because their problem was insoluble. When one begins to talk about all-inclusive wholes, one lands inevitably either in the negative theology or in allegory. And that is well illustrated by the works of Ficino, Pico, and Bruno, just as it is by those of Fichte, Schelling, and even Hegel. To talk of the love that binds the elements together or of the music of the spheres is not in essence less literal than talking of the masculine and feminine forces in Nature or of the absorption of all negation in the Absolute. But there comes a moment in philosophy when one yearns to go beyond the obvious, to think in terms that transcend logic, to paint a picture, if need be, rather than to reason. This may be reprehensible, but if so, then certain conclusions had best be drawn with precision, even if they kill all aspiration. One is that science can proceed only by abstracting from the mass of possible experiences a selected few, which was the technique of a man like Telesio and the so-called British Empiricists. This accomplished, one proceeds to use them as the basis for further constructions. A second is that the various groups selected need not be logically connected, a conclusion that has been utilized by those who speak of various kinds of truth, of the fundamental difference between faith and reason or of a pluralistic universe. A third is that we must abandon the idea of a universe which would include all things and be satisfied with knowledge about small areas, whether because of the existence of something called The Unknowable, something like Kant's things-in-themselves, or because of practical difficulties. A fourth is that the fine arts convey no information whatsoever which can reasonably be called true or false. A fifth is that what we think of as human values are created by human beings and not by the natural world and that consequently we must think of nature as simply a barrier to our imagination This is not unlike, but certainly not identical with, the conclusion of some Existentialists. Let me now leave these conclusions as my valedictory. They combine to suggest an alternative to the philosophic structure of the Florentine Neo-Platonists. There may be other possibilities too of which I have not thought, but these will suffice both to indicate what Neo-Platonism contributed to European philosophy and why our contemporaries find it so difficult to accept it *in toto.* Yet historians of ideas can see its basic premises passing into English philosophy through Lord Herbert of Cherbury to the English Platonists, from them to the Vicaire Savoyard, and from him to Kant. Whether that was a noble or a humble career, I leave to you.

THE HERMETIC TRADITION IN RENAISSANCE
SCIENCE 🙰 FRANCES A. YATES

f there is any characteristic by which the Renaissance can be recognised it is, I believe, in the changing conception of Man's relation to the Cosmos."[1] That is a quotation from a fairly recent book on *Science and the Renaissance*, the writer of which proceeds to inquire where we should look for the origins of a change in the climate of opinion in western Europe which could have produced this changed relation to the cosmos. He looks, naturally, first of all in the movement known as "Renaissance Neo-Platonism," originating in the renewed study of Plato and the Platonists in the Florentine circle of Marsilio Ficino, but he dismisses this movement as useless for his search. There is no evidence, he thinks, that the Florentine academicians had any but an incidental interest in the problem of knowledge of the external world or of the structure of the cosmos.[2] Yet the movement loosely known as "Renaissance Neo-Platonism" is the movement which—coming in time between the Middle Ages and the seventeenth century—ought to be the originator of the changed climate of opinion, the change in man's attitude to the cosmos, which was to be fraught with such momentous consequences. The difficulty has been, perhaps, that historians of philosophy may have somewhat misled us as to the nature of that movement. When treated as straight philosophy, Renaissance Neo-Platonism may dissolve into a rather vague eclecticism. But the new work done in recent years on Marsilio Ficino and his sources has demonstrated that the core of the movement was Hermetic, involving a view of the cosmos as a network of magical forces with which man can operate. The Renaissance magus had his roots in the Hermetic core of Renaissance Neo-Platonism, and it is the Renaissance magus, I believe, who exemplifies that changed attitude of man to the cosmos which was the necessary preliminary to the rise of science.

The word "Hermetic" has many connotations; it can be vaguely used as a generic term for all kinds of occult practices, or it can be used more particularly of alchemy, usually thought of as the Hermetic science *par excellence*. This loose use of the word has tended to obscure its historical meaning—and it is in the historical sense alone that I use it. I am not an

[1] W. P. D. Wightman, *Science and the Renaissance* (Aberdeen, 1962), I, 16.
[2] *Ibid.*, p. 34.

occultist, nor an alchemist, nor any kind of a sorceress. I am only a humble historian whose favorite pursuit is reading. In the course of this reading and reading, I came to be immensely struck by the phenomenon—to which scholars in Italy, in the United States, and in my own environment in the Warburg Institute had been drawing attention, namely the diffusion of Hermetic texts in the Renaissance.[3]

I must very briefly remind you that the first work which Ficino translated into Latin at the behest of Cosimo de' Medici was not a work of Plato's but the *Corpus Hermeticum,* the collection of treatises going under the name of "Hermes Trismegistus." And I must also remind you that Ficino and his contemporaries believed that "Hermes Trismegistus" was a real person, an Egyptian priest, almost contemporary with Moses, a Gentile prophet of Christianity, and the source—or one of the sources with other *prisci theologi*—of the stream of ancient wisdom which had eventually reached Plato and the Platonists. It was mainly, I believe, in the Hermetic texts that the Renaissance found its new, or new-old, conception of man's relation to the cosmos. I illustrate this very briefly from two of the Hermetic texts.

The "Pimander,"[4] the first treatise of the *Corpus Hermeticum,* gives an account of creation which, although it seems to recall Genesis, with which Ficino of course compared it,[5] differs radically from Genesis in its account of the creation of man. The second creative act of the Word in the "Pimander," after the creation of light and the elements of nature, is the creation of the heavens, or more particularly of the Seven Governors or seven planets

[3] The fundamental bibliographical study of Ficino's translation of the *Corpus Hermeticum* and its diffusion is P. O. Kristeller's *Supplementum Ficinianum* (Florence, 1937), I, lvii–lviii, cxxix–cxxxi; see also Kristeller's *Studies in Renaissance Thought and Letters* (Rome, 1956), pp. 221ff. The Hermetic movement is studied by E. Garin in his *Medioevo e Rinascimento* (Bari, 1954), pp. 150ff., and in his *La cultura filosofica del Rinascimento italiano* (Florence, 1961). The volume *Testi umanistici su l'ermetismo,* ed. by E. Garin (Rome, 1955), publishes some Renaissance texts containing Hermetic influence. The importance of the *prisca theologia* tradition in establishing Hermetic influence in the Renaissance is brought out by D. P. Walker in his article "The *Prisca Theologia* in France," *Journal of the Warburg and Courtauld Institutes,* XVII (1954), 204–59. D. P. Walker's book *Spiritual and Demonic Magic from Ficino to Campanella* (London, 1958), analyzes Renaissance magic particularly in relation to Ficino. In the first ten chapters of my book *Giordano Bruno and the Hermetic Tradition* (Chicago, 1964), I have endeavored to give an outline of the Hermetic tradition in the Renaissance before Bruno.

The best modern edition of the *Corpus Hermeticum* and the *Asclepius* is that by A. D. Nock and A.-J. Festugière, with French translation (Paris, 1945 and 1954).

[4] *Corpus Hermeticum,* ed. Nock and Festugière, I, 7–19. A précis of this work is given in my *Giordano Bruno and the Hermetic Tradition,* pp. 22–25.

[5] In the *Argumentum* before his Latin translation of the *Corpus Hermeticum* (*Opera omnia* [Bâle, 1576], pp. 1837–39). Ficino gave his translation the collective title of *Pimander* though this is really the title of only the first treatise.

on which the lower elemental world was believed to depend. Then followed the creation of man who "when he saw the creation which the demiurge had fashioned . . . wished also to produce a work, and permission to do this was given him by the Father. Having thus entered into the demiurgic sphere in which he had full power, the Governors fell in love with man, and each gave to him a part of their rule. . . ."

Contrast this Hermetic Adam with the Mosaic Adam, formed out of the dust of the earth. It is true that God gave him dominion over the creatures, but when he sought to know the secrets of the divine power, to eat of the tree of knowledge, this was the sin of disobedience for which he was expelled from the Garden of Eden. The Hermetic man in the "Pimander" also falls and can also be regenerated. But the regenerated Hermetic man regains the dominion over nature which he had in his divine origin. When he is regenerated, brought back into communion with the ruler of "the all" through magico-religious communion with the cosmos, it is the regeneration of a being who regains his divinity. One might say that the "Pimander" describes the creation, fall, and redemption not of a man but of a magus—a being who has within him the powers of the Seven Governors and hence is in immediate and most powerful contact with elemental nature.

Here—in the Hermetic core of Ficinian Neo-Platonism—there was indeed a vast change in the conception of man's relation to the cosmos. And in the Hermetic *Asclepius*,[6] the work which had been known all through the Middle Ages but which became most potently influential at the Renaissance through the respect accorded to the Egyptian Hermes Trismegistus and all his works, the magus man is shown in operation. The Egyptian priests who are the heroes of the *Asclepius* are presented as knowing how to capture the effluxes of the stars and through this magical knowledge to animate the statues of their gods. However strange his operations may seem to us, it is man the operator who is glorified in the *Asclepius*. As is now well known, it was upon the magical passages in the *Asclepius* that Ficino based the magical practices which he describes in his *De vita coelitus comparanda*.[7] And it was with a quotation from the *Asclepius* on man as a great miracle that Pico della Mirandola opened his "Oration on the Dignity of Man." With that oration, man as magus has arrived, man with powers of operating on the cosmos through magia and through the numerical conjurations of cabala.[8]

---

[6] *Corpus Hermeticum*, ed. Nock and Festugière, II, 296–355. Précis in Yates, *Giordano Bruno and the Hermetic Tradition*, pp. 35–40.
[7] As demonstrated by Walker, *Spiritual and Demonic Magic*, pp. 40ff.
[8] On Pico's yoking together of Magia and Cabala, see Yates, *Giordano Bruno and the Hermetic Tradition*, pp. 84ff.

I believe that the tradition which has seen in Pico della Mirandola's oration and in his nine hundred theses a great turning point in European history has not been wrong, though sometimes wrongly interpreted. It is not as the advocate of "humanism" in the sense of the revival of classical studies that he should be chiefly regarded but as the spokesman for the new attitude to man in his relation to the cosmos, man as the great miracle with powers of acting on the cosmos. From the new approach to them, Ficino and Pico emerge not primarily as "humanists," nor even primarily, I would say, as philosophers, but as magi. Ficino's operations were timid and cautious; Pico came out more boldly with the ideal of man as magus. And if, as I believe, the Renaissance magus was the immediate ancestor of the seventeenth-century scientist, then it is true that "Neo-Platonism" as interpreted by Ficino and Pico was indeed the body of thought which, intervening between the Middle Ages and the seventeenth century, prepared the way for the emergence of science.

While we may be beginning to see the outlines of a new approach to the history of science through Renaissance magic, it must be emphasized that there are enormous gaps in this history as yet—gaps waiting to be filled in by organized research. One of the most urgent needs is a modern edition of the works of Pico della Mirandola, an edition which should not be merely a reprint but which would trace the sources of, for example, the nine hundred theses. Though laborious, this would not be an impossible task, and until it is done, the historian of thought lacks the foundation from which to assess one of its most vital turning points.

It is convenient to consult the practical compendium for a would-be magus compiled by Henry Cornelius Agrippa as a guide to the classifications of Renaissance magic.[9] Based on Ficino and the *Asclepius*, and also making use of one of Ficino's manuscript sources, the *Picatrix*,[10] and based on Pico and Reuchlin for Cabalist magic, Agrippa distributes the different types of magic under the three worlds of the Cabalists. The lowest or elemental world is the realm of natural magic, the manipulation of forces in the elemental world through the manipulation of the occult sympathies running through it. To the middle celestial world of the stars belongs what Agrippa calls mathematical magic. When a magician follows natural philosophy and mathematics and knows the middle sciences which come from them—arithmetic, music, geometry, optics, astronomy, mechanics—he can do marvelous things. There follow chapters on Pythagorean numerology and on world

[9] H. C. Agrippa, *De occulta philosophia* (1533); see Yates, *Giordano Bruno and the Hermetic Tradition*, pp. 130ff.
[10] The *Picatrix* is a treatise on talismanic magic, originally written in Arabic, a Latin translation of which circulated in the Renaissance in manuscript.

harmony, and on the making of talismans. To the highest or supercelestial world belongs religious magic, and here Agrippa treats of magical rituals and of the conjuring of angels.

The magical world view here expounded includes an operative use of number and regards mechanics as a branch of mathematical magic. The Hermetic movement thus encouraged some of the genuine applied sciences, including mechanics, which Campanella was later to classify as "real artificial magic."[11] Many examples could be given of the prevalent confusion of thought between magic and mechanics. John Dee, for example, branded as the "great conjuror" for his angel-summoning magic, was equally suspect on account of the mechanical Scarabaeus which he constructed for a play at Trinity College, Cambridge.[12] In his preface to Henry Billingsley's translation of Euclid, Dee bitterly protests against the reputation for conjuring which his skill in mechanics has brought him:

And for . . . marueilous Actes and Feates, Naturally, Mathematically, and Mechanically wrought and contriued, ought any honest Student and Modest Christian Philosopher, be counted & called a Coniuror?"[13]

Yet there is no doubt that for Dee his mechanical operations, wrought by number in the lower world, belonged into the same world view as his attempted conjuring of angels by Cabalist numerology. The latter was for him the highest and most religious use of number, the operating with number in the supercelestial world.

Thus the strange mental framework outlined in Agrippa's *De occulta philosophia* encouraged within its purview the growth of those mathematical and mechanical sciences which were to triumph in the seventeenth century. Of course it was through the recovery of ancient scientific texts, and particularly of Archimedes, that the advance was fostered, but even here the Hermetic outlook may have played a part which has not yet been examined. Egypt was believed to have been the home of mathematical and mechanical sciences. The cult of Egypt, and of its great soothsayer, Hermes Trismegistus, may have helped to direct enthusiastic attention toward newly recovered scientific texts. I can only give one example of this.

In 1589 there was published in Venice a large volume by Fabio Paolini entitled *Hebdomades*. D. P. Walker has said of this work that it contains

---

[11] Tommaso Capanella, *Magia e Grazia*, ed. R. Amerio (Rome, 1957), p. 180; see Yates, *Giordano Bruno and the Hermetic Tradition*, pp. 147–48.

[12] See Lily B. Campbell, *Scenes and Machines of the English Stage during the Renaissance* (Cambridge, 1923), p. 87.

[13] H. Billingsley, *The Elements of Euclid* (London, 1570), Dee's preface, sig. A i *verso*.

"not only the theory of Ficino's magic but also the whole complex of theories of which it is a part: the Neo-Platonic cosmology and astrology on which magic is based, the *prisca theologia* and *magia*"[14] and so on. It represents the importation of the Florentine movement into Venice and into the discussions of the Venetian academies. The movement has not yet been adequately studied in its Venetian phase, in which it underwent new developments. When speaking of the magical statues of the Hermetic *Asclepius*, Paolini makes this remark: "we may refer these to the mechanical art and to those machines which the Greeks call *automata*, of which Hero has written."[15] Paolini is here speaking in the same breath of the statues described by Hermes Trismegistus in the *Asclepius*, which the Egyptian magicians knew how to animate, and of the work on automata by Hero of Alexandria which expounds mechanical or pneumatic devices for making statues move and speak in theaters or temples. Nor is he intending to debunk the magic statues of the *Asclepius* by showing them up as mere mechanisms, for he goes on to speak with respect of how the Egyptians, as described by Trismegistus, knew how to compound their statues out of certain world materials and to draw into them the souls of demons. There is a basic confusion in his mind between mechanics as magic and magic as mechanics, which leads him to a fascinated interest in the technology of Hero of Alexandria. Such associations may also account for passages in the *Hebdomades*, to which Walker has drawn attention, in which Paolini states that the production of motion in hard recalcitrant materials is not done without the help of the *anima mundi*, to which he attributes, for example, the invention of clocks.[16] Thus even the clock, which was to become the supreme symbol of the mechanistic universe established in the first phase of the scientific revolution, had been integrated into the animistic universe of the Renaissance, with its magical interpretations of mechanics.

Among the great figures of the Renaissance who have been hailed as initiators of modern science, one of the greatest is Leonardo da Vinci. We are all familiar with the traditional reputation of Leonardo as a precursor, throwing off the authority both of the schools and of rhetorical humanism to which he opposed concrete experiment integrated with mathematics. In two essays on Leonardo, published in 1965, Professor Eugenio Garin argues,

[14] Walker, *Spiritual and Demonic Magic*, pp. 126–27.

[15] Fabio Paolini, *Hebdomades* (Venice, 1589), p. 208. See also Agrippa's listing of the "speaking statues of Mercurius" among mechanical marvels, quoted in Yates, *Giordano Bruno and the Hermetic Tradition*, p. 147; and Dee's citation of the works of Hero followed by mentions of the brazen head made by Albertus Magnus and of the "Images of Mercurie" (preface to the Euclid, sigs. A i *recto* and *verso*).

[16] Paolini, *Hebdomades*, p. 203, quoted in Walker, *Spiritual and Demonic Magic*, p. 135, n. 1.

with his usual subtlety, that Vasari's presentation of the great artist as a magus, a "divine" man, may be nearer the truth.[17] Garin points to Leonardo's citation of "Hermes the philosopher" and to his definition of force as a spiritual essence. According to Garin, Leonardo's conception of spiritual force "has little to do with rational mechanics but has a very close relationship to the Ficinian-Hermetic theme of universal life and animation."[18] If as Garin seems to suggest, it is after all within the Renaissance Hermetic tradition that Leonardo should be placed, if he is a "divine" artist whose strong technical bent is not unmixed with magic and theurgy, whose mechanics and mathematics have behind them the animist conception of the universe, this would in no way diminish his stature as a man of genius. We have to get rid of the idea that the detection of Hermetic influences in a great Renaissance figure is derogatory to the figure. Leonardo's extraordinary achievements would be, on the hypothesis put forward by Garin, one more proof of the potency of the Hermetic impulses toward a new vision of the world, one more demonstration that the Hermetic core of Renaissance Neo-Platonism was the generator of a movement of which the great Renaissance magi represent the first stage.

In the case of John Dee, we do not have to get rid of a reputation for enlightened scientific advance, built up by nineteenth-century admirers, in order to detect the Hermetic philosopher behind the scientist. Dee's reputation has not been at all of a kind to attract the enlightened. The publication in 1659 of Dee's spiritual diaries, with their strange accounts of conferences with the spirits supposedly raised by Dee and Kelly in their conjuring operations, ensured that it was as a conjuror, necromancer, or deluded charlatan of the most horrific kind that Dee's reputation should go down to posterity. Throughout the nineteenth century this image of Dee prevailed, and it warned off those in search of precursors of scientific enlightenment from examining Dee's other works. Though Dee's reputation as a genuine scientist and mathematician has been gradually growing during the present century, some survival of the traditional prejudice against him may still account for the extraordinary fact that Dee's preface to Billingsley's translation of Euclid (1570), in which he fervently urges the extension and encouragement of mathematical studies, has not yet been reprinted. While I suppose that practically every educated person either possesses one of the many modern editions of Francis Bacon's *Advancement of Learning* or has had easy access to them in some library, Dee's mathematical preface can still only be read in the rare early editions of the Euclid. (Fortunately this

---

[17] Eugenio Garin, *Scienza e vita civile nel Rinascimento italiano* (Bari, 1965), pp. 57–108.
[18] *Ibid.*, p. 71. See also Garin's *Cultura filosofica del Rinascimento italiano*, pp. 397ff., for a similar approach to Leonardo.

situation will not long continue, for a long-awaited edition of the Euclid and its preface is planned for publication in the near future.) Yet Dee's preface is in English, like Bacon's *Advancement*, and in a nervous and original kind of English; and as a manifesto for the advancement of science it is greatly superior to Bacon's work. For Dee most strongly emphasizes the central importance of mathematics, while the neglect or relative depreciation of mathematics is, as we all know, the fatal blind spot in Bacon's outlook and the chief reason why his inductive method did not lead to scientifically valuable results.

It is not for me here to go through the mathematics of the preface nor to discuss Dee's work as a genuine scientist and mathematician, consulted by technicians and navigators. The work done on these matters by E. G. R. Taylor[19] and F. R. Johnson[20] is well known, and there is a remarkable thesis on Dee by I. R. F. Calder[21] which is unfortunately still unpublished. My object is solely to emphasize the context of Dee's mathematical studies within the Renaissance tradition which we are studying. That Dee goes back to the great Florentine movement for his inspiration is suggested by the fact that he appeals, in his plea for mathematics, to the "noble Earle of Mirandula" and quotes from Pico's nine hundred theses the statement in the eleventh mathematical conclusion that "by numbers, a way is to be had to the searching out and understanding of euery thyng, hable to be knowen."[22] And it was certainly from Agrippa's compilation with its classification of magical practices under the three worlds that he drew the discussion of number in the three worlds with which the preface opens. It may be noticed, too, that it is with those mathematical sciences which Agrippa classifies as belonging to the middle celestial world that the preface chiefly deals,[23] though there are many other influences in the preface, particularly an important influence of Vitruvius. This may raise in our minds the curious thought that it was *because*, unlike Francis Bacon, he was an astrologer and a conjuror, attempting to put into practice the full Renaissance tradition of Magia and Cabala as expounded by Agrippa, that Dee, unlike Bacon, was imbued with the importance of mathematics.

---

[19] *Late Tudor and Early Stuart Geography* (London, 1934), pp. 75ff.

[20] *Astronomical Thought in Renaissance England* (Baltimore, 1937), pp. 135ff.

[21] I. R. F. Calder, "John Dee Studied as an English Neoplatonist" (unpublished Ph.D. thesis, University of London, 1952).

[22] Dee's preface to the Euclid, sig. *i verso. See Yates, *Giordano Bruno and the Hermetic Tradition*, p. 148; also my note in *L'Opera e il Pensiero di Giovanni Pico della Mirandola nella storia dell'umanesimo* (Convegno internazionale, Mirandola 15–18 Settembre 1963), Instituto Nazionale e di Studi sul Rinascimento (Florence, 1965), I, 152–54.

[23] "Thynges Mathematicall," he says in the preface (sig. *verso), are "middle betwene thinges supernaturall and naturall."

I should like to try to persuade sensible people and sensible historians to use the word *Rosicrucian*. This word has bad associations owing to the uncritical assertions of occultists concerning the existence of a secret society or sect calling themselves Rosicrucians, the history and membership of which they claim to establish. Though it is important that the arguments for and against the existence of a Rosicrucian society should be carefully and critically sifted, I should like to be able to use the word here without raising the secret society question at all. The word *baroque* is used, rather vaguely, of a certain style of sensibility and expression in art without in the least implying that there were secret societies of baroquists, secretly propagating baroque attitudes. In a similar way the word *Rosicrucian* could, I suggest, be used of a certain style of thinking which is historically recognizable without raising the question of whether a Rosicrucian style of thinker belonged to a secret society.

It would be valuable if the word could be used in this way as it might come to designate a phase in the history of the Hermetic tradition in relation to science. A very generalized attempt to define two such phases might run somewhat on the following lines. The Renaissance magus is very closely in touch with artistic expression; the talisman borders in this period on painting and sculpture; the incantation is allied to poetry and music. The Rosicrucian type, though not out of touch with such attitudes, tends to develop more in the direction of science, mixed with magic. Thus though the Rosicrucian type comes straight out of the Renaissance Hermetic tradition, like the earlier magi, he may orientate it in slightly different directions or put the emphasis rather differently. The influx of Paracelsan alchemy and medicine, itself originally stimulated by Ficinian influences, is important for the latter or Rosicrucian type, who is often, perhaps always, strongly influenced by Paracelsus. The tradition in its later or Rosicrucian phase begins to become imbued with philanthropic aims, possibly as a result of Paracelsan influence. Finally, the situation of the Rosicrucian in society is worse and more dangerous than that of the earlier magi. There were always dangers, which Ficino timidly tried to avoid and from which Pico della Mirandola did not escape. But as a result of the worsening political and religious situation in Europe, and of the strong reactions against magic in both Catholic and Protestant countries, the Rosicrucian seems a more hunted being than the earlier magi, some of whom seem able to expand quite happily in the atmosphere of the early Renaissance Neo-Platonism, feeling themselves in tune with the age. The artist Leonardo or the poet Ronsard might be examples of such relatively happy expansion of great figures who are not untinctured with the Hermetic core of Neo-Platonism. The Rosicrucian, on the other hand, tends to have persecution mania. Though usually of an intensely religious temper, he avoids identifying himself with any of the religious parties and hence is suspected as an atheist by them all, while his reputation as a magician inspires fear and hatred. Whether or not he belongs to a secret society, the Rosicrucian is a secretive type, and has to be. His experience of life has confirmed him in the Hermetic belief that the deepest truths cannot be revealed to the multitude.

John Dee seems obviously placeable historically as a Renaissance magus of the later Rosicrucian type. Paracelsist and alchemist, a practical scientist who wished to develop applied mathematics for the advantage of his country-men, full of schemes for the advancement of learning, branded in the public eye as conjuror and atheist, Dee felt himself to be an innocent and a per-secuted man. "O unthankfull Countrey men," he cries in the preface to the Euclid, "O Brainsicke, Rashe, Spitefull, and Disdainfull Countrey men. Why oppresse you me, thus violently, with your slaundering of me. . . ." And he goes on to compare himself, significantly, with "Ioannes Picus, Earle of Mirandula," who also suffered from the "raging slander of the Malicious ignorant against him."[24]

In the so-called Rosicrucian manifesto published in Germany in 1614 in the name of the Fraternity of the Rosy Cross,[25] the characteristics of what I have called the Rosicrucian type of thinking are perceptible. The brethren are said to possess the books of Paracelsus, and the activity to which they are said to bind themselves is the philanthropic one of healing the sick, and that gratis. The manifesto states that the founder of the society based his views and activities on "Magia and Cabala," a mode of thinking which he found agreeable to the harmony of the whole world. It expresses a wish for closer collaboration between magician-scientists. The learned of Fez, says the writer, communicated to one another new discoveries in mathematics, physics, and magic, and he wishes that the magicians, cabalists, physicians, and philosophers of Germany were equally co-operative. Thus whether or not this manifesto really emanates from a secret society, it sets forth a Rosi-crucian type of program, with its devotion to Magia and Cabala, its mixed scientific and magical studies, its Paracelsan medicine.

The utopias of the Renaissance show many traces of Hermetic influences which can even be discerned, I believe, in Thomas More's foundation work. Campanella's *City of the Sun*, which he first wrote in prison in Naples in the early years of the seventeenth century, is a utopian city governed by priests skilled in astral magic who know how to keep the population in health and happiness through their understanding of how to draw down beneficent

---

[24] Dee's preface to the Euclid, sig. A ii *recto*.

[25] *Fama Fraternitas, dess Löblichen Ordens des Rosencreutzes etc.* (Cassel, 1614). See Yates, *Giordano Bruno and the Hermetic Tradition*, pp. 410–11.

astral influences.[26] This is after all a philanthropic use of magical science, though somewhat arbitrarily applied. And the Solarians were in general greatly interested in applied magic and science; they encouraged scientific inventions, all inventions to be used in the service of the community. They were also healthy and well skilled in medicine, that is in astral medicine of the Ficinian or Paracelsan type. I would classify the City of the Sun as belonging to the later or Rosicrucian phase of the Hermetic movement.

The Rosicrucian flavor is also clearly discernible in a less well-known work, the description of the ideal city of Christianopolis by Johann Valentin Andreae, published at Strasbourg in 1619.[27] Andreae's Christianopolis is heavily influenced by Campanella's City of the Sun. Its inhabitants, like Campanella's Solarians, are practicers of astral magic and at the same time are deeply interested in every kind of scientific research. Christianopolis is busy with the activity of scientists who are applying their knowledge in inventions which are to improve the happiness and well-being of the people.

When, after a course of reading of this type, one returns once again to the so much more famous *New Atlantis* of Francis Bacon (written in 1624), it is impossible not to recognize in it something of the same atmosphere. The New Atlantis is ruled by mysterious sages who keep the citizens in tune with the cosmos; and in this late utopia the wisdom tradition is turning ever more and more in the direction of scientific research and collaboration for the betterment of man's estate. Yet there are significant differences as compared with the earlier Rosicrucian utopias which I have mentioned; the priests of the New Atlantis do not practice astral magic and are not exactly magi; its scientific institutions are drawing closer to some future Royal Society. But to me it seems obvious that the New Atlantis has its roots in the Hermetic-Cabalist tradition of the Renaissance, though this is becoming rationalized in a seventeenth-century direction. The magus had given place to the Rosicrucian, and the Rosicrucian is giving place to the scientist, but only very gradually.

Francis Bacon is, in my opinion, one of those figures who have been misunderstood and their place in history distorted by those historians of science and philosophy who have seen in them only precursors of the future without examining their roots in the past. The only modern book on Bacon which makes, or so it seems to me, the right historical approach is Paolo Rossi's *Francesco Bacone,* published in Italian in 1957[28] and now trans-

---

[26] See Yates, *Giordano Bruno and the Hermetic Tradition*, pp. 370ff.

[27] J. V. Andreae, *Reipublicae Christianopolitanae Descriptio* (Strasbourg, 1619); English translation by F. E. Held, *Christianopolis, an Ideal State of the Seventeenth Century* (New York, 1916). Andreae was the author of the *Chemical Wedding of Christian Rosencreutz* [*Chymische Hochzeit Christiani Rosencreutz*], (Strasbourg, 1616).

[28] English translation to be published shortly by Routledge and Kegan Paul, London.

lated into English. The significant subtitle of Rossi's book is *Dalla Magia alla Scienza* [*From Magic to Science*]. Rossi begins by outlining the Renaissance Hermetic tradition, pointing out that Bacon's emphasis on the importance of technology cannot be disentangled from the Renaissance Hermetic tradition in which magic and technology are inextricably mingled. He emphasizes those aspects of Bacon's philosophy which show traces of Renaissance animism, and he argues that the two main planks of the Baconian position—the conception of science as power, as a force able to work on and modify nature, and the conception of man as the being to whom has been entrusted the capacity to develop this power—are both recognizably derivable from the Renaissance ideal of the magus. While urging that the approach to Bacon should take full cognizance of his roots in the Renaissance Hermetic tradition, Rossi emphasizes that such an approach does not diminish Bacon's great importance in the history of thought but should enable the historian to analyze and bring out his true position. In Rossi's opinion, Bacon's supreme importance lies in his insistence on the co-operative nature of scientific effort, on the fact that advance does not depend on individual genius alone but in pooling the efforts of many workers. He emphasizes, and this second point is related to the first one, Bacon's polemic against the habit of secrecy which was so strongly ingrained in the older tradition, his insistence that the scientific worker must not veil his knowledge in inscrutable riddles but communicate it openly to his fellow workers. And finally he draws attention to Bacon's dislike of illuminism and of the pretensions of a magus to knowledge of divine secrets, his insistence that it is not through such proud claims but through humble examination and experiment that nature is to be approached.

I believe that Rossi has indicated the right road for further research on Bacon, who should be studied as a Rosicrucian type but of a reformed and new kind, reformed on the lines indicated by Rossi, through which the Rosicrucian type abandons his secrecy and becomes a scientist openly co-operating with others in the future Royal Society, and abandons also his pretensions to illuminism, to being the "divine" man admired in the Hermetic tradition, with its glorification of the magus, for the attitude of a humble observer and experimentalist. The interesting point emerges here that the humble return to nature in observation and experiment advocated by Bacon takes on a moral character, as an attitude deliberately opposed to the sinful pride of a Renaissance magus with his claims to divine insights and powers.

Yet Bacon's reactions against the magus type of philosopher or scientist themselves belong into a curious context. Rossi has emphasized that Bacon regarded his projected *Instauratio Magna* of the sciences as a return for man of that dominion over nature which Adam had before the Fall but which he lost through sin. Through the sin of pride, Aristotle and Greek philosophers generally lost immediate contact with natural truth, and in a significant passage Bacon emphasizes that this sin of pride has been repeated in recent times in the extravagances of Renaissance animist philosophers. The proud fantasies of the Renaissance magi represent for Bacon something like a second Fall through which man's contact with nature has become even more distorted than before. Only by the humble methods of observation and experiment in the Great Instauration will this newly repeated sin of pride be redeemed, and the reward will be a new redemption of man in his relation to nature.[29] Thus Bacon's very reaction against the magi in favor of what seems a more modern conception of the scientist contained within it curious undercurrents of cosmic mysticism. Though Bacon's attitude would seem to dethrone the Hermetic Adam, the divine man, his conception of the regenerated Mosaic Adam, who is to be in a new and more immediate and more powerful contact with nature after the Great Instauration of the sciences, seems to bring us back into an atmosphere which is after all not so different from that in which the magus lived and moved and had his being. In fact, Cornelius Agrippa repeatedly asserts that it is the power over nature which Adam lost by original sin that the purified soul of the illuminated magus will regain.[30] Bacon rejected Agrippa with contempt, yet the Baconian aim of power over nature and the Baconian Adam mysticism were both present in the aspirations of the great magician. Though for Bacon, the claim of the magus to Illuminism would itself constitute a second Fall through pride.

Bacon's reaction against the animist philosophers as proud magi who have brought about a second Fall is extremely important for the understanding of his position as a reformed and humble scientific observer, and I would even go further than Rossi and suggest that some of Bacon's mistakes may have been influenced by his desire to rationalize and make re-

[29] See F. Bacon, *Historia naturalis et experimentalis, quae est Instaurationis Magna pars tertia* (London, 1622) in the *Works*, ed. Spedding et al., 1857 edition, II, 13–16. Bacon constantly repeats the statement that it was not his pure and direct knowledge of nature which caused Adam's fall, but his proud judging of good and evil; see *Advancement of Learning, ibid.*, III, 264–65; *Instauratio Magna, Praefatio, ibid.*, I, 132, etc. See Rossi, *Francesco Bacone*, pp. 321ff., 392ff., etc.

[30] See *De occulta philosophia*, III, 40; and see C. G. Nauert, *Agrippa and the Crisis of Renaissance Thought* (Urbana, 1965), pp. 48, 284.

Nauert in this book makes interesting comparisons between Agrippa's theory of the magus as possessing power through his magical knowledge and Bacon's promises that man will be lord and master of nature; but he does not know of Rossi's book with its analysis of the difference between the outlook of the magus and that of Bacon.

spectable a tradition which was heavily suspected by its opponents, by the Aristotelians of the schools and by the humanists of the rhetoric tradition. Bacon's admirers have often been puzzled by his rejection of Copernican heliocentricity and of William Gilbert's work on the magnet. I would like to suggest, though there is hardly time to work this out in detail, that these notions might have seemed to Bacon heavily engaged in extreme forms of the magical and animist philosophy or like the proud and erroneous opinions of a magus.

In the sensational works published by Giordano Bruno during his visit to England, of which Bacon must have been well aware, Bruno had made use of heliocentricity in connection with the extreme form of religious and magical Hermetism which he preached in England. Bruno's Copernicanism was bound up with his magical view of nature; he associated heliocentricity with the Ficinian solar magic and based his arguments in favor of earth movement on a Hermetic text which states that the earth moves because it is alive.[31] He had thus associated Copernicanism with the animist philosophy of an extreme type of magus. When Bacon is deploring the sinful pride of those philosophers who have brought about the second Fall, who, believing themselves divinely inspired, invent new philosophical sects which they create out of their individual fantasy, imprinting their own image on the cosmos instead of humbly approaching nature in observation and experiment, he mentions Bruno by name as an example of such misguided Illuminati, together with Patrizi, William Gilbert, and Campanella.[32] Is it possible that Bacon avoided heliocentricity because he associated it with the fantasies of an extreme Hermetic magus, like Bruno? And is it further possible that William Gilbert's studies on the magnet, and the magnetic philosophy of nature which he associated with it, also seemed to Bacon to emanate from the animistic philosophy of a magus, of the type which he deplored?

The magnet is always mentioned in textbooks of magic as an instance of the occult sympathies in action. Giovanni Baptista Porta, for example, in his chapters on the occult sympathies and how to use them in natural magic constantly mentions the loadstone.[33] The animist philosophers were equally fond of this illustration; Giordano Bruno when defending his animistic version of heliocentricity in the *Cena de le ceneri* brings in the magnet.[34] I think that it has not been sufficiently emphasized how close to

[31] See Yates, *Giordano Bruno and the Hermetic Tradition*, pp. 241–43.
[32] *Historia naturalis; Works*, ed. Spedding et al., II, 13.
[33] G. Porta, *Natural Magick*, ed. D. J. Price (New York, 1957), (reprint of the English translation of Porta's *Magia Naturalis*), pp. 10, 14, etc. As is well known, this book was the source of a large part of Bacon's *Sylva sylvarum*.
[34] G. Bruno, *Cena de le ceneri*, dialogue III; see G. Bruno, *Dialoghi italiani*, ed. G. Aquilecchia (Florence, 1957), p. 109.

Bruno's language in the *Cena de le ceneri* is Gilbert's defense of heliocentricity in the *De magnete*. Gilbert, like Bruno, actually brings in Hermes and other *prisci theologi* who have stated that there is a universal life in nature when he is defending earth movement.[35] There are passages in the *De magnete* which sound almost like direct quotations from Bruno's *Cena de le ceneri*. The magnetic philosophy which Gilbert extends to the whole universe is, it seems to me, most closely allied to Bruno's philosophy, and it is therefore not surprising that Bacon should list Gilbert with Bruno as one of the proud and fantastic animist philosophers[36] or that notions about heliocentricity or magnetism might seem to him dangerous fantasies of the Illuminati, to be avoided by a humble experimentalist who distrusts such proud hypotheses.

Finally, there is the suggestion at which I hinted earlier. Is it possible that the reputation of John Dee, the conjuror, conjuring angels with number in the supercelestial world with a magus-like lack of humility of the kind which Bacon deplored, might have made the Lord Verulam suspicious also of too much operating with number in the lower worlds? Was mathematics, for Bacon, too much associated with magic and with the middle world of the stars, and was this one of the reasons why he did not emphasize it in his method? I am asking questions here, obviously somewhat at random, but they are questions which have never been asked before, and one object in raising them is to try to startle historians of science into new attitudes to that key figure, Francis Bacon. To see him as emerging from the Renaissance Hermetic tradition and as anxious to dissociate himself from what he thought were extreme and dangerous forms of that tradition may eventually lead to new adjustments in the treatment both of his own thought and of his attitude toward contemporaries. It would be valuable if careful comparisons could be organized between the works of Dee, Bacon, and Fludd. The extreme Rosicrucian types, Dee and Fludd, might come out of such an examination with better marks as scientists than Bacon. Dee certainly would, and even Fludd might do better than expected.

Nevertheless, all this does not do away with Bacon's great importance.

---

[35] Against the "monstrous" opinion of Aristotle that the earth is dead and inanimate, Gilbert cites "Hermes, Zoroaster, Orpheus," who recognize a universal life; see W. Gilbert, *On the Magnet*, ed. D. J. Price (New York, 1958), p. 209.

E. Zilsel, "The origins of William Gilbert's scientific method," *Journal of the History of Ideas*, II (1941), 4ff., emphasizes that Gilbert's philosophy of magnetism is animistic and belongs into the same current as that of Bruno.

[36] Marie Boas, "Bacon and Gilbert," *Journal of the History of Ideas*, XII (1951), 466–67, has suggested that it was primarily Gilbert's expansion of his work on the magnet into a magnetic philosophy of nature to which Bacon took exception, having studied these ideas, not only in *De magnete*, but perhaps primarily in Gilbert's posthumous work, *De mundo nostro sublunari philosophia nova*.

As compared with Dee and Fludd, Bacon has unquestionably moved into another era in his conception of the role of the scientist and of the character of the scientist. Though Bacon descends from the magus in his conception of science as power and of man as the wielder of that power, he also banishes the old conception of the magus in favor of an outlook which can be recognized as modern, if the Adamic mysticism behind the Great Instauration is not emphasized. Bacon obviously qualifies as a member of the future Royal Society, though one with surviving affiliations with the occult tradition—as was the case with many early members of the Society. The figure of Bacon is a striking example of those subtle transformations through which the Renaissance tradition takes on, almost imperceptibly, a seventeenth-century temper and moves on into a new era.

I would thus urge that the history of science in this period, instead of being read solely forwards for its premonitions of what was to come, should also be read backwards, seeking its connections with what had gone before. A history of science may emerge from such efforts which will be exaggerated and partly wrong. But then the history of science from the solely forward-looking point of view has also been exaggerated and partly wrong, misinterpreting the old thinkers by picking out from the context of their thought as a whole only what seems to point in the direction of modern developments. Only in the perhaps fairly distant future will a proper balance be established in which the two types of inquiry, both of which are essential, will each contribute their quota to a new assessment. In the meantime, let us continue our investigations in which the detection of Hermetic influences in some great figure and acknowledged precursor should be a parallel process to the detection of genuine scientific importance in figures who have hitherto been disregarded as occultists and outsiders.

And we must constantly beware of giving an impression of debunking great figures when we expose in them unsuspected affiliations to the Hermetic tradition. Such discoveries do not make the great figures less great; but they demonstrate the importance of the Renaissance Hermetic tradition as the immediate antecedent of the emergence of science. The example of this which I made the subject of a book is Giordano Bruno. Long hailed as the philosopher of the Renaissance who burst the bonds of medievalism and broke out of the old world view into Copernican heliocentricity and a vision of an infinitely expanded universe, Bruno has turned out to be an "Egyptian" magus of a most extreme type, nourished on the Hermetic texts. Bruno's vision of an infinite universe ruled by the laws of magical animism with which the magus can operate is not a medieval or a reactionary vision. It is still the precursor of the seventeenth-century vision, though formulated within a Renaissance frame of reference. As I have tried to suggest in this

270

paper, even the mathematical and mechanical progress which made possible the seventeenth-century advance may have been encouraged by Hermetic influences in the earlier movement. The emergence of modern science should perhaps be regarded as proceeding in two phases, the first being the Hermetic or magical phase of the Renaissance with its basis in an animist philosophy, the second being the development in the seventeenth century of the first or classical period of modern science. The two movements should, I suggest, be studied as inter-related; gradually the second phase sheds the first phase, a process which comes out through the double approach of detecting intimations of the second phase in the first and survivals of the first phase in the second. Even in Isaac Newton, as is now well known, there are such survivals, and if Professor Garin is right, even in Galileo,[37] while Kepler provides the obvious example of a great modern figure who still has one foot in the old world of universal harmony which sheltered the magus.

Renaissance and early seventeenth-century literature abounds in vast tomes which it is beyond the power of any one scholar to tackle unaided. They sleep undisturbed on library shelves or are only dipped into at random, while people turn to the easier and more lucrative occupation of writing little books about the Renaissance and seventeenth century and the great names—Kepler, Newton, Galileo—run easily off all our pens. Yet do we really understand what happened? Has anyone really explained where Kepler, Newton, Galileo, came from? I wish that a concerted effort could be made, less on the published writings of the great in their modern and accessible editions than on the vast sleeping tomes. I think of two in particular with which I have often tried to struggle: Francesco Giorgi's *De harmonia mundi* and Marin Mersenne's *Harmonie universelle*. Giorgi's *Harmony of the World* is full of Hermetic and Cabalist influences; the Franciscan friar who wrote it was a direct disciple of Pico della Mirandola. This tome represents the Renaissance Hermetic-Cabalist tradition working on the ancient theme of world harmony. Mersenne is a seventeenth-century monk, friend of Descartes. And just as Bacon does in his sphere, Mersenne attacks and discards the old Renaissance world; his *Universal Harmony* will have nothing to do with the *anima mundi* and nothing to do with Francesco Giorgi, of whom he sternly disapproves. Mathematics replaces numerology in Mersenne's harmonic world; magic is banished; the seventeenth century has arrived. The emergence of Mersenne out of a banished Giorgi seems somehow a parallel phenomenon to the emergence of Bacon out of the magus. It is perhaps somehow in these transitions from Renaissance to

[37] On survivals in Newton, see J. E. McGuire and P. M. Rattansi, "Newton and the Pipes of Pan," *Notes and Records of the Royal Society of London*, XXI (1966), 108–43; on survivals in Galileo, see Garin, *Scienza e vita civile nel Rinascimento*, p. 157.

seventeenth century that the secret might be surprised, the secret of how science happened. But to understand Mersenne and Mersenne's rejection of Giorgi, one must know where Giorgi came from. He came out of the Pythagoro-Platonic tradition plus Hermes Trismegistus and the Cabala.

In a review of my book on Bruno,[38] Allen G. Debus has suggested that I have overemphasized the importance of the dating of the Hermetic writings by Isaac Casaubon in 1614 as weakening the influence of the Hermetic writings after that date. He points out that "the first half of the seventeenth century saw an increased interest in the occult approach to nature which parallels the contemporary rise of mechanical philosophy. The real collapse of the Renaissance magical science only occurs in the period after 1660. Until then it remained a positive force stimulating some scientists to a new observational approach to nature."[39] I would accept this criticism as valid; I think that I may have overestimated the importance of Casaubon's dating, which was totally ignored by, for example, Fludd and Kircher, and I also believe, as indeed I have suggested in this paper, that the late Renaissance movement which I would like to label "Rosicrucian" does continue to exert a strong influence through the seventeenth century. Nevertheless I still think that Casaubon's dating does, as it were, mark a historical term which helps to define and delimit the Hermetic movement. Though the importance of Ficino's propagation of the Hermetic writings and his adoption of Hermetic philosophy and practice must not be exaggerated to the exclusion of the many other influences fostering the movement, yet it was basic, and the Hermetic attitude toward the cosmos and toward man's relation to the cosmos which Ficino and Pico adopted was, I believe, the chief stimulus of that new turning toward the world and operating on the world which, appearing first as Renaissance magic, was to turn into seventeenth-century science. And it was the sanction which the misdating of the *Hermetica* gave to these writings that sanctioned procedures and attitudes which St. Augustine had severely condemned and which were prohibited by the Church. If, as Ficino believed, the *Hermetica* were all written many centuries before Christ by a holy Egyptian who foresaw the coming of Christ, this encouraged him and other Christian souls to embark on the Hermetic magic. Casaubon's dating of the *Hermetica* as written after Christ destroyed an illusion without which the movement might not have gained its original momentum, though it could not stop the movement after it had gained such force and influence. That is perhaps a better way of putting it.

[38] In *Isis*, LV, No. 180, (1964), 389–91.
[39] See also the many observations in Allen G. Debus' book *The English Paracelsians* (London, 1965), confirming the connections between Renaissance magic and Neo-Platonism and the rise of science.

It would be absurd of course to suggest that the Hermetic texts and Ficino's interpretation of them were the only causes of the movement. These were only factors, though important ones, in disseminating a new climate of opinion through Europe which was favorable to the acceptance of magico-religious and magico-scientific modes of thinking. Neo-Platonism itself was favorable to this climate, and medieval traditions of the same type revived. If one includes in the tradition the revived Platonism with the accompanying Pythagoro-Platonic interest in number, the expansion of theories of harmony under the combined pressures of Pythagoro-Platonism, Hermetism, and Cabalism, the intensification of interest in astrology with which genuine astronomical research was bound up, and if one adds to all this complex stream of influences the expansion of alchemy in new forms, it is, I think, impossible to deny that these were the Renaissance forces which turned men's minds in the direction out of which the scientific revolution was to come. This was the tradition which broke down Aristotle in the name of a unified universe through which ran one law, the law of magical animism. This was the tradition which had to contend with the so much more prominent and successful disciplines of rhetorical and literary humanism. This was the tradition which prepared the way for the seventeenth-century triumph. But it must be emphasized that the detailed work, the great body of research, necessary for tracing this movement is not yet done. It lies in the future.

There is yet another way of regarding this strange history of the Renaissance Hermetic tradition in its relation to science. We may ask whether the seventeenth century discarded notions from the earlier tradition which may have been actually nearer to the views of the universe unfolded by the science of today than the movement which superseded it. Was the magically animated universe of Bruno, so close to the magnetic universe of Gilbert, a better guess about the nature of reality than those seemingly so much more rational universes of the mechanistic philosophers?

It may be illuminating to view the scientific revolution as in two phases, the first phase consisting of an animistic universe operated by magic, the second phase of a mathematical universe operated by mechanics. An enquiry into both phases and their interactions, may be a more fruitful line of approach to the problems raised by the science of to-day than the line which concentrates solely on the seventeenth-century triumph.

Professor Debus quotes these words of mine in his review,[40] adding, "I heartily agree with this opinion, and in essence it is the approach which I

[40] In *Isis*, LV, 390, quoting *Giordano Bruno and the Hermetic Tradition*, p. 452.

have been taking in my own courses on Renaissance science." It is most gratifying to me to learn that a point of view which I put forward in some fear and trembling is actually already the basis of teaching in the United States. I must, however, not come before you on false pretenses, and I must emphasize that, just as I was careful to state in the beginning that I am no magician, so I must be even more careful to state at the end that I am no scientist. Though when I read in the *Observer* for September 26, 1965, that five hundred of the world's most expensive scientists, gathered at Oxford, were in a mood of breathless expectation because they believed that high-energy physics, burrowing ever deeper into matter, may be about to break into "quite a new level of reality," it seemed to me that I had heard something like this before. In the Rosicrucian manifesto of 1614 it is announced that some great aurora is at hand in the light of which man is about "to understand his own nobleness and worth, and why he is called Microcosmus, and how far his knowledge extendeth into nature." Perhaps these words are not so much a prophecy of the limited vision of the seventeenth-century revolution as of yet another aurora. And perhaps the view of nature of a Rosicrucian like John Dee as a network of magical forces which can be dealt with by mathematics is nearer to the new aurora—notwithstanding his belief in talismans and in the conjuring of angels—than an ignorant person like myself can understand.

SOME NOVEL TRENDS IN THE SCIENCE OF THE FOURTEENTH CENTURY ❦ MARSHALL CLAGETT

t seems evident that the announced topics of the two papers that Mr. Drake and I have prepared represent the upper and lower limits of Renaissance science. In fact, while much of what I have to say has some bearing on science in the fifteenth and sixteenth centuries, not all of the trends here singled out as being novel in fourteenth-century speculation were actually of much substantial influence in the succeeding centuries, even though they do represent a changing tone by bringing cosmological theory closer to the rather more impersonal approach of terrestrial physics. Let me say further that what I have to say in the second half of the paper concentrates on efforts at quantification in the fourteenth century since out of these efforts arise some concepts that are rather more significant to the later Scientific Revolution.

Even the most cursory glance at the rich manuscript holdings in European libraries that bear on fourteenth-century science will convince the student of this period that, aside from the extraordinary increase in the volume of scientific material, there is a perfect explosion of philosophical speculation beyond the confines of Aristotelian philosophy. In part, this may be associated with the check put by the Condemnations of 1277 at Paris to considering the Aristotelian philosophy, with its unique cosmos, as the only possible natural explanation of the way things are. At least this is the theory held by Pierre Duhem in his celebrated studies completed some two generations ago.[1] And while this thesis has not been tested in any extensive way (and does not seem of great significance in some of the the cases discussed by Duhem) so that we cannot affirm it to be the principal cause of the varied speculation of the fourteenth century, it may have played some role in certain cosmological speculations like those connected with the possibility of the existence of a plurality of worlds,[2] certainly not a new topic but one handled with considerable skill and novelty by the schoolmen of

---

[1] P. Duhem, *Études sur Léonard de Vinci*, Vol. 3 (Paris, 1913; new printing, 1955), vii. See *Le Système du monde*, Vol. 6 (Paris, 1954), in general; see particularly p. 80: ". . . l'Université parisienne ne cessa de maintenir avec fermeté les condamnations portées par Étienne Tempier; en fait, le décret de 1277 resta, pendant toute la durée du xive siècle, une code révéré des maîtres parisiens; nous en aurons maintes fois la preuve."
[2] A. Koyré was highly skeptical of the significance of the Condemnations on the develop-

the fourteenth century. For example, let us look at the manner in which Nicole Oresme, whose ideas will provide us with some of our exemplary material, takes up this problem. Before turning to Oresme's bold speculations, let me remark on one general device that catches on in scientific speculation in the fourteenth century in an all-pervasive manner (in fact, enough so to constitute a trend in itself); this is the device of the *ymaginatio*, i.e., a kind of thought experiment or imaginative scheme constructed as if it existed in nature, with the purpose of illustrating basic theoretical ideas. Such schemes were usually not thought to be possible (and indeed I would say that many were clearly invented with the understanding that certain features of the real world were to be set aside). The main point to emphasize in connection with the *ymaginatio* is that it was the chief instrument of the speculative science of the fourteenth century, and this is particularly true of the schoolmen at the University of Paris. One such *ymaginatio* is the detailed picture of a possible plurality of worlds as visualized by Oresme. The account I wish to quote is from his *Livre du ciel et du monde*, written in 1377 at the end of his brilliant career:

All heavy things of this world tend to be conjoined in one mass (*masse*) such that the center of gravity (*centre de la pesanteur*) of this mass is in the center of this world, and the whole constitutes a single body in number. And consequently they all have one [natural] place according to number. And if a part of the [element] earth of the other world was in this world, it would tend towards the center of this world and be conjoined to its mass. . . . But it does not accordingly follow that the parts of the [element] earth or heavy things of the other world (if it exists) tend to the center of this world, for in their world they would make a mass which would be a single body according to number, and which would have a single place according to number, and which would be ordered according to high and low [in respect to its own center] just as is the mass of heavy things in this world. . . . I conclude then that God can and would be able by His omnipotence (*par toute sa puissance*) to make another world other than this one, or several of them either similar or dissimilar, and Aristotle offers no sufficient proof to the contrary. But as it was said before, in fact (*de fait*) there never was nor will there be any but a single corporeal world. . . .[3]

I perhaps should comment in passing that this passage has some importance in reflecting the growing interest in the use of the concept of

---

ment of scientific thought, see "Le Vide et l'espace infini au XIVe siècle," in *Études d'histoire de la pensée philosophique* (Paris, 1961), pp. 33–41.

[3] M. Clagett, *The Science of Mechanics in the Middle Ages* (Madison, 1959), pp. 592–93. For the original text, see Nicole Oresme, *Le Livre du ciel et du monde*, Bk. I, Chap. 24 (ed. A. D. Menut and A. J. Denomy in *Mediaeval Studies*, Vol. 3 [1941], 243–44).

center of gravity in large bodies, an interest which in its mature development in the seventeenth century was to be of considerable importance to the marriage in astronomy of force physics and kinematics.

While the plurality of worlds was clearly rejected by Oresme, certainly that idea was to grow in the Renaissance, where, joined with Copernican astronomy and the substitution of Euclidian infinite space for the finite cosmos, it was to have some success with Bruno and others.

A less successful novel speculation concerned with the wider field of the supposed structure of this one world, but like it, a speculation that seemed to shake the Aristotelian dichotomy of two physics—a physics of the heavens and that of the earth—was that concerned with the celestial movers themselves. A standard view in the Middle Ages, and this was equally true of the philosophers of the fourteenth century, was that the movers of the heavens were the intelligences, moving the heavenly bodies as they willed. Hence, forces like these natural forces of the terrestrial realm were not involved. This was spelled out in great detail on several occasions by Oresme. For example, listen to him distinguish between natural and voluntary force in his *Questiones de spera:*

I posit three distinctions. . . . The first distinction is that certain force is natural and certain voluntary and free, so that I distinguish natural in opposition to voluntary. And they differ in two ways: (1) Natural force moves with some exertion or effort as some horse strives and exerts itself to draw a cart, but the force which is a pure, voluntary one does not move with some effort but by volition alone, so that these forces are of differing natures. (2) The second difference is that natural force is designatable by numbers and ratios. . . . But the force which is will is not designatable by numbers or ratios. Therefore, one such voluntary force ought not be described as twice or triple another. And accordingly we say that the first kind (i.e., natural force) is a certain quality which is receptive to increase and decrease. But the second (i.e., voluntary force) is pure substance itself which is not susceptible to comparison [by ratios]. The second distinction concerns resistance, which can be understood in two ways. The first and proper way is as a certain difficulty or quality contrary to some force. Spoken of in this first way, the velocity of motion increases or decreases according to its (i.e., the resistance's) decrease or increase. . . . In the second and improper way, resistance is nothing but a certain impossibility of moving faster or slower. And thus it can be said that a heaven resists being moved faster. The third distinction is that sometimes a certain effect or motion arises from a natural force, such is the motion of a stone downward, while sometimes a motion arises primarily from a force which is an intelligence . . . , such is the motion of a heaven. . . . [Hence,] I say that between an intelligence and a [celestial] orb there is no ratio, but only something analogous, in a way similar to that of things [here] below. For we see that a certain velocity arises from a definite ratio, and the cause is that every ratio,

277

if it is rational, is signifiable by numbers. But an intelligence cannot be signified by numbers.[4]

The point to underline in this passage is that the action of intelligences is not quantifiable in terms of proportionality theorems and no force mechanics seems allowable. Hence, this speculative effort seems far removed from early modern mechanics. But there were speculations on the part of some natural philosophers that throw some doubt on the role of intelligences and seem to offer an opening wedge for force physics. One of the most interesting is that of Oresme's supposed master, Jean Buridan. He makes a suggestion that impetuses (like the impetuses found in this world that move projectiles after the initial force introducing the impetus is no longer in contact with the body) move the heavenly bodies. He was led to his suggestion by the idea that such an impetus tends to permanence, although in this terrestrial realm impetus is never permanent because of the continual presence of resistance and contrary force, which destroy the impetus. But if such impetuses were implanted in heavenly bodies where there is no resistance and no contrary force, then there is no reason why those bodies would not continue to move indefinitely with a uniform speed. Hence, since impetus was ranked by Buridan as a force, he is clearly, in this one instance, at least suggesting the possibility that the motion of heavenly bodies is explicable by forces akin to natural forces rather than by intelligences. Listen to Buridan:

. . . it does not appear necessary to posit intelligences of this kind [as celestial movers], because it could be said that when He created the world God moved each of the celestial orbs as He pleased, and in moving them impressed in them impetuses which moved them without His having to move them any more. . . . And these impetuses which He impressed in the celestial bodies were not decreased or corrupted afterwards because there was no inclination of the celestial bodies for other movements. Nor was there resistance which would be corruptive or repressive of that impetus.[5]

And in another place, Buridan specifically ties the kind of impetus present in the terrestrial realm to the suggested celestial impetuses:

Many posit that the projectile after leaving the projector is moved by an impetus given [it] by the projector, and it is moved so long as the impetus remains stronger

[4] I have published the Latin text of this passage in my "Nicole Oresme and Medieval Scientific Thought," *Proceedings of the American Philosophical Society*, Vol. 108 (1964), 300, n. 20.

[5] Clagett, *The Science of Mechanics*, pp 524–25. (I have here changed the translation slightly.)

than the resistance. And it would last indefinitely (*in infinitum*) if it were not diminished or corrupted by a resisting contrary or by a contrary motion; and in celestial motion there is no resisting contrary. . . .[6]

But Buridan's disciple, Oresme, in this case clearly would not go along with the master, primarily because he viewed impetus in a different way. For him it was expendable as it acted, and thus it could not be conceived of as tending toward permanence.[7] Furthermore, he seemed to connect impetus with acceleration. Hence, he considered it inapplicable to the uniform motion of the heavenly bodies. As we have already seen, he tended to accept the intelligences as movers. But he was never one for complete consistency; he liked to follow the argument where it led him, and so we see that he too made suggestions in the *Livre du ciel et du monde* that seemed to contradict or modify conclusions stated elsewhere in that and other works. Two of his suggestions are of particular interest. In the first he says:

And, according to truth, an intelligence is simply immobile, and it is not logical that it exists throughout the heaven which it moves nor that it exists in a part of such a heaven, it having been posited that the heavens are moved by intelligences. For, perhaps, when God created them (i.e., the heavens), He placed in them motive qualities and powers, just as He placed weight into terrestrial things, and He put in them resistances [to] counter the motive powers. And these powers and resistances are of a nature and a matter differing from any sensible thing or quality which exists here below. And these powers are so moderated, tempered and accorded to their opposing resistances that the movements take place without violence. And with violence excepted, it is somewhat like a man making a [mechanical] clock and letting it go and be moved by itself. And so God let the heavens be moved continually according to the ratios that the motive powers have to the resistances and according to the established order.[8]

It is clear that here Oresme's suggestion is completely contradictory to what he has previously said about the inapplicability of ratios and numbers to the heavenly movers. A further passage elaborates and repeats this suggestion of implanted powers and resistances (if not impetuses):

It is not impossible . . . that the heaven be moved by an inherent corporeal quality or power [and that it be moved] without violence and without work (*travail*), for the resistance which is in the heavens inclines it to no other move-

[6] *Ibid.*, p. 524, n. 38.
[7] *Ibid.*, p. 552.
[8] *Le Livre du ciel et du monde*, Bk. II, Chap. 2 (*ed. cit. Mediaeval Studies*, Vol. 4 [1942], 170).

ment nor rest but only so that it does not move more rapidly [or, we could add, more slowly].[9]

In other words, Oresme has joined to his suggestion of implanted forces and resistances the idea of improper resistance, spelled out in our earlier quotation. While one might be inclined to see some inertial adumbration in Oresme's suggestion, I think it must be observed that it had very little, if any, direct influence on early modern authors so far as inertial ideas are concerned, although it certainly was one of several efforts to make celestial motion open to force analysis.

It can be pointed out further that Oresme seemed to feel no unease in transferring the basic proportionality theorem describing the relationship of velocity and force as present in natural motions to the heavens in a work entitled *Ratios of Ratios*[10] in spite of his oft-repeated assertion that the intelligences as celestial movers are not susceptible to treatment by ratios and numbers. If challenged, he would perhaps have answered that, although he believed intelligences to be movers, this is not certain and it might well be argued (as in fact he did in the previous quotations) that there are analogous (but essentially different) implanted forces and resistances in the celestial bodies, which forces and resistances might be related to the velocities of celestial motions by the same basic formula embracing movements arising from natural forces. At any rate, this tendency to take positions that seem contrary to most of the tenets of natural philosophy that he supports gives a strongly tentative and speculative tone to his natural philosophy that certainly appears to distinguish his approach from that of Galileo or Copernicus. In fact, the more one reads of his rather extensive bibliography, the surer he becomes that Oresme has been significantly influenced by the skeptical tendencies so evident in the philosophy of the first half of the century.[11] Further, in his *Quodlibeta* we hear him twice say that when it comes to natural knowledge (as distinguished from the true knowledge of faith), the only thing he knows is that he knows nothing.[12] It would appear to me, then, that such skeptical tendencies nurtured the speculative approach

[9] *Ibid.*, Chap. 3, p. 175.
[10] See the new edition by E. Grant: Nicole Oresme, *The "De proportionibus proportionum" and "Ad pauca respicientes"* (Madison, 1966).
[11] The classic work on the skepticism of the fourteenth century is that of K. Michalski, "Le Criticisme et le scepticisme dans le philosophie du XIVe siècle," *Bulletin international de L'Académie Polonaise des Sciences et des Lettres. Classe de philologie. Classe d'histoire et de philosophie*, L'Année 1925, Part I (1926), pp. 41–122; Part II (1927), pp. 192–242.
[12] *Quodlibeta*, MS Paris, BN lat. 15126, 98v: "Ideo quidem nichil scio nisi quia scio me nichil scire." See also 118v: "Et quamvis multis appareant faciles, mihi tamen difficiles videntur. Ideo nichil scio nisi quia me nichil scire scio."

that would just as well consider (and even support) one view as another. But lest I oversell this idea and leave you with the impression that Oresme everywhere introduces and supports opposing views of the basic problems of natural philosophy, may I say that, whether skeptical or not in some instances, Oresme was highly rational in his approach to natural phenomena and he often did take determinate positions. He did, as far as was possible for a believing Christian of the fourteenth century, seek only natural causes for the events of this world. In this connection, he was highly critical of the tendency on the part of many to seek supernatural causes for remarkable natural events. In the *Contra divinatores* which precedes the *Quodlibeta* we read: "From the said arguments it is evident that the diversity and plurality of effects here below arises by reason of matter and passive and immediate causes rather than from superior things."[13] And further, he says: "Every diversity here below can be saved by natural dispositions. Therefore, such particular [celestial] influences are posited in vain."[14] And still further:

Whence certain people attribute unknown causes to God immediately, certain to the heavens, and certain to the Devil. But an expert ought never to draw a conclusion from pure ignorance . . . and there is great doubt on the part of any expert as to whether demons exist.[15]

In the *Quodlibeta* attached to the *Contra divinatores*, Oresme repeats this theme of explaining terrestrial effects without recourse to the heavens, to God, or to demons.[16] In one place he highlights the difficulty of proving from natural effects the existence of demons in a question entitled "Whether Demons exist naturally." He admits that they do exist, but only because faith says so and, in fact, their existence cannot be proved naturally: "In regard to the question, I say that, speaking truly, demons exist, as must be believed from faith. But I assert that this cannot be proved naturally. . . . If

[13] *Ibid.*, 5ᵛ: "Ex dictis rationibus patet quod diversitas et pluralitas effectuum hic inferius plus provenit ratione materie et passivarum et causarum immediatarum magis quam superiorum."

[14] *Ibid.*, 7ᵛ: ". . . et dispositionibus naturalibus potest salvari omnis diversitas hic inferius. Igitur frustra ponuntur tales influentie particulares. . . ."

[15] *Ibid.*, 33ᵛ: "Unde quidam causas ignotas attribuunt Deo immediate, quidam celo, quidam dyabolo. Homo autem peritus nunquam ex pura ignorantia debet concludere seu incurrere maiorem nec ex parvo inconsequenti maius concedere, nunquam tibi, et cuicumque perito est magis dubium demones esse. . . ."

[16] *Ibid.*, 60ᵛ: "Ad aliud, ut superius dixi, illi qui nesciunt causas immediatas et naturales fugiunt ad demones, alii ad celum, alii ad deum, et quia talia videntur mirabilia, ideo attribuunt etc., sed hoc est falsum etc." See 80ʳ, 95ʳ, and *passim*. Incidentally, on fol. 96ᵛ he indicates that it seems to him "to believe easily is and was the cause of the destruction of natural philosophy" (". . . videtur mihi quod faciliter credere est et fuit causa destructionis philosophie naturalis").

the faith were not to pose their existence, I would say that they cannot be proved to exist from any natural effect, because all things [supposedly arising from them] can be saved naturally. . . ."[17] Although I have stressed the sharpness of the attacks of Nicole Oresme on the justice of seeking celestial influences on natural events, certainly he was not alone in his attacks, as is apparent in the equally strong attack of his younger associate Henry of Hesse (see Chapter Two, Part C, of my edition of Oresme's *De configurationibus qualitatum et motuum*); and there are no doubt others. And, in fact, Oresme's attack in the *Quodlibeta* on the concept of celestial influences was summarized in the fifteenth century (about 1478) by one Claudius Coelestinus in a work entitled *De his quae mundo mirabiliter eveniunt*, a work that was later published in 1542 by the well-known mathematician Oronce Finé. As Thorndike has pointed out (*Osiris*, Vol. 1 [1936], 631): " . . . it must have been largely through Oronce Finé's printing of the summary of Coelestinus that the views of Oresme continued to exert influence in early modern times." Indeed, this summary was translated into French at Lyon in 1557.

Before leaving our brief consideration of heavenly motions, three things ought to be remarked. The first is that formal astronomy with its Ptolemaic models and general kinematic results as inherited from antiquity continued in the fourteenth century. Hence, the *Almagest* of Ptolemy as well as the popular works of Arabic astronomy such as those of al-Farghānī and al-Battānī were studied, as were two of the best-known medieval works. These were both entitled *Theorica planetarum*: the more elementary one attributed to Gerard of Cremona (probably falsely) and the other to Campanus of Novara, the celebrated commentator on the *Elements* of Euclid. The first was subjected to commentary in the fourteenth century. And in one such commentary by an Oxford schoolman, Walter Brytte, we find the kinematic ideas of Oxford which were developed in connection with qualitative analysis, joined with the purely Greek astronomy.[18] One should further point out that there was also a sprightly development in trigonometry in the same circle, as evidenced in the works of John Maudith, Richard Wallingford, and Simon Bredon, and in France with Levi ben Gerson.[19] And, finally, there was considerable activity in the preparation of astronomical tables, much of which remains to be investigated.

[17] *Ibid.*, 127^v: "Ad propositum dico quod veraciter loquendo demones sunt, sicut credendum est ex fide. Sed dico quod non potest probari naturaliter, ut dixi. . . . Nisi autem fides poneret eos esse dicerem quod ex nullo effectu possent probari esse, quia naturaliter omnes possunt salvari. . . ."

[18] O. Pedersen, "The Theorica planetarum—Literature of the Middle Ages," *Classica et Mediaevalia*, Vol. 23 (1962), 228.

[19] G. Sarton, *Introduction to the History of Science*, Vol. 3 (Baltimore, 1947), 598–602, 660–61, 662–68, 673.

I am rather more interested in efforts at quantification that arose in the treatment of qualities and terrestrial motions; and we have now moved down from the heavens. I hardly need point out that quantification has a basic twofold aspect: the first is in the development of a language of quantity that can be applied to phenomena; the second is in the development of theories and techniques of measurement to produce numerical values. In its latter aspect of quantification as measurement, the fourteenth-century scientists contributed almost nothing (except perhaps in the above mentioned preparation of astronomical tables and the increasing development of practical handbooks of mensuration). In the former, of theoretical quantification, the contributions were of considerable importance, and in one way or another will occupy the remainder of this paper.

One of the initial points to mention in connection with the development of a language of quantity as applied to nature is that the medieval natural philosophers inherited the theory and substance of Euclidian proportions as the language of science, and they scarcely, if ever, transcended this language in their quantified statements of the functions they saw in nature. Hence, definitions of velocity and the like were always expressed as the ratios of like quantities, and not as metrical statements couched in algebraic form and involving proportionality constants.

Now in concentrating on the period of our discussion, we should note that within the confines of this proportionality language the fourteenth-century schoolmen made some interesting contributions. The first of these was in the development of the so-called Bradwardine rule for representing the supposed relations of force and resistance to velocity in the production of motion. The so-called Bradwardine rule or function can be expressed as follows:[20]

$$\frac{F_2}{R_2} = \left(\frac{F_1}{R_1}\right)^{\frac{V_2}{V_1}}.$$

In discussing this rule, Bradwardine made an effort to quantify more specifically than it had been done in the past the rule usually attributed by modern authors (but not by Bradwardine) to Aristotle and which he (Bradwardine) rejects. This can be written as follows:

$$\frac{V_2}{V_1} = \frac{F_2}{F_1} \text{ when } R_2 = R_1 \text{ and } \frac{V_2}{V_1} = \frac{R_1}{R_2} \text{ when } F_2 = F_1.$$

He also more neatly and mathematically expressed another rule that perhaps had a pre-history in late antiquity in a work of John Philoponus and in

---

[20] For a discussion of Bradwardine's contribution to dynamics, see my *The Science of Mechanics*, pp. 437–40.

Islam in one of Ibn Bājja. Bradwardine for the first time unambiguously gives it this form:

$$\frac{V_2}{V_1} = \frac{F_2 - R_2}{F_1 - R_1}.$$

This is equally rejected by Bradwardine in favor of his exponential law. Where did Bradwardine's curious "law" arise? I suggested some years ago in *The Science of Mechanics* that it had an analogue in medical or pharmacological treatments and particularly in the treatment of the degrees in medicinal compounds and simples.[21] Thus in the basic work of Alkindi *On the Degrees of Medicinal Compounds*, the author related an arithmetic ordering of degrees of intensity with a geometrical increase of the determinate qualitative powers. Hence, a medicament or a compound was "in equality" when its frigidity (i.e., its power of frigidity) was equal to that of its calidity. It was hot in the first degree when its calidity was twice its frigidity, hot in the second degree when its calidity was four times its frigidity, hot in the third degree when its calidity was eight times is frigidity, and finally, hot in the fourth degree when its calidity was sixteen times its frigidity. Hence, I proposed earlier that it was not much of a step to go from (1) an exponential relationship between qualitative degrees of intensity and qualitative powers of compounds to (2) an exponential relationship between degrees of the intensity of motion and the ratio of motive and resistive powers, particularly in view of the growing tendency from the early years of the fourteenth century to treat the intensities of qualities and the intensities of velocities in a parallel manner. My earlier suggestions have been rendered more probable by a thesis done at Princeton by Michael McVaugh,[22] except that the route of influence was more circuitous than I had imagined. It shows the influence of Alkindi, particularly on Arnald of Villanova, who appears to have conceived of Alkindi's pharmacological "law" as but one manifestation of a more general law relating intensities and powers. As he states it, however, his supposedly general law is rather a special case of Bradwardine's law (the case where the ratio of powers is continually doubled). If I spend considerable time on this clearly erroneous law, I do so only because of its widespread influence in the fourteenth and fifteenth centuries (with only occasional dissenting voices raised against it: Blasius of Parma, Giovanni Marliani, and later Alessandro Achellini). Interestingly enough, its development led to a whole new branch of mathematical proportionality dealing with fractional exponents (called by the

---

[21] P. 439, n. 35.
[22] "The Mediaeval Theory of Compound Medicines" (1956), pp. 228–33.

schoolmen soon after Bradwardine, "ratios of ratios"). And the language of this branch of proportionality theory as developed particularly by Oresme in his *De proportionibus proportionum* continued into the sixteenth and seventeenth centuries to the time of Kepler and Newton (Kepler's Third Law is expressed in the medieval exponential language[23]). Two aspects of Bradwardine's law make it particularly important. First, it was applied to all aspects of motion in the Aristotelian sense; to local motion, motion of alteration, and motion of augmentation. It thus had a surprisingly universal appeal. In the second place, it focused attention on instantaneous changes and led to a remarkable development in kinematics at Merton College.

In Chapters Four, Five and Six of my *Science of Mechanics in the Middle Ages* I have treated the crucial aspects of this development at Merton. It is associated with the names of four natural philosophers present at Merton between 1328 and 1350: Thomas Bradwardine, William Heytesbury, Richard Swineshead, and Thomas Dumbleton. In brief, its essential contributions were: (1) a careful distinction of kinematics from dynamics; (2) the elaboration of precise definitions (indeed every bit as precise as those of Galileo later) of uniform speed, uniform acceleration, and the like; (3) a growing appreciation of the concept of instantaneous velocity; and (4) the elaboration and proof of a theorem describing uniform acceleration in terms

---

[23] In the "ratio of ratios" vocabulary developed in the fourteenth century, when one ratio is said to be a certain ratio of another, we are to understand that the "certain ratio" is the exponent to which the second ratio is to be raised rather than merely the arithmetic ratio in which the two compared ratios stand. Thomas Bradwardine in his *Tractatus de proportionibus*, ed. H. L. Crosby, Jr. (Madison, 1955; 2d printing, 1961), p. 126, for example, speaks of "the ratio of any two spheres as three-halves the ratio of their surfaces" ("Quarumlibet duarum spherarum proportio ad proportionem superficierum suarm eodem ordine sesquialtera comprobatur"), when he obviously intends that

$$\frac{V_1}{V_2} = \left(\frac{S_1}{S_2}\right)^{\frac{3}{2}}.$$

Similarly, Kepler, in his third law (which is, of course, entirely original with him), says that "the ratio between the periodic times of any two planets is precisely three-halves the ratio of their mean distances" (*Harmonices mundi*, Bk. V, ed. C. Frisch, Joannis Kepleri, *Opera omnia*, Vol. 5 [Frankfurt, 1964], 279; ". . . proportio, quae est inter binorum quorumcunque planetarum tempora periodica, sit praecise sesquialtera proportionis mediarum distantiarum . . ."), by which he obviously means that

$$\frac{T_1}{T_2} = \left(\frac{D_1}{D_2}\right)^{\frac{3}{2}}$$

and so he is still using the medieval exponential language. See the new edition of Oresme's *De proportionibus proportionum* (ed. Grant), where the exponential language is used throughout.

of its mean velocity. Let me say a word about the last of these contributions, since the invention of the mean speed theorem is one of the true glories of fourteenth-century science.

While we are in doubt as to the exact origin of the mean speed theorem, the earliest statement of it that we can pin down to a specific date occurs in the *Regule solvendi sophismata* of William Heytesbury, dated in 1335. In that statement of the theorem, we see the law given as an obvious truth without proof, as follows:

For whether it commences from zero degree or from some [finite] degree every latitude (i.e., increment of velocity or velocity difference), as long as it is terminated at some finite degree (i.e., of velocity), and as long as it is acquired or lost uniformly, will correspond to its mean degree. Thus the moving body, acquiring or losing this latitude (increment) uniformly during some assigned period of time, will traverse a distance exactly equal to what it would traverse in an equal period of time if it were moved uniformly at its mean degree of velocity.[24]

Other early statements of the mean speed theorem without proofs are found in the two short treatises on motion by Swineshead, which are attached to the *Liber calculationum* in a Cambridge manuscript.

However the mean speed law might have arisen at Merton, a number of interesting proofs of the theorem were given in the various works attributed to Heytesbury, Swineshead, and Dumbleton. While the relative chronological order of these proofs is by no means certain, we should not be too far wrong in supposing that the theorem was discovered some time during the early 1330's and that all of these proofs were completed before 1350.

All of these Merton proofs are of an arithmetical and logical character, while Galileo's proof of the mean speed theorem is strictly geometrical, as we shall observe. Where does the latter come from? The answer is also from the fourteenth century. The kinematic studies of the English spread by 1350 to France. And at Paris there developed a system of graphing movements by the use of a kind of co-ordinate geometry. The most complete treatment of this system—which came to be known as "latitudes of forms" or, better, "configurations of forms"—was that developed by the famous Parisian schoolman, Nicole Oresme, in two treatises, the first entitled the *Questions on the Geometry of Euclid* and the second *De configurationibus qualitatum et motuum*.

The system is interesting enough to describe briefly, even though its

[24] Clagett, *The Science of Mechanics*, pp. 262–63. The next several paragraphs depend heavily on my account of kinematics in Chapter Five of this work.

outlines are fairly well known. It was equally applied to qualities and to motions. In the case of qualities, the abscissa or base line represented the extension of the quality in a subject, while perpendiculars erected on the line of extension represented the intensities of the qualities at the various points of the subject or extension line. (See Fig. 1.) Thus, the whole two-

Fig. 1.

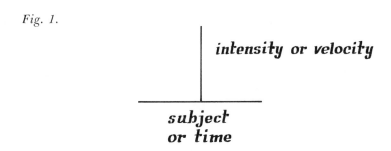

intensity or velocity

subject
or time

dimensional figure was used to represent the quantity of a quality. Hence, a rectangle represented a uniformly intense quality, while a right triangle represented a uniformly non-uniform quality. (See Figs. 2–4.) It should be obvious that when applied to motion the rectangle figures a uniform motion and the right triangle a uniform acceleration from rest. Now Oresme in the third part of his *De configurationibus* gave a geometrical proof of the Merton uniform acceleration theorem of the same kind as given by Galileo much later. The rectangle is employed to represent the uniform motion at the mean degree of velocity and the right triangle to represent the uniform acceleration. (See Fig. 5.) Oresme then showed by simple plane geometry that the two areas were equal. By the way, the quantity of velocity (equivalent to the area of the whole figure, or as we should say "the area under the curve") was definitely related by Oresme to the distance traversed, as it was later by Galileo in his explanation of the corollary to the theorem.

Its ultimate influence on Galileo's treatment of the law of free fall can, I believe, be shown without any doubt. I shall not try to do so here, for the details of the argument would unduly extend this paper and they are given in my forthcoming edition of the *De configurationibus*, where I have attempted to trace the step-by-step passage of the configuration doctrine from the fourteenth through the sixteenth century.[25] I should like, however, to say something here about the origins of the system with Oresme. I sug-

[25] *Nicole Oresme and the Medieval Geometry of Qualities and Motions* (Madison, in press), Chap. Two, Part B.

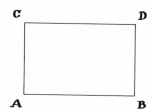

Fig. 2.—*A uniform linear quality or uniform velocity*

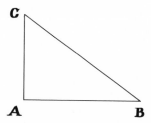

Fig. 3.—*Uniformly difform quality beginning at zero degree or uniform acceleration from rest*

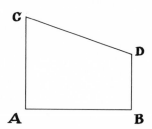

Fig. 4.—*Uniformly difform quality from a certain degree or uniform acceleration beginning at a certain velocity*

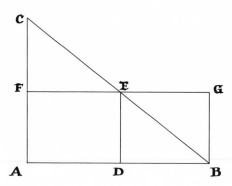

Fig. 5.—*Oresme's geometric proof of the Merton rule of uniformly difform*

288

gested some years ago that the system essentially grew out of an analogy drawn between the quantity of quality or motion and the area and dimensions of a surface.[26] At that time I pointed to the analogy as given in the brilliant *Liber calculationum* of Richard Swineshead, perhaps written a few years prior to the two main works of Oresme. Now recently I have discovered the presence of the analogy in Oresme's mind in two of his own early works: *The Questions on the Generation and Corruption of Aristotle* and *The Questions on the Physics of Aristotle*, works that are extant only in manuscript. The pertinent passage in the first work occurs in the midst of a question as to whether an indivisible could be altered:

Fourth corollary: if something were to be difform in quality, there would be more quality in the whole than in some part of it. *Whence quality is to be imagined to have two dimensions: longitude according to the extension of the subject and latitude according to intensity in degree.* Then there is a second conclusion, that if the whole quality of one body were in a point it would be infinitely intense. This is proved, for the subject is uniform. By the preceding conclusion the quality of the whole is double the quality of the half. Therefore, if by imagination the whole were placed in one half of the subject, it would be twice as intense as before. . . . It is obvious, if one uses a surface as an example, that if it is two feet long and one foot high and if it is made half as long while the total quantity of the surface remained as before, then it would be twice as high. And in the same way, if the whole quality were placed in a third part, it would be triply intense; and if in a fourth part, quadruply intense, and so on without end. Therefore, if the whole were placed in a point, it would be infinitely intense. . . . A third conclusion is that it does not follow from this that that quality would be infinite [in quantity]. This is evident in the first place because it is the same as before; it was not augmented, for, although infinitely intended, its extension has been proportionally divided to infinity, as is imaginable in regard to a surface.[27]

The basic idea, presented here so lucidly, is that we can imagine a quality as having two dimensions, longitude and latitude, the one associated

[26] Clagett, *The Science of Mechanics*, pp. 335–36.
[27] Bk. I. Quest. 20, MS Florence, Bibl. Naz. Centr., Conv. Soppr. H.ix.1628, 40ᵛ–41ʳ: "4ᵐ corollarium: quod, si esset aliquid difformiter quale, tunc esset maior qualitas in toto quam in aliqua eius parte. Unde qualitas ymaginatur habere duas dimensiones, scilicet longitudinem secundum extensionem subiecti et latitudinem secundum intensionem in gradu. Tunc est secunda conclusio, quod si tota qualitas unius corporis esset in puncto, illa esset in infinitum intensa. Probatur quia subiectum est uniforme. Per precendentem conclusionem qualitas totius est dupla ad qualitatem medietatis. Igitur si per ymaginationem tota poneretur in una medietate subiecti, esset in duplo intensior quam ante. Patet statim quia aliter non esset tanta qualitas sicud ante, quia si ipsa esset equaliter (41ʳ)

with the extension of the subject of the quality and the other with its intensity expressed in degrees. These, of course, are the very names that Oresme adopts in his fully developed graphing system. Because of this imagined two-dimensionality of the quality, we can accordingly imagine its quantity by means of a surface, with the quantity of the surface varying in the same way as the quantity of the quality. That Oresme also thought of the surface analogy as applying to motion is clear from a passage in his *Questions on the Physics*:

This then is the first conclusion concerning a motion qualifiedly infinite: A motion infinite in velocity can be produced in a finite time. Proof: It is possible for something to be moved during the proportional parts of an hour, first with some velocity, then twice as fast, then three times, four times, and so on, this being accomplished by the diminution of resistance to infinity. Then I infer by way of corollary that such a motion would be qualifiedly infinite, for it would only traverse a finite space, namely four times the space traversed in the first [proportional] part [of the hour]. One could concede this to be demonstrable as follows. It having been posited that in the first part it traverses the space of a foot, then if it were moved with such a degree through the remainder [of the hour] it would traverse two feet in the whole time. Again, if with the latitude acquired in the second part it would be moved through that second part and the rest [of the hour], it would traverse one foot. And if it were moved with the latitude of the third part through that part and the rest [of the hour], it would traverse one half foot, and then one half of one half, and so on in this way continually; it traverses [all together] precisely as much as that posited in the case; hence it traverses four feet. . . . And if upon a line of two feet in length, which is divided into proportional parts, one proceeded in that way so that upon the first part there would be a surface one foot in altitude, and [on the second one of two feet, and] on the third one of three feet, and so on continually, then this [total] surface would be equal to a surface four feet in length and one foot wide.[28]

intensa, cum ipsa sit in duplo minus extensa, sequitur quod esset in duplo minor, quod est contra ypothesim. Patet in exemplo de superficie, quod si sit longa duobus pedibus, alta uno pede, et si fiat minus longa in duplo et maneat tanta sciud ante, esset in duplo magis alta. Et eodem modo si qualitas totius poneretur in parte ⟨tertia⟩, esset in triplo plus intensa; et si in 4ᵃ, in quadruplo, et sic sine fine. Igitur si tota poneretur in puncto, ipsa esset in infinitum intensa, quod est propositum. Et eodem modo argueretur, si subiectum esset difforme et per ymaginationem posset reduci ad uniformitatem. . . . Tertia conclusio est quod ex hoc non sequitur quod illa qualitas esset infinita. Patet primo quia est eadem quam prius; erat non augmenta quia licet sit in infinitum intensa tamen eius extensio est in infinitum divisionata proportionaliter, sicud patet ymaginari de superficie."

[28] *Questiones super libros physicorum*, Bk. VI, Quest. 8, Seville, Bibl. Col. 7.6.30, 71ʳ, c. 2: "Tunc de motu infinito secundum quid est prima conclusio, quod infinitus motus in velocitate potest fieri tempore finito. Probatur, quia possible est quod secundum partes

This application of surfaces to illustrate a line of reasoning about velocities varying in time reappears in more formal dress in Chapter III. viii of the *De configurationibus*, which I shall discuss later. But in the *Questions on the Physics*, it is being given as a geometrical illustration or analogy to a preceding verbal argument. I should also point out there seems to be no further use or treatment of the nascent configuration doctrine in the *Questions on the Physics*. I am thus reasonably confident that these questions as well as those on the *De generatione* date from a period prior to the composition of the *De configurationibus* and even prior to his *Questions on the Geometry of Euclid*, where he first outlines the configuration doctrine in detail. This expression of the analogy as present in the earlier works of Oresme seems then to be the last step prior to a full exposition of the system, and, as I have said, Oresme's first full exposition came in the *Questions on the Geometry of Euclid*, written some time close to 1350, I would suppose as a part of his arts teaching assignments. While I shall not give here a lengthy exposition of the doctrine as outlined in the *Questions* or distinguish the treatment adequately from his later exposition in the *De configurationibus qualitatum et motuum*, I should like to share with you my discovery of a rather remarkable coincidence in Oresme's treatment of the rule of uniformly difform with that of Galileo in the *Discorsi*. In order to do this, let me say a word about the use of the configuration doctrine by Galileo.

Galileo's use of the configuration doctrine is most fruitfully present in his proof of the fundamental uniform acceleration theorem (Theorem I, Proposition I) in the "Third Day" of his *Discorsi e dimonstrazioni matematiche intorno a due nuove scienze*,[29] where Galileo states the Merton College acceleration theorem in slightly different form and gives a geometric proof employing a right triangle to represent uniform acceleration from rest and a rectangle to represent uniform motion at the speed of the middle

---

proportionales hore aliquid moveatur aliqua velocitate, deinde duplo velocius, deinde triplo, et quadruplo, et cetera, et hoc propter diminutionem in infinitum resistentie. Tunc infero corollarie quod talis motus esset infinitus secundum quid quia non pertransiret nisi spacium finitum, videlicet quadruplum ad pertransitum in prima parte; sicut posset concedere demonstrari, quia posito quod in prima parte pertranseat spacium pedem, tunc sit tali gradu moveretur per residuum pertransiret duplos in toto tempore; et iterum si latitudine asquisita in secunda parte moveretur per ipsam et per residuum pertransiret unum pedem; et si latitudine tertie moveretur per ipsam et per residuum pertransiret dimidium pedem, deinde medietatem medietatis, et sic semper et continue, precise tantum pertransit in casu posito; quare pertransit 4^or pedes. . . . Et si super lineam bipedalem divisam per partes proportionales fieret illo modo: super primam partem esset superficies pedalis altitudinis [et super secundam bipedalis] et super tertiam tripedalis et sic semper, tunc ista superficies equivaleret uni superficiei 4^or pedum in longum et unius in latum."

[29] *Le Opere, ed. naz.*, Vol. 8 (Florence, 1898), 208–12, where the first two theorems and the first corollary to the second theorem are given. For an English translation, see

instant of the period of acceleration. Theorem I states: "The time in which a certain space is traversed by a moving body uniformly accelerated from rest is equal to the time in which the same space would be traversed by the same body traveling with a uniform speed whose degree of velocity is one-half of the maximum, final degree of velocity of the original uniformly accelerated motion." This theorem is the basis of the proof of his celebrated second theorem: "If some body descends from rest with a uniformly accelerated motion, the spaces traversed in any times at all by that body are related to each other in the duplicate ratio of these same times, that is to say, as the squares of these times." This theorem was used for the proof of its first corollary, which held that in uniform acceleration from rest the spaces traversed in any number of equal and consecutive time periods starting from the first instant of motion "will be related to each other as the odd numbers beginning with unity, i.e., 1, 3, 5, 7. . . ." And so we see that Galileo has given three forms of the acceleration theorem. Now let us step back to Oresme's *Questions on the Geometry of Euclid*. The *Questions* first includes the conventional form of the Merton rule which measured a uniformly difform quality or motion by its middle degree:

The penultimate [conclusion] is that from this latter together with the aforesaid it can be proved that a quality uniformly difform is equal to the middle degree, i.e., that it would be just as great in quantity as if it were uniform at the middle degree. And this can be proved as for a surface.[30]

Although Oresme probably gave no proof of the theorem here, there is one manuscript which includes a garbled proof.[31] The proposition is repeated in a somewhat different form in Question 15:

Then the first proposition is that it is impossible for *b*, which is uniformly difform to no degree, to have as much quality as *a* [which is equal in subject and is

Clagett, *The Science of Mechanics,* pp. 409–16. A discussion of the Merton theorem with the appropriate literature has been given in the same work, Chaps. 5 and 6, and pp. 630–31n, 646–47, 649, 654–55. Note that the medieval form of the theorem usually emphasized that the spaces traversed in equal times were equal, while Galileo's theorem states that the times for the traversal of the equal spaces are equal, a change in wording without great mathematical significance.

[30] Nicole Oresme, *Quaestiones super Geometriam Euclidis,* Quest. 10, ed. H. L. L. Busard (Leiden, 1961), p. 28, ll. 8–11. (Note: I have given a new edition and English translation of Questions 10–14, and the beginning of 15, in Appendix I of my *Nicole Oresme and the Medieval Geometry of Qualities.* There are no significant changes in any of the passages quoted below, but in the full proofs there are numerous changes and the new text should be consulted, if possible.)

[31] See n. 6 of my English translation of Question 10 in the above cited Appendix, where I discuss the corrupt proof appearing in MS Seville, Bibl. Colomb. 7.7.13.

uniform in quality], unless it begins from a degree double [the intensity of *a* and ends with no degree].[32]

For this proposition, one manuscript gives the well-known diagram of a right triangle and its equivalent rectangle constructed on the same base (Seville, Bibl. Colomb, 7.7.13, 109ʳ), i.e., the figure that appears in *De con-figurationibus*, Chapter III.vii. (See Fig. 5.) Oresme in the *Questions* also recognized that the proposition held for uniform deceleration to rest:

If *a* is moved uniformly for an hour and *b* is uniformly decelerated in the same hour by beginning from a degree [of velocity] twice [that of *a*] and terminating at no degree, then they will traverse equal distances, as can be easily proved.[33]

Since Oresme stresses here that it is equality of distance (in the same time) that is involved, it is evident that when talking about velocities, he conceives of the areas of the surfaces as representing total distances traversed. It is then clear that Oresme, even in the earlier work, understood what was to be the substance of Galileo's first theorem and, as we pointed out, Oresme later added a formal geometric proof in his *De configurationibus*, Chapter III. vii.

But in addition to the Merton rule for uniformly difform qualities and motions, Oresme has given in his early work a proposition that is formally equivalent to Galileo's second theorem relating the distances to the squares of the times, a proposition usually considered as entirely original with Galileo. It is true, however, that Oresme's proposition is applied to qualities rather than directly to velocities. Oresme asserts:

The second is that, in the case of a subject uniformly difform [in quality] to no degree, the ratio of the whole quality to the quality of a part terminated at no degree is as the square of the ratio of the whole subject to that part [of the sub-ject]. This is evident in the first place because, by the preceding [proposition] such triangles and such qualities [which the triangles represent] are similar. But the ratio of similar triangles is as the square of the ratio of a side to a corresponding side, by VI.17 [of the *Elements*], but the extension of the subject is as the side of the triangle designating the quality. From this, the proposition is evident. . . . A third argument for it [i.e., the proposition] is, that by the first question [i.e., Question 10], it is obvious that any such quality would be equal to its middle degree with respect to intensity. Therefore, the [equivalent] intensity of the whole [uniformly difform quality] is twice the [equivalent] intensity of its half, and also the extension [of the whole is twice] the extension [of its half].

[32] Ed. Busard, p. 42, ll. 6–8.
[33] *Quaestiones,* ed. Busard, Quest. 13, p. 37, ll. 17–20.

Hence, by the second question [i.e., Question 11, giving a rule for comparing the quantities of two uniform qualities one of which is both twice as intense and twice as extended as the other], the ratio of the qualities is as the square of the ratio of subject to subject.[34]

Oresme later in the same question notes that one can speak in the same way "of velocities with respect to time." Hence, if such a transfer to velocities is made so that the quantities of the velocities, i.e., their total distances traversed (e.g. represented by triangles *ABC* and *ADE* in Fig. 6), are considered in the place of quantities of qualities, and their times (e.g. represented by lines *AB* and *AD*) are considered in place of subjects of qualities, we will have Galileo's second theorem. Furthermore, one of Oresme's arguments in proof of the proposition is based on the Merton rule for uniformly difform, just as Galileo's proof of his second theorem is based on his expression of the Merton rule found in his first theorem.

Fig. 6.

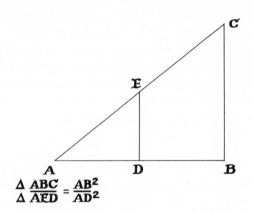

$$\frac{\triangle\ ABC}{\triangle\ AED} = \frac{AB^2}{AD^2}$$

Finally, it should be remarked that Oresme also has given in the *Questions* a proposition equivalent to Galileo's corollary to Theorem II holding that in the case of uniform acceleration from rest the distances traversed in equal, consecutive time periods are as the odd numbers: 1, 3, 5, 7, . . . Again, it should be noted that Oresme has framed the proposition in terms of qualities:

The second conclusion is that, with a subject so divided [into equal parts] and with the most remiss part designated as the first part, the ratio of the partial

[34] *Ibid.*, Quest. 13, p. 36, ll. 20–27; l. 32 to p. 37, l. 4.

294

qualities, i.e., their mutual relationship, is as the series of odd numbers where the first is 1, the second 3, the third 5, etc., as is evident in the figure.[35] (See Fig. 7.)

Fig. 7.

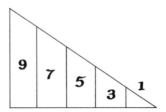

Oresme's proof, it will be clear to the reader consulting the full text, goes back ultimately to the proposition relating the quantities of the qualities to the squares of the subject lines. Similarly, Galileo proved his corollary by means of his Theorem II relating distances to the squares of the times. One must presume that Oresme would also have recognized that his proposition held for the distances traversed, since throughout the work he stresses the applicability of his propositions to velocities as well as qualities. And, in fact, the first step toward Oresme's odd-number theorem was taken by the Merton College authors in the context of velocities when they proved that a body uniformly accelerated from rest traverses three times as much space in the second half of the time as in the first.[36]

In view of the fact that Oresme drew much the same consequences from the Merton rule as did Galileo, and in view of the further fact that the order and substance of proofs in the two authors is essentially the same, we might well ask whether Galileo read Oresme's *Questions*. For the present, the answer certainly must be in the negative. There is no evidence of Galileo's having used any Oresme manuscripts, and it must be remembered that Oresme's *Questions on the Geometry of Euclid* was not published until the twentieth century. I hasten to add, however, that while I cannot connect this particular work with Galileo, there is no doubt in my mind that Galileo was the heir, in at least an indirect way, of the medieval configuration doctrine. And, in fact, I believe that one can make a fair case for identifying the particular published works from which he drew his general knowledge of the system and his particular knowledge of the mean speed theorem with its geometric proof.

[35] *Ibid.*, Quest.. 14, p. 38, ll. 29–32.
[36] Clagett, *The Science of Mechanics*, p. 266.

In centering my attention on the development of kinematics with its geometric representation by Oresme, I should also point out that as a part of this evolution of kinematics there appears a remarkable treatment of the summation of infinite series. This appears to be almost wholly a medieval development and largely a product of the fourteenth century. Again, the key names are those at Merton College, such as, Swineshead, and Oresme at Paris. It may surprise some of you that the treatment of the summation of infinite series had its beginning in the consideration of Aristotle's views of the infinite rather than in Greek mathematical works. That is to say, it arose in a strictly philosophical context. The point of departure seems to have been Book III, Chapter 6 (206^b, 3–12), of the *Physics* of Aristotle, where Aristotle says that in a certain way an infinite according to addition is the same as an infinite according to division.[37] The example given is of a finite magnitude, which is divided successively according to a given ratio (in medieval terminology "divided according to proportional parts"). And so the parts of the line produced by this division when added together produce the finite magnitude. It is but a small jump from what the passage says to the conclusion that, if the division is completed to infinity, the whole original finite quantity is produced by the addition of an infinity of parts. In his exposition of this passage, Aquinas gives, as a specific example of this division, a cubit line divided first into halves, then one of the halves into halves, one of the quarters into halves, and so on to infinity.[38] Hence, if one thought of the infinite division as completed, we would then have by addition the following infinite series:

$$(1) \qquad \tfrac{1}{2} + \tfrac{1}{4} + \tfrac{1}{8} + \ldots + \tfrac{1}{2^n} + \ldots = 1$$

I should add, however, that neither Aristotle nor Aquinas seems to have actually thought of the infinite series as completable, since the division to

[37] See the Medieval Latin translation of Moerbeke accompanying Thomas Aquinas, *In octo libros physicorum Aristotelis expositio*, ed. P. M. Maggiòlo (Turin, 1954), p. 183 (Text. No. 59): "Quod autem secundum appositionem idem quodammodo est et quod est secundum divisionem. Infinitum enim secundum appositionem fit e contrario: secundum quod enim divisum videtur in infinitum sic appositum videbitur ad determinatum. In finita enim magnitudine si accipiens aliquis determinatum, accipiet eadem ratione, non eandem aliquam magnitudinem ratione accipiens, non transibit finitum. Sin vero sic augmentet rationem, ut semper eandem aliquam sit accipere magnitudinem, transibit finitum, propter id quod omne finitum absumitur quolibet finito. Aliter quidem igitur non est, sic autem est infinitum, potentia et divisione. Et actu autem est, sicut diem esse dicimus et agonem; et potentia sic sicut materiam, et non per se sicut finitum. (Text No. 60) Et secundum appositionem igitur sic infinitum potentia est, quod idem dicimus quodammodo esse ei quod est secundum divisionem: semper quidem enim aliquid ipsius extra est accipere."

[38] *Ibid.*, p. 185, ". . . puta si a linea cubitali accipiat medietatem, et iterum a residuo medietatem; et sic in infinitum procedere potest. . . ."

infinity was only potential rather than actual. However, by the time of the fourteenth century, the schoolmen were blithely assuming the summation of this and other more complicated series. Furthermore, it is important to realize that most of the treatment of the more complicated series rests on the assumption of the summation of this first simple series. To show this to be so, let us take the most popular of the series summed by the schoolmen of the fourteenth, fifteenth, and sixteenth centuries:

$$(2) \qquad 1 + \tfrac{1}{2} \cdot 2 + \tfrac{1}{4} \cdot 3 + \ldots + \tfrac{1}{2}^{n-1} \cdot n + \ldots = 4$$

It should be noted that this was the series which we have already mentioned as appearing in geometric form in Nicole Oresme's *Questions on the Physics of Aristotle*. Let us now look at the more complete geometric treatment of it given by Oresme in his *On the configurations of qualities and motions* (see Fig. 8):

Fig. 8.

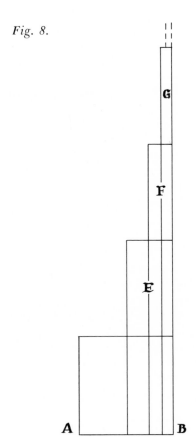

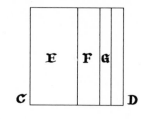

III.viii. On the measure and intension to infinity of certain difformities. A finite surface can be made as long as we wish, or as high, by varying the extension without increasing the size. For such a surface has both length and breadth and it is possible for it to be increased in one dimension as much as we like without the whole surface being absolutely increased so long as the other dimension is diminished proportionally, and this is also true of a body. For example, in the case of a surface, let there be a surface of one square foot in area whose base line is *AB*; and let there be another surface, similar and equal to it, whose base line is *CD*. Let the latter surface be imagined to be divided on line *CD* to infinity into parts continually proportional according to the ratio of 2 to 1, with its base divided in the same way. Let *E* be the first part, *F* the second, *G* the third, and so on for the other parts. Therefore, let the first of these parts, namely *E*, which is half the whole surface, be taken and placed on top of the first surface towards the extremity *B*. Then upon this whole let the second part, namely *F*, be placed, and again upon the whole let the third part, namely *G*, be placed, and so on for the others to infinity. When this has been done, let base line *AB* be imagined as being divided into parts continually proportional according to the ratio 2 to 1, proceeding toward *B*. And it will be immediately evident that on the first proportional part of line *AB* there stands a surface one foot high, on the second a surface two feet high, on the third one three feet high, on the fourth four feet high, and so on to infinity, and yet the whole surface is only the two [square] feet [in area] previously given, without augmentation. And consequently the whole surface standing on line *AB* is precisely four times its part standing on the first proportional part of the same line *AB*. Therefore that quality or velocity which would be proportional in intensity to this figure in altitude would be precisely four times the part of it which would be in the first part of the time or the subject so divided. For example, let the first part (towards extreme *A*) of the proportional parts divided along *AB* according to the ratio 2 to 1 be a certain amount white or hot, the second twice as white [intensively], the third three times as white, the fourth four times, and so on to infinity on both sides according to the [natural] series of [whole] numbers. Then from the prior statements it is apparent that the total whiteness of line *AB* is precisely four times the whiteness of the first part; and it would be the same for a surface [whiteness], or for a corporeal whiteness, if it were increased in intensity in a similar fashion. In the same way, if some mobile were moved with a certain velocity in the first proportional part of some period of time, divided in such a way, and in the second part it were moved twice as rapidly, and in the third three times as fast, in the fourth four times, and increasing in this way successively to infinity, the whole velocity would be precisely four times the velocity of the first part, so that the mobile in the whole hour would traverse precisely four times what it traversed in the first half of the hour; e.g., if in the first half or proportional part it would traverse one foot, in the whole remaining period it would traverse three feet and in the total time it would traverse four feet. . . .[39]

[39] *The Science of Mechanics*, pp. 380–81. A slightly different new edition of this appears in my *Nicole Oresme and the Medieval Geometry of Qualities and Motions*, and it is from

It is evident that what Oresme has done by piling up the surfaces is to indicate that the original series (2) is equivalent to the following series:

$$(3) \qquad 2 + [1 + \tfrac{1}{2} + \tfrac{1}{4} + \ldots + \tfrac{1}{2}^{n-1} + \ldots] = 4$$

And indeed the subseries in brackets is not "proved" in this work, it being merely stated that one of his square feet is divided into proportional parts according to the ratio of 2 to 1. But he has given a proof of this in his earlier *Questions on the Geometry of Euclid*,[40] where a proof is expounded for the general series:

$$(4) \qquad \frac{a}{b} + \frac{a}{b}\left(1 - \frac{1}{b}\right) + \frac{a}{b}\left(1 - \frac{1}{b}\right)^2 + \ldots + \frac{a}{b}\left(1 - \frac{1}{b}\right)^{n-1} + \ldots = a$$

$\dfrac{a}{b}$ being the first aliquot part of a to be removed; and series (1), i.e.,

$$\tfrac{1}{2} + \tfrac{1}{4} + \tfrac{1}{8} + \ldots + \tfrac{1}{2}^{n} + \ldots = 1,$$

is shown to be a corollary of the general series.

Now series (2) summing at 4, with which I opened this discussion, was also given by Richard Swineshead in his *Liber calculationum*, with the objective of determining the average intensity of a quality whose subject has been divided into proportional parts of increasing intensity.[41] And

---

this rather more complete text that my translation has been made. See H. Wieleitner, "Über den Funktionsbegriff und die graphische Darstellung bei Oresme," *Bibliotheca Mathematica*, 3. Folge, Vol. 14 (1913–14), 231–33.

[40] *Quaestiones*, ed. Busard, Quest. 1, p. 2, ll. 23–34; p. 3, ll. 1–3; Quest. 2, p. 5, ll. 9–15. See the review of Busard's edition of the *Quaestiones* by John Murdoch in *Scripta mathematica*, Vol. 27 (1964), 68.

[41] "De difformibus," in the *Calculationes* (Pavia, 1498), pp. 14–15 (cf. MSS Cambridge, Gonville and Caius 499/268, 168r–v; Paris, BN lat. 6558, 6r–v; Pavia, Bibl. Univ. Aldini 314, 5v–6r): "Prima tamen opinio de qualitate difformi, cuius utraque medietas est uniformis, potest sustineri, scilicet quod corresponderet (*C*, respondeat *Ed*) gradiu medio inter illas qualitates. Et fundatur argumentum super illo: in duplo plus facit qualitas extensa per totum subiectum ad totius denominationem quam si sola (*C*, tota *Ed*) per medietatem extenderetur, quod arguitur sic. Signetur *a* quod habeat caliditatem ut 4 per totum. Tunc totum erit (*C*, est *Ed*) calidum per totam caliditatem ut 4. Sed una medietas illius tantum facit ad denominationem totius (*C, om. PEd* subiecti) sicut alia medietas. Igitur tota illa qualitas in duplo plus denominat totum quam una eius medietas (*C*, pars sive medietas *Ed*) totum denominat, quod fuit probandum. Ex quo sequitur quod denominatio totius subiecti per qualitatem extensam per medietatem subiecti (*C, om. Ed*) solum est subdupla ad illam qualitatem, quia in duplo minus dominat totum quam medietatem illam per quam extenditur, et illam medietatem denominat gradu suo (*C*, summo *Ed*). Igitur totum per illam qualitatem denominatur gradu subduplo ad illam qualitatem. Et si extenderetur (*C*, extendatur *Ed*) per quartam totius, tantum tunc denominaret totum gradu subquadruplo ad illam qualitatem. Et sic correspondenter sicut proportionaliter extenditur per minorem partem quam est totum, ita totum denominat (*CP*, nominatur *Ed*) gradu remissiori quam partem (*C*, pars *Ed*) per quam illa (*C, om. Ed*) extenditur. . . . Contra quam positionem et eius fundamentum arguitur sic, quia sequitur quod si prima pars proportionalis alicuius esset aliqualiter intensa, et secunda in duplo intensior, et

Swineshead's treatment was made the object of a Renaissance commentary by the Portuguese master at Paris, Alvarus Thomas, in his *Liber de triplici motu* (Paris, 1509), where the series is proved in a more general form,[42] with the division into proportional parts made according to any ratio and not just the ratio of 2 to 1. Incidentally, this brilliant tract illustrates exceedingly well the continuity of the discussion of infinite series between the Middle Ages and the Renaissance.

But let us turn away from the Swineshead type of expression and proof to another fourteenth century proof appearing in one copy of a short philosophical tract entitled from its incipit *A est unum calidum*,[43] and perhaps composed by Johannes Bode. The author states the conclusion of the series as applied to velocities in almost exactly the same way as Oresme did in the last part of Chapter III.viii, which I have just quoted. In his proof the author ingeniously transforms series (2) into the following series:

$$(5) \qquad 1 + 1 + 1 + [\tfrac{1}{2} + \tfrac{1}{4} + \tfrac{1}{8} + \ldots + \tfrac{1}{2^n} + \ldots] = 4$$

---

tertia in triplo, et sic in infinitum, totum esset eque intensum precise sicut est secunda pars proportionalis, quod tamen non videtur verum. Nam apparet (*add. C ex illa conclusione*) quod illa qualitas est infinita, ergo si sit sine contrario infinite denominabit suum subiectum. . . . (15, c. 2) . . . Ad argumentum in oppositum negatur consequentia: hec est qualitas infinite intensa, ergo infinite totum subiectum denominat, quia illo modo extenditur quod infinite modicum faciet qualitas illa infinita respectu illius subiecti. Nam qualitas quarte partis proportionalis est in duplo intensior quam qualitas secunde partis proportionalis et est per (*CP, om. Ed*) subiectum subquadruplum (*PEd, subduplum C*) et ideo in duplo facit minus (*add. Ed* scilicet quam secunda *quod om. CP*). Si enim quarta pars proportionalis esset in octuplo (*EdP, duplo C*) intensior quam prima (*EdP, secunda pars C*), sicut est in octuplo (*EdP, quadruplo A*) minor illa [in extensione], tunc tantum faceret ad totius denominationem sicut prima (*EdP, secunda C*). Sed quarta pars, ut constat, est nunc in duplo minus intensa quam tunc esset, quia nunc est in duplo remissior quam tunc esset. Ergo nunc quarta pars proportionalis in duplo minus facit in comparatione ad totum quam facit prima pars proportionalis et prima tantum facit comparatione ad totum ut secunda, ut patet. Igitur quarta pars proportionalis in duplo minus facit quam secunda ad totius denominationem (*C, intensionem APEd*), et tamen sua qualitas est in duplo intensior. Et sic discurrendo, quelibet qualitas extensa per partem posteriorem minus facit quam qualitas extensa per partem priorem, vocando partes priores que sunt propinquiores extremo ubi partes maiores terminantur. Et hoc est verum de omnibus partibus proportionalibus *a* nisi de prima et secunda que equaliter faciunt ad totius denominationem." I have made a few corrections from the three manuscripts (*C*-Cambridge; *P*-Paris; *A*-Pavia, Aldini) but only those which help to emend the edition so that it would make better sense. I have altered the punctuation freely.

[42] This edition is without pagination, but see Sig. p. 4v, c. 2–q 1r, c. 2. The full Latin text is given in my *Nicole Oresme and the Medieval Geometry*, Commentary to Chap. III.viii.

[43] The Latin text is given by H. L. L. Busard, "Unendliche Reihen in *A est unum calidum*," *Archive for History of Exact Sciences*, Vol. 2, No. 5 (1965), 394–95. I have also given the text independently with a somewhat different reading in several places in the *Nicole Oresme and the Medieval Geometry*, Commentary to Chap. III.viii.

Hence, since the subseries sums at 1, the total sums at 4, and he has his proof. The ingenious method of transformation is first applied to the velocity of the third proportional part of the time and then successively to each succeeding part. It is first noted that in the third part, and indeed in every succeeding part, the time span of that part is equal to the time remaining. The implication of this remark is that there are just as many instantaneous velocities in the third part as in the time remaining. And so, if we take from all the remaining velocities one degree and add it to velocities in the third proportional part, then the velocity during the third proportional part will be uniform at 4 (instead of at 3), and the uniform velocity of each of the remaining velocities will be reduced by one. Since the uniform velocity in the third proportional part is now 4 and so twice the velocity of the second part, and since it is half as extended in time, it is evident that in the third part of the time just as much space will be traversed as in the second; namely, a space of 1. Similarly the velocity of the fourth proportional part, having been reduced from its original 4 to a velocity of 3 by the increase of the velocity of the third proportional part, is now restored to 4 by borrowing a degree from all of the succeeding velocities. Hence, with its velocity now restored to 4 (and, hence, equal in velocity to that of the third part) and with a time period half that of the third part, the mobile in the fourth part traverses a distance half of that traversed in the third part. Similarly, the velocity of the fifth part, having been reduced by two borrowings to 3, is now raised to 4 by borrowing one degree from all the succeeding velocities. Hence, with a velocity now equal to that of the fourth part and with a time period half that of the fourth, the mobile will traverse in the fifth part half as much space as that traversed in the fourth. In the same manner, all of the succeeding velocities are, one by one, brought up to a velocity of 4 by successive borrowings, and since each part is one half that of the preceding one in time span, the space traversed in each proportional part of the time will be half that traversed in the preceding part. Hence, the transformation of the series is complete and the proof is obvious.

This proof was not without its influence in the Renaissance, since it was repeated (although with some minor differences) by the fifteenth-century Florentine Bernardus Torni, in his commentary to Heytesbury's *Regule solvendi sophismata*, published with that latter text in Venice, 1494.[44] In fact, Torni praises Oresme as the one who stimulated his consideration of the conclusion, but it was the proof from the *A est unum calidum* that he takes over. He has made one addition, namely, a proof of the summation

[44] Folios 76ᵛ–77ʳ. The full text is given in *Nicole Oresme and the Medieval Geometry*, Commentary to Chap. III.viii.

of series (1); that is, the subseries in the transformed series of the *A est unum calidum*: $\frac{1}{2} + \frac{1}{4} + \frac{1}{8} + \ldots + \frac{1}{2^n} + \ldots = 1$. His proof is based on Euclid V.13 (Greek text, V.12). Incidentally, Torni's treatment was also influential on the group of schoolmen at Paris in the early fifteenth century, perhaps on Alvarus Thomas, but certainly on Juan de Celaya, Luiz Coronel, and Jean Dullaert.[45]

Before leaving this rather exciting development of infinite series, let me underline that the discussion of all of these series occurs as part of the discussions of qualities and velocities and so as part of the treatment of hypothetical physical situations. Furthermore, it is worth pointing out that Swineshead and others actually employed the summation of infinite series in a very subtle way to prove such crucial theorems as the mean speed theorem and, hence, the use of such series was already improving the mathematical facility of the natural philosophers.[46]

In my brief search for trends in the fourteenth century—those influential and those not so influential—I have left some glaring omissions. Hence, I have not even mentioned the interesting progress made in statics.[47] This omission is partly justified in that the main medieval contributions to a quantified statics were made in the thirteenth rather than the fourteenth century and in essentially Hellenistic terms, although the author of one commentary on Jordanus' *Elementa de ponderibus* known as the *Aliud commentum* showed, in a way more clear than any before him, that Jordanus' proof of the law of the lever rested on the principle of virtual displacements. I have also barely touched upon the dynamical discussion of *impetus* and its rudimentary quantification in the hands of Jean Buridan, with the impetus described as being proportional to prime matter and velocity.[48] Its ultimate influence on sixteenth-century authors[49] is much more debatable than the kinematic doctrines here emphasized. Furthermore, it has been so often treated that I forbear further discussion. Nor have I commented on certain developments in mathematics in the fourteenth century, such as the increasing use of the Moerbeke translations of Archimedes at Paris and, in particular, their influence on the French mathematician Johannes de Muris.[50] (Incidentally, this same author also produced the most mature handbooks of mensuration and calculation in the fourteenth cen-

[45] Duhem, *Léonard de Vinci*, Vol. 3, 546–48.
[46] *The Science of Mechanics*, pp. 295–96.
[47] *Ibid.*, Chap. 2.
[48] *Ibid.*, pp. 522–25.
[49] *Ibid.*, pp. 653–57.
[50] For a preliminary summary of this growing use of Archimedes, see my *Archimedes in the Middle Ages*, Vol. 1 (Madison, 1964), Chap. I. Volume 2 of this work will treat the matter in more detail.

tury.)[51] My only hope is that I have left you with the impression of a fertile upheaval in natural philosophy in the fourteenth century that augured well for the scientific quickening that followed. And if, like Swineshead, you are inclined to see the clarification of change in the finding of means, perhaps you will be successful in assaying the scientific thought of the Renaissance in terms of a mean between the trends I have stressed for the fourteenth century and those which Mr. Drake will discuss as representing Galileo.

[51] These are his *Quadripartitum numerorum,* which gave numerous formulas that apply to physical science, all with numerical examples, and his *De arte mensurandi.* Both deserve critical editions.

MATHEMATICS, ASTRONOMY, AND PHYSICS IN THE WORK OF GALILEO ❧ STILLMAN DRAKE

I

ntil the present century it was customary to call Galileo the founder of modern physical science. Ancient science was thought of as having ended with the decline of Greek civilization, and no real contribution to scientific thought was known to have been made during the long ensuing period to the late Renaissance. The seemingly abrupt emergence of many recognizably modern scientific concepts early in the seventeenth century thus appeared to have been a true revolution in human thought. In that scientific revolution, Galileo appeared as the prime mover. The persecution that he suffered as a result of his active propagation of new ideas lent color to the idea that his had been a totally new and revolutionary kind of science.

As historians of ideas gave more careful attention to the treatment of scientific questions in medieval and early Renaissance times, the traditional role that had rather romantically been assigned to Galileo was critically re-examined. Many early anticipations of modern science or its fundamental concepts were found in manuscript commentaries on the philosophy of Aristotle and in treatises on statics forming what is known as the medieval science of weights. Anti-Aristotelian traditions were shown to have existed perennially in the universities, particularly those of Oxford, Paris, and Padua —traditions which had long been obscured by an overwhelming ascendancy of the peripatetic philosophy in the official curricula at most times. Study of works by such men as John Philoponus, Jordanus Nemorarius, Thomas Bradwardine, Walter Burley, Robert Grosseteste, Nicole Oresme, and Jean Buridan, coupled with evidence that their writings had circulated widely among scholars, first in manuscript and later in printed form, suggested that modern science was not suddenly born with Galileo, but rather emerged about that time after a long period of incubation. Many scholars now question whether it is proper to speak of a scientific revolution as having occurred in the late Renaissance or at any other time and whether it would not be more accurate to characterize the emergence of modern science as a gradual event in a continuous process of thought which has merely had periods of slower and of more rapid development.

Though much may be said for this sophisticated modern viewpoint, it does tend to obscure a striking historical fact which the older, more naïve conception recognized and at least attempted to explain. The fact is that man's attitude toward the world about him and his control of natural phenomena have altered more in the four centuries that have elapsed since the birth of Galileo than in as many millennia before that time. However many of the fundamental ideas underlying modern science may have been in some sense anticipated by his predecessors, their insights had no marked effect on the pursuits of other men, as did those of Galileo. Whether one wishes to call his undoubted effectiveness revolutionary or not is a matter of taste; but whatever one calls it, the manner in which it is to be explained is still deserving of serious study.

The principal sources of Galileo's effectiveness as I see them were intimately related to the temper of the time in which he lived. But in saying that, I do not mean that the emergence of a Galileo at that period was in any way inevitable. Such social-deterministic views are coming to be more and more widely held, but to me they seem too smug, implying as they do that we now know so much that we can retroactively predict the appearance of a genius and that the spirit of an age is bound to produce one.

Reflecting on what I know of Galileo's contemporaries, I find it hard to select one who could have filled his place. Those who took up his work did not hesitate to credit him with having placed them on the path to further achievement. Those who opposed his work rarely even understood what Galileo was telling them, despite his clarity of expression; hence I think it unlikely that they would have found out the same things for themselves. Two outstanding candidates, Kepler and Descartes, each excelled him in one field—Kepler in astronomy and Descartes in mathematics—but they were both woefully deficient as physicists. And this leads me to the first point that I wish to make in explanation of Galileo's effectiveness.

Galileo was born into a world that already had a highly developed and technically advanced mathematical astronomy, but it had no coherent mathematical physics and no physical astronomy at all.[1] It was Galileo who, by consistently applying mathematics to physics and physics to astronomy, first brought mathematics, physics, and astronomy together in a truly significant and fruitful way. The three disciplines had always been looked upon as

---

[1] In the discussion following the presentation of this paper, the work of Simon Stevin was adduced as an example of sound mathematical physics prior to the work of Galileo. Galileo was born in 1564, and Stevin's first book on physics appeared in 1586, so the statement in question remains literally true. More important is the fact that Stevin first published in Dutch, so that his work remained relatively unnoticed until 1605, when a Latin translation appeared. As will be seen, Galileo was by that time far along with his own mechanical investigations. Similarly, Johannes Kepler contributed to physical astronomy, but his contributions appeared still later than those of Stevin and had little effect on Galileo's program of unification of mathematics, physics, and astronomy. The work of René Descartes began to appear only at the close of Galileo's life.

essentially separate; Galileo revealed their triply-paired relationships, and thereby opened new fields of investigation to men of widely divergent interests and abilities. Mathematical astronomy, mathematical physics, and physical astronomy have ever since constituted an inseparable triad of sciences at the very base of modern physical science. Therein, I think, lies the primary explanation of Galileo's effectiveness.

The inner unity of mathematics, astronomy, and physics is more often implied than overtly stated in the work of Galileo. It is to me doubtful whether a philosophical theory of that unity would have sufficed to produce the effects here under consideration. A contrary view may be taken by those historians who see in Galileo's work little more than a loyal carrying-out of the philosophical program of Plato, through showing that geometry indeed governs the physical universe. But there are difficulties with that view. Galileo did not acknowledge any leadership but that of Archimedes; he spent most of his life trying to persuade people not to swear by the words of any master; his notion of the fundamental role of geometry seems quite different from that of Plato; and eminent Platonists of the time were much better informed in that philosophy than Galileo, yet had no influence on the study of physical science. In any event, Galileo, rather than offering a general statement from outside concerning the relationship of mathematics, astronomy, and physics, set forth particulars within each pairing of those sciences which established their interconnections. And that suggests a second source of his effectiveness: he was an acknowledged expert in each of the three separate disciplines. In view of the recognized merit of his own contributions to each of them, it was evident that he knew what he was talking about, even when what he said appeared paradoxical or absurd. Astronomers were drawn to his physics by his eminence in astronomy; mathematicians took up his mechanics because he was a respected professor of mathematics.

One may contrast the case of Galileo with that of his contemporary Francis Bacon, who likewise formulated a program for the reform of knowledge. That program included three basic ideas which are also found in Galileo's works; namely, recognition of the inadequacy of traditional knowledge, an understanding of the obstacles standing in the way of any departure from it, and the suggestion of methods by which real advances in knowledge might be made. But Bacon, unlike Galileo, was no expert in mathematics or science; hence, though he could discern the goal and suggest a path, he did not actually lead others along that path to the goal. As he himself wrote in a letter to Dr. Playfere: "I have only taken upon me to ring a bell to call other wits together."[2] Bacon's program stressed the accumulation of obser-

[2] Cited from R. F. Jones, *Ancients and Moderns* (Berkeley, 1965), p. 41.

vational data, a valuable antidote to authority and a most useful procedure in the natural sciences. But it did not clearly open a new path to physical science, as did the bringing together of mathematics, astronomy, and physics. Bacon's bell served its purpose; but—at least so far as physical science was concerned—when the wits assembled, it was in Galileo's study.

Of course, to be an expert in as many as three separate fields was not uncommon for men of the Renaissance. That is a part of what I meant when I said that Galileo's effectiveness was closely related to the temper of his time. But to be an expert in each of the three disciplines now under discussion was not a likely combination at his time, for reasons which will presently appear. It was not unusual in 1600 to be an expert in mathematics, astronomy, and music, for example, or in astronomy, physics, and metaphysics. But Galileo's expertness in mathematics, astronomy, and physics was exceptional and in certain curious ways was largely accidental. Indeed, I believe that it was only gradually that Galileo himself fully discerned their essential interrelationships. Eventually he did so, and as a result he became a critic and a reformer of, as well as a contributor to science; that is, a man with a definitely formulated program for the study of natural phenomena, not just a technical specialist. That was, in my opinion, the third major source of his effectiveness.

Certain personal traits of Galileo's deserve to be mentioned, at least in passing, which also contributed to his effectiveness. One group of such traits comprised his scientific temperament; another aided him in getting a hearing for his views. First, he had extraordinary powers of observation and memory; even small effects seldom escaped his notice. Second, he was deeply interested in detecting relationships among seemingly scattered facts, though he was not inclined to multiply examples for their own sake. Third, his concern for precision was sufficient to protect him from indulging in rash conjectures, but not so great as to hinder him from declaring certain kinds of observed discrepancies or apparent contradictions to be irrelevant or negligible. This last trait is essential in a discoverer of truly fundamental laws, just as an inflexible and uncompromising precision is necessary for their refinement and extension. It is interesting that most of Galileo's scientific contemporaries tended to fret over detail, which fitted them admirably for the next step in physical science, but that is just what makes it hard for me to see any of them in Galileo's role, as pioneers in perceiving and putting faith in the existence of essential mathematical regularities in nature.

Among the personal traits of Galileo which heightened his effective presentation of the new sciences was a ready wit, which he used to enliven his writings and to explode any appeal made by his adversaries to authority or tradition. Also, he wrote particularly well in the vernacular tongue, enabling him to convey his program effectively to men of intelligence whether or not they were imbued with the prejudices then associated with academic learning. Finally, he was able to form lasting friendships with men of high station and to interest them personally in his campaign.

II

Let us now examine the origin and development of Galileo's program for the reform of physical science, describing first the period of his life that immediately preceded his full comprehension of the magnitude of his task. In so doing we shall be better able to see how his work contrasted with that of his predecessors and what traditions and prejudices he was obliged to break down.

For a period of more than twenty years, from 1589 to 1610, Galileo was a professor of mathematics, first at the University of Pisa and then at Padua. No record remains of his lectures at Pisa. At Padua, however, records show his assigned subjects for various years.

1593—leget Sphaeram et Euclidem
1594—leget quintum librum Euclidis et theoricas planetarum
1598—leget Euclidis Elementa et Mechanicas Aristotelis Quaestiones
1599—Leget Sphaeram et Euclidem
1603—leget librum De sphera et librum Elementorum Euclidis
1604—leget theoricem planetarum

By *librum De sphera* in the entry for 1603 we may assume that the text of Sacrobosco was meant and that the same text was explained in 1593 and 1599. The syllabus of Galileo's public lectures on that subject is preserved in five closely corresponding copies in manuscript, one of which was published in 1656 by Urbano d'Aviso as the *Trattato della Sfera di Galileo Galilei*. Apart from its clarity of style, it differs little from dozens of similar treatises of the epoch.

Galileo's lectures on Euclid's *Elements* probably also represented the usual course in geometry; if he wrote a syllabus for them, it has been lost. The fifth book of Euclid, however, was of particular interest to Galileo, and we have what is probably an edited version of his course of 1594 in a book published by Vincenzio Viviani in 1674 under the title, "The fifth book of Euclid or the universal science of proportion explained by the teaching of Galileo, arranged in a new order." Galileo's special interest in the theory of proportion had an important bearing on his method of applying mathematics to physics, but it is unlikely that such applications were discussed in his public lectures at Padua.

There is reason to believe that Galileo also wrote a syllabus for his course on the treatise *Questions of Mechanics*, then attributed to Aristotle though now ascribed to one of his disciples. That syllabus, now lost, must not be confused with Galileo's own treatise *On Mechanics*, drafted about 1593 and successively revised, which seems to have been used by Galileo in his private lessons to special pupils. At the end of that treatise and in a letter written in 161C, Galileo refers specifically to another, on questions of mechanics, which has not survived.

The basis of Galileo's lectures on planetary theory is conjectural, as again no syllabus has survived. Probably they constituted an exposition of Ptolemy's *Almagest*, for Galileo mentioned that he had composed a commentary on Ptolemy about 1590, which he intended to publish. It is improbable that his public lectures included an exposition of the Copernican theory, though the heliocentric theory was probably mentioned in them, as (contrary to the assertion of many modern historians) it was in his lectures on the sphere.

In short, Galileo's public teaching had the general characteristics of most courses offered in mathematics at any leading university around 1600. Even at the enlightened University of Padua, Galileo appears to have departed only to a small degree from conventional routine instruction. In the six years for which we have a definite record, he lectured only for one term on anything that might today be considered as related to physics—and even that course was designed as a commentary on an ancient treatise.

Physics at that time belonged to the department of philosophy rather than mathematics and was taught in universities as an exposition of Aristotle's *Physica*. Such a course was probably taken by Galileo as a student at Pisa under Francesco Buonamico or Girolamo Borro. Since the chair of mathematics at Pisa was vacant during Galileo's student days, his only formal instruction in astronomy was also associated with philosophy and not with mathematics. From the so-called *Juvenilia* preserved among his papers, it appears that he heard the lectures of Buonamico on Aristotle's *De caelo*, a purely speculative treatment of astronomy devoid not only of what we should call physics, but even of the mathematical theory of astronomical calculation which had already been carried to an impressive degree of accuracy.

It should be remembered that the entire tradition in professional astronomy up to that time was essentially technical rather than scientific. The task of astronomers was to improve the methods of describing and calculating the observed positions and motions of heavenly bodies, rather than to explain such motions physically. That tradition found strong support in the Aristotelian philosophy, which made a fundamental distinction between terrestrial and celestial matter and motions. Such a distinction had long been incorporated into Christian theology. Thus astronomy, like physics, had evolved along two distinct paths in formal education. The philosophical part of each science remained quite foreign to the mathematical part, in the minds of even the best-educated men. Astronomical calculations, moreover, which fell to the mathematicians, remained strictly kinematic. The science of dynamics had not yet been born, and even if terrestrial dynamics had existed, it would not have been applied to astronomy so long as celestial objects were regarded as pure lights constituted of quintessential material. But not even terrestrial dynamics existed; it was effectively precluded in scholarly circles by Aristotle's dictum that every body moved must be in contact with a separate mover. With respect to physics, all that was traditionally open to mathematical treatment was a small branch of mechanics, namely statics.

Galileo, however, even in his earliest studies, defied Aristotle and attempted to expand the applications of mathematics to physics. Since his achievements along that line gave him the first essential clue to his wider program, we should consider some of the ways in which his methods differed from those of his predecessors. An excellent example is provided by comparing his treatment of a celebrated theorem in statics with the analyses of the same proposition by the best medieval writer on statics, Jordanus Nemorarius, and by an older contemporary of Galileo, Simon Stevin. Here is the theorem of Jordanus:

IF TWO WEIGHTS DESCEND ALONG DIVERSELY INCLINED PLANES, AND IF THE INCLINATIONS ARE DIRECTLY PROPORTIONAL TO THE WEIGHTS, THEN THEY WILL BE OF EQUAL FORCE IN DESCENDING.

Fig. 1.

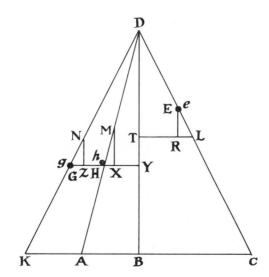

Let there be a line ABC parallel to the horizon, and let BD be erected vertically to it; and from D draw the lines DA and DC, with DC the more oblique. I then mean by proportion of inclinations, not the ratio of the angles, but of the lines taken to where a horizontal line cuts off an equal segment of the vertical.

Let the weight e be on DC, and the weight h on DA, and let e be to h as DC is to DA. I say that those weights are of the same force in this position.

For let DK be a line of the same obliquity as DC, and let there be on it a weight g, equal to e. If then it is possible, suppose that e descends to L and draws h up to M. Now let GN be equal to HM, which is equal to EL. Draw a perpendicular on DB through G to H and Y, and another from L, which will be TL. On GHY erect the perpendiculars NZ and MX, and on LT erect the perpendicular ER.

Then since the proportion of NZ to NG is that of DY to DG, and hence as that of DB to DK; and since likewise MX is to MH as DB is to DA, MX is to NZ as DK is to DA; that is, as the weight g is to the weight h.

But because e does not suffice to lift g to N, it will not suffice to lift h to M. Therefore they will remain as they are.[3]

In this demonstration the physical element is, as you see, brought in as an afterthought to the geometry, by main force as it were. That the weights e and g are in equilibrium perhaps requires no proof; but the relation of that fact to the situation of the weight h does demand some further explanation than a mere reiteration of the theorem itself. The theorem happens to be true, but what Jordanus has to say in support of it does not afford any hint of a method by which other physical theorems might be developed.

Next, consider the ingenious treatment given to the same proposition by Simon Stevin in 1586. Here the previous situation is reversed; geometry is eliminated in favor of pure mechanical intuition. Here is a paraphrase of Stevin's proof of the theorem:[4]

Imagine a circular necklace of equal-spaced heavy balls, draped over two diversely inclined planes. (See Fig. 2.) The chain will not move, there being no more reason for it to rotate in one direction than the other. Now, the bottom part of the chain is perfectly symmetrical, so it cannot affect the balancing. Cut it off, and the remainder of the chain will still not move. But since the total weight now resting on either plane is in proportion to the length of that plane, such weights are in equilibrium.

[3] Slightly modified from E. A. Moody and M. Clagett, eds., *The Medieval Science of Weights* (Madison: The University of Wisconsin Press, 1952), p. 191, with permission of the copyright owners, The Regents of the University of Wisconsin.
[4] See *The Principal Works of Simon Stevin*, ed. E. J. Dijksterhuis, I (Amsterdam, 1955), 177ff.

Fig. 2.

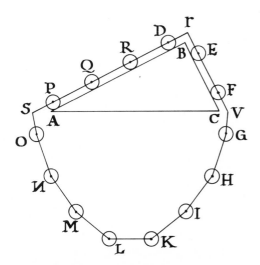

Stevin's physical reasoning is superb, but the theorem still remains isolated; no further physical application of the method suggests itself, this time because geometry is left out. In the attempted proof by Jordanus, geometry ruled and physics was left out.

Galileo's way of merging geometry and physics in his proof of the same theorem became apparent in his early treatise on motion, dating from 1590. The method itself suggested to him not only many corollaries, but successive improvements of the proof itself and further physical implications of it. Here is Galileo's original proof:

Fig. 3a.

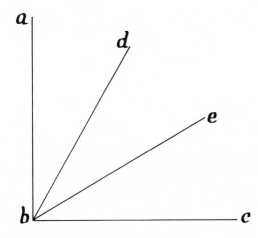

. . . A heavy body tends downward with as much force as is necessary to lift it up. . . . If, then, we can find with how much less force the heavy body can be drawn up on the line *bd* than on the line *ba*, we will have found with how much greater force the same heavy body descends on line *ab* than on line *db*. . . . We shall know how much less force is required to draw the body upward on *bd* than on *be* as soon as we find out how much greater will be the weight of that body on the plane . . . along *bd* than on the plane along *be*.

Fig. 3b.

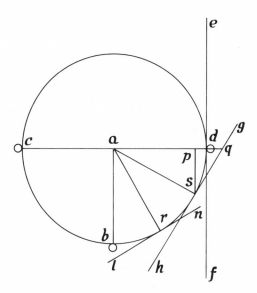

Let us proceed then to investigate that weight. Consider a balance *cd*, with center *a*, having at point *c* a weight equal to another weight at point *d*. Now, if we suppose that the line *ad* moves toward *b*, pivoting about the fixed point *a*, then the descent of the body *at the initial point d* will be as if on the line *ef*. Therefore the descent of the body on line *ef* will be a consequence of the weight of the body at point *d*. Again, when the body is at *s*, its descent *at the initial point s* will be as if on line *gh*, and hence its motion on *gh* will be a consequence of its weight at point *s*. . . .

Now it is clear that the body exerts less force at point *s* than at *d*. For the weight at point *d* just balances the weight at point *c*, since the distances *ca* and *ad* are equal. But the weight at point *s* does not balance that at *c*. For if a line is drawn from point *s* perpendicular to *cd*, the weight at *s*, as compared to the weight at *c*, is as if it were suspended from *p*. But a weight at *p* exerts less force than at *c*, since the distance *pa* is less than the distance *ac*. . . . It is obvious, then, that the body will descend on line *ef* with greater force than on line *gh*. . . .

But with how much greater force it moves on *ef* than on *gh* will be made clear as follows. Extend line *ad* beyond the circle to meet line *gh* at point *q*. Now since the body descends on line *ef* more readily than on *gh* in the same ratio as the body is heavier at point *d* than at point *s*, and since it is heavier at *d* than at *s* in proportion as line *da* is longer than *ap*, it follows that the body will descend on line *ef* more readily than on *gh* in proportion as line *da* is longer than *pa*. . . . And as *da* is to *pa*, so *qs* is to *sp*; that is, as the length of the oblique descent is to the vertical drop. And it is clear that the same weight can be drawn up an inclined plane with less force than vertically in proportion as the vertical ascent is shorter than the oblique. Consequently the same heavy body will descend vertically with greater force than on an inclined plane in proportion as the length of the descent is greater than the vertical fall. . . .[5]

First, note the physical rule stated at the outset, that the downward tendency is measured by the upward resistance, suggesting Newton's law of action and reaction; this device was repeatedly used by Galileo as a step to mathematical analysis in physics. Next, you see how the entire demonstration constitutes a reduction of the problem of equilibrium on inclined planes to the lever, which in itself removes the theorem from the isolation in which it stood before. Galileo applied the idea of virtual displacements as Jordanus had intended to do; but unlike Jordanus, Galileo assumed that the force acting on a body at the initial point of movement alone would determine its mode of descent along a given plane. In place of the macroscopic intervals geometrically equated by Jordanus, Galileo's method assumes a determinate tendency to motion at a point, or along an infinitesimally small distance. That form of analysis is all the more noteworthy if we recall that when Galileo composed this first demonstration, he still believed that the motion of descent was essentially uniform and that acceleration was only an accidental and temporary effect. Probably his words were governed by physical intuition that the force of descent or tendency to move is not the same for a body at rest as for one already in motion, or his choice of terminology may have been governed by his feeling for mathematical precision of statement. In any event, his choice of a correct form of expression at the outset helped him later to discover, and perhaps even led him directly to, the essential and fundamental role of acceleration in free fall. A further significant feature of Galileo's analysis here is that he made vertical motion the common measure of all motions on various inclined planes, instead of attempting a direct comparison of different planes as Jordanus and Stevin had done. That approach enhanced the fruitfulness of his demonstration by suggesting to him several corollaries and later helped him to generalize correctly when comparing the inclined plane to other simple machines.

But when Galileo framed the foregoing demonstration, he was attempting to analyze the *speeds* of bodies on inclined planes. He mistakenly

[5] I. E. Drabkin and S. Drake, *Galileo Galilei on Motion and on Mechanics* (Madison, 1960), pp. 64–65. Emphasis added.

supposed the speed for any plane to be constant, and to be related to the force or effective weight that he had correctly analyzed. Because of that misapprehension, several corollaries he deduced from his theorem were illusory. Luckily, it is characteristic of scientific reasoning, especially when mathematics is applied, that errors and inconsistencies are readily detected and may even aid in the discovery of truth. Galileo himself pointed this out in his writings. In the present instance, he deduced the ratios of speeds that would hold under his assumption for a given body on different planes, and observed that those ratios were contradicted by experiment. Very likely it was that troublesome fact that caused him to withhold from publication his early treatise on motion. He also deduced that a body on a horizontal plane could be moved, at least in theory, by a force smaller than any previously assigned force. This terminology is arresting—it is that of an Archimedean concept in pure mathematics, later to become the foundation of the theory of limits, applied here for the first time to a purely physical concept. The fruitfulness of that application of mathematics to physics was enormous; in Galileo's hands it led quickly to a limited inertial concept and to the first bridge between statics and dynamics. But let us now examine Galileo's development of the same theorem in his next composition, his treatise on mechanics.

Fig. 4.

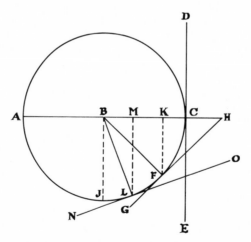

Consider the circle AJC and in it the diameter ABC with center B, and two weights at the extremities A and C, so that the line AC being a lever or balance, movable about the center B, the weight C will be sustained by the weight A. Now if we imagine the arm of the balance as bent downward along the line BF . . . then the moment of the weight C will no longer be equal to the moment of the weight A, since the distance of the point F from the perpendicular line BJ has been diminished.

Now if we draw from the point F a perpendicular to BC, which is FK, the moment of the weight at F will be as if it were hung from the line KB; and as the distance KB is made smaller with respect to the distance BA, the moment of the weight F is accordingly diminished from the moment of weight A. Likewise, as the weight inclines more, as along the line BL, its moment will go on diminishing, and it will be as if it were hung from the distance BM along the line ML, in which point L it may be sustained by a weight placed at A that is as much less than itself as the distance BA is greater than the distance BM.

You see, then, how the weight placed at the end of the line BC, inclining downward along the circumference CFLJ, comes gradually to diminish its moment and its impetus to go downward, being sustained more and more by the lines BF and BL. But to consider this heavy body as descending and sustained now less and now more by the radii BF and BL, and as constrained to travel along the circumference CFL, is not different from imagining the circumference CFLJ to be a surface of the same curvature placed under the same movable body, so that this body, being supported upon it, would be constrained to descend along it. For in either case the movable body traces out the same path, and it does not matter whether it is suspended from the center B and sustained by the radius of the circle, or whether that support is removed and the body is supported by and travels upon the circumference CFLJ. Whence we may undoubtedly affirm that the heavy body descending from the point C along the circumference CFLJ, its moment of descent *at the first point C* is total and integral, since it is in no way supported by the (corresponding) circumference; and at this first point C it has no disposition to move differently from what it would freely do in the perpendicular tangent line DCE. But if the movable body is located at the point F, then its heaviness is partly sustained by the circular path placed under it, and its moment downward is diminished in that proportion by which the line BK is exceeded by the line BC. Now when the movable body is at F, *at the first point of its motion* it is as if it were on an inclined plane according with the tangent line GFH, since the tilt of the circumference at the point F does not differ from the tilt of the tangent FG, apart from the insensible angle of contact. . . .[6]

The balance of this revised demonstration need not detain us here; it suffices to note how fruitful Galileo's approach had already been for him. By his original restriction of virtual motions to the initial point, he was led to perceive on the one hand the analogy between a series of tangential tendencies and the motion of a pendulum; and on the other hand, the analogy between those tendencies and the descent of a body along a circular path supported from below. From that, he arrived at a number of new and

[6] *Ibid.,* pp. 173–74. Emphasis added.

interesting corollaries concerning descent along arcs and chords of circles, and wrote to Guido Ubaldo del Monte about these in 1602. Probably it was in the course of those investigations that he was led to his discovery that the spaces traversed in free fall, whether vertical, inclined, or tangential, were proportional to the squares of the times of descent. That conclusion was communicated to Paolo Sarpi in 1604, though at that time Galileo had not yet satisfactorily defined uniform acceleration or derived his result correctly from fundamental assumptions. The patient ordering of his definitions, assumptions, theorems, and observations took many years. But enough has now been said to illustrate the novelty and effectiveness of Galileo's method in applying mathematics to the problems of mechanics.

III

Galileo's assigned university courses gave him little or no scope for the communication of his researches in mathematical physics. He went on with them, but one might say that he did so almost despite his official position rather than in pursuance of it. From a letter written to him by Luca Valerio, a Roman mathematician best known for his work on centers of gravity, it is evident that by June, 1609, Galileo had arrived at two fundamental propositions on which he was prepared to found the science of mechanics. But in that same month, his attention was diverted from that project by news of the telescope, and his energies were applied immediately to its improvement and to the astronomical discoveries which soon made him famous throughout Europe.

Galileo's sudden celebrity quickly suggested to him a way in which he might free himself from routine teaching and find time to write and publish the books which had been evolving in his mind. It was mainly for that purpose that Galileo, after twenty years of university teaching, applied for a court position with the Grand Duke of Tuscany. In making his application he included a significant and unusual request: ". . . As to the title of my position," he wrote, "I desire that in addition to the title of 'mathematician,' his Highness will annex that of 'philosopher,' for I may claim to have studied for a greater number of years in philosophy than months in pure mathematics."[7]

To the historian, Galileo's request is striking in two respects: first, because he sought the unusual title of court philosopher, and second, because he did *not* seek the title of court astronomer. That title was common throughout Europe; Johannes Kepler, for example, was imperial astronomer

[7] S. Drake, *Discoveries and Opinions of Galileo* (New York, 1957), p. 64.

at the court of Rudolph II at Prague. That Galileo did not mention, let alone demand, such a title is made still more striking by the fact that his chief claim to fame in 1610 lay in the astronomical discoveries he had just published in his *Sidereus Nuncius* and dedicated to his proposed employer, the Grand Duke of Tuscany.

Galileo's failure to seek the title of court astronomer is striking, but it is not difficult to explain. In the first place, a court astronomer at that period was in fact an astrologer, or at least his primary value to his employer lay in that capacity. Any contributions he might make to theoretical or observational astronomy were the merest by-products, so far as the sovereign was concerned. As Kepler once ruefully remarked, it was the wayward daughter astrology who had to support the honest dame astronomy. Galileo was perfectly competent to cast horoscopes, but he did not enjoy doing so; on the contrary, he was openly critical of both astrology and alchemy. In the second place, Galileo does not seem to have cared much about mathematical astronomy outside his routine public lectures, which in fact were offered by the university chiefly because of their astrological applications, deemed necessary for medical students especially.

As we have seen, the subject to which Galileo *had* devoted his principal researches during twenty years of teaching was the application of mathematics to problems of motion and mechanics. But in applying for employment by the Grand Duke, he could scarcely have described that specialty as physics, nor could he have asked for the title of "physicist" with any hope of being understood. Physics was no more a subject in its own right than metaphysics or ethics. It was merely a branch of philosophy—and that is what throws light on Galileo's otherwise odd request for the title of "philosopher." That was, in fact, the only appropriate word, when we consider that the term *mechanic* was entirely undignified and unthinkable as a court title. The historic appropriateness of Galileo's choice of title is shown by the fact that long after his time, the term *natural philosopher* served as the official name for a physicist in England, the word *physics* itself being no more than a synonym for *nature*.

I think it is safe to assume that the philosophy which Galileo had in mind when he asked for the title of philosopher was that which he later described in one of his most frequently cited passages, which begins: "Philosophy is written in that grand book of nature that stands forever open to our eyes. . . ."[8] I think it is not too strained to say that what Galileo was seeking, when he asked to be made mathematician and philosopher to the Tuscan court, was the world's first post as a mathematical physicist.

[8] *Ibid.*, p. 237.

Support for this viewpoint is afforded by a difficulty that arose in Galileo's securing of the title he had asked for. The Grand Duke, doubtless in the belief that he was granting the request, sent to Galileo a document naming him chief philosopher to the court and chief mathematician at the University of Pisa. But Galileo was not satisfied with this, which is puzzling unless it was because he wanted the two titles definitely combined. Eventually he had his way and was named chief mathematician and philosopher to the Grand Duke and chief mathematician at the University of Pisa without obligation to reside or teach there. It may seem that he had made a major issue of a minor point, but I believe that even before Galileo moved from Padua to Florence, he had in mind the conception that completes the famous quotation of which I cited the opening words a moment ago. To paraphrase the rest of it: ". . . that grand book of nature . . . is written in the language of mathematics, without a knowledge of which one cannot understand a word of it, but must wander about forever as in a maze."[9] What Galileo wanted, and could get only from a powerful patron, was a position in which he could openly expound the unity of mathematics and physics—and probably astronomy also, though without any astrological connotations. The Grand Duke could grant him this, but the departments of philosophy were too strong and the departments of mathematics too weak to accommodate such a program in the universities.

Galileo's first publication in his new position was a book on the behavior of bodies placed in water. In that book he used the principle of Archimedes and the principle of virtual velocities to refute the qualitative ideas of the Aristotelians, and he also introduced physical experiments to combat tradition and authority. His earlier use of virtual displacements had by now led him on to a true principle of virtual velocities, however primitive it may seem from a modern standpoint. Observing that the unequal arms of a steelyard in equilibrium would necessarily move their unequal weights through inversely proportional distances *in equal time*, he stated the general rule that equal weights are of equal *moment* when given inversely proportional *speeds*. As the source of this proposition, he credited the Aristotelian treatise on *Questions of Mechanics*, where indeed it had been adumbrated in the first "question," dealing with circular motions in general. But the author of that ancient treatise preferred to use the geometrical relation between forces and their displacements and failed to note that the significant physical factor was the relation of masses to velocities. Thus in the third "question," dealing with the explanation of the lever, the Aristotelian writer reverted to displacements, and even those he treated as circu-

[9] *Ibid.*, p. 238.

lar rather than perpendicular motions. Galileo proceeded to remedy this. His recognition of the general roles of force and speed enabled him to extend the principle of virtual velocities from statics to hydrostatics, as exemplified in this theorem in his book on bodies in water:

Fig. 5.

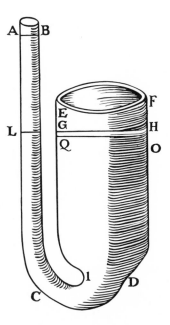

Consider the above figure (which, if I am not mistaken, may serve to reveal the errors of some practical mechanicians who on false premises sometimes attempt impossible tasks) in which, to the large vessel EIDF, the narrow pipe ICAB is connected, and suppose water in them to the level LGH, which water will rest in that position. This astonishes some people, who cannot conceive how it is that the heavy weight of water GD, pressing downwards, does not lift and push away the small quantity contained in the pipe CL, which nevertheless resists and hinders it. But the wonder will cease if we suppose that the water GD goes down only to QD, and ask what the water CL has done. It, to make room for the other, descending from GH to QO, would have in the same time to ascend to the level AB. And the rise LB will be greater in proportion to the descent GD as the width of the vessel GD is to that of the pipe IC, or as the water GD is to the water CL. But since the moment of the speed of motion in one vessel compensates that of the weight in the other, what is the wonder if the swift rise of the lesser water CL shall resist the slow descent of the greater amount GD?[10]

[10] See *Galileo Galilei. Discourse on Bodies in Water*, ed. S. Drake (Urbana, 1960), p. 17.

Not only did this constitute an important addition to what Archimedes had written on hydrostatics; it also opened the door to the creation of hydrodynamics. Another of Galileo's insights in his earlier treatise *On Mechanics*, in which he pointed out that the force required to disturb a system in equilibrium is negligible, had similarly constituted a bridge from statics to dynamics, as mentioned a moment ago. In later years, as you know, Galileo applied mathematics to the laws of falling bodies and to the analysis of strength of materials. But we have now sufficiently illustrated the fruitfulness of his applications of mathematics to physics as compared with the attempts of his predecessors, and it is time to turn to his novel applications of physics to astronomy.

As mentioned before, the Aristotelian philosophy insisted on a complete dichotomy between terrestrial and celestial substances and motions. Astronomers were not concerned with the nature of heavenly bodies, but only with purely mathematical descriptions of the observed motions and predictions of positions. Since Christian theology had adopted the Aristotelian separation of the base earth from the noble heavens, the belief was one that was very difficult to attack. Nevertheless, the opening sections of Galileo's *Dialogue* were devoted to a refutation of the Aristotelian assumption, as a point of departure for his pioneer attempt to unify astronomy and physics.

A complete chronological account of Galileo's efforts to bring physics and astronomy together would take us back to 1597, when he wrote to Kepler that he believed certain physical events on earth to be explicable only by its motions. But his first public attempts to relate physics and astronomy are found in the *Sidereus Nuncius* of 1610.

In announcing his first telescopic observations, Galileo declared the moon to be like another earth because of its rough, mountainous surface. Applying geometric methods familiar to land surveyors, he calculated the heights of the lunar mountains from their shadows. Denying the peripatetic contention that all heavenly bodies were perfect spheres, he accounted in later years for the moon's roughness by assuming its material to be similar to that of the earth and similarly drawn to its common center of gravity. The darker surfaces on the moon he declared on earthly analogies to be relatively smooth as compared to the brighter parts. Conversely, he deduced from the moon's rough surface and its bright reflection of sunlight that the earth, with a similar surface, must reflect sunlight to the moon; and in that way he correctly explained the secondary light on the moon, seen when it is thinly crescent. At first he suggested also that the moon had an atmosphere, but he later withdrew that further analogy of the moon with the earth. However, he postulated an atmosphere for Jupiter in order to account for changes in the appearances of its satellites in their various positions. All Galileo's physical analogies between the earth and any heavenly body, even the lowly moon, aroused strong opposition from philosophers and some protest from theologians.

When Galileo wrote the *Starry Messenger*, he was rapidly approaching his ultimate firm conviction that astronomy must be completely integrated with physics. But his interest in astronomy was then still subsidiary to his absorption in the application of mathematics to physics on which he had labored for so many years. The telescope was for a time principally a means to him for gaining fame and improving his position. Thus he first observed sunspots not later than the spring of 1611, but he paid little attention to them except as a curiosity to show his friends. He did not seize immediately on them as a ready basis for expansion of his analogies between terrestrial and celestial phenomena, as one would expect him to do had he already completely formulated his program to unify mathematics, physics, and astronomy. But the sunspots were destined to be decisive in that regard when a rival astronomer proposed a theory about the spots.

Galileo's treatise on bodies placed in water was about to go to press when he received from Mark Welser three printed letters on sunspots written by a German Jesuit, Christopher Scheiner. Scheiner argued that the spots were only apparently on the sun and were in reality varying clusters of small opaque bodies rotating about it at some distance. His opinion probably had its origin in a desire to maintain, for religious and philosophical reasons, the doctrine of incorruptibility of heavenly bodies, especially the sun. Welser wrote to ask for Galileo's opinion on the whole subject, and Galileo undertook a series of careful observations, after which he replied in refutation of the Jesuit's theory.

By precise mathematical reasoning, Galileo demonstrated that the sunspots must be located either on the surface of the sun or at a negligible distance from it. He went on to say that the only proper method available for assigning the causes of distant or unfamiliar events was to apply our experience of things near at hand and suggested that the sunspots might better be explained as vast clouds or smokes than as stars or planets. He noted that terrestrial clouds often cover a whole province and that when a terrestrial cloud happens to be near the sun and is compared with a sunspot, it generally appears much darker than the spot. Still more significant was his explanation of the rotation of the sunspots, which he attributed to an axial rotation of the sun itself, for that in turn led him to a first attempt in the direction of celestial dynamics. He wrote:

I seem to have observed that physical bodies have a natural inclination to some motion (as heavy bodies downward) . . . without need of an external mover, whenever they are not impeded by some obstacle. And to some other motion they have a repugnance (as have the same heavy bodies to motion upward), and therefore they never move in that manner unless thrown violently by an external mover. Finally, to some movements they are indifferent, as are these same heavy bodies to horizontal motion. . . . And therefore, all external impediments being removed, a heavy body on a spherical surface concentric with the earth will be indifferent to rest and to movement toward any part of the horizon. And it will maintain itself in that state in which it has once been placed; that is, if placed in a state of rest, it will conserve that; and if placed in movement toward the west, for example, it will maintain itself in that movement. Thus a ship, for instance, having once received some impetus through the tranquil sea, would move continually around our globe without ever stopping, and placed at rest it would perpetually remain at rest, if all extrinsic impediments could be removed and no external cause of motion were added.

Now if this is true, as indeed it is, what would an inert body do if continually surrounded with an ambient that moved with a motion to which it was indifferent? I do not see how one can doubt that it would move with the motion of the ambient. [Recall Galileo's old proof that such a body would be set in motion by a force less than any previously assigned force.] And the sun, a body of spherical shape suspended and balanced on its own center, cannot fail to follow the motion of its ambient, having no intrinsic repugnance or external impediment to rotation. It cannot have an internal repugnance because by such a rotation it is neither moved from its place, nor are its parts permuted among themselves. Their natural arrangement is not changed in any way, so that as far as the constitution of its parts is concerned, such movement is as if it did not exist. . . . This may be further confirmed, as it does not appear that any movable body can have a repugnance to a movement without having a natural propensity to the opposite motion, for in indifference no repugnance exists. . . .[11]

Here Galileo's inertial concept and his knowledge of the conservation of angular momentum were brought to bear on a problem of celestial motion—a real step toward the unification of physics and astronomy.

Mathematics having already been linked firmly to both the latter sciences, Galileo's program was now essentially complete. That program was presented successively in *Il Saggiatore*, or *The Assayer* (1623), Galileo's scientific manifesto (as it was called by Leonardo Olschki); in the *Dialogo* or *Dialogue* (1632), dealing primarily with astronomical arguments but heavily weighted with physics; and in the *Discorsi* or *Two New Sciences* (1638), dealing exclusively with physics (since Galileo was forbidden to deal with astronomy after writing the *Dialogue*) but extending beyond mechanics to

[11] Drake, *Discoveries*, pp. 113–14.

include speculations on sound, light, and other physical topics. *The Assayer* and the *Dialogue* offer many illustrations of Galileo's program of linking physics with astronomy and of his simultaneous linkage of all three sciences. We shall consider a few of these here, neglecting the *Two New Sciences*, since examples have already been given of Galileo's methods of applying mathematics to physics.

In *The Assayer*, Galileo had occasion to discuss the so-called third motion attributed by Copernicus to the earth in order to maintain the earth's axis continually parallel throughout the year. Of this he says:

This extra rotation, opposite in direction to all other celestial motions, appeared to many people to be a most improbable thing. . . . I used to remove the difficulty by showing that such a phenomenon was far from improbable . . . for any body resting freely in a thin and fluid medium will, when transported along the circumference of a large circle, spontaneously acquire a rotation in the directions contrary to the larger movement. This is seen by taking in one's hand a bowl of water, and placing in it a floating ball. Then, turning about on one's toe with the hand holding the bowl extended, one sees the ball turn on its axis in the opposite direction, and completing its revolution in the same time as one's own. [Really] . . . this would not be a motion at all, but a kind of rest. It is certainly true that to the person holding the bowl, the ball appears to move with respect to himself and to the bowl, turning on its axis. But with respect to the wall . . . the ball does not turn, and any point on its surface will continue to point at the same distant object.[12]

Galileo's appeal to familiar terrestrial observations in explanation of heavenly events was designed to further the idea of a physical astronomy, which would eliminate the clumsy hypothetical apparatus of crystalline spheres which earlier astronomers had introduced as a means of accounting for the motions of heavenly bodies. The idea of relative motion, introduced in the preceding passage, became pivotal in the later *Dialogue*. Astronomers, it is true, had long been aware of the importance of optical relativity, but even Tycho Brahe had been unable to conceive of a literal physical relativity of motion. Galileo, armed with that concept and with his principles of the composition of motions, inertia, and conservation of angular momentum, was able effectively to meet many common-sense objections against motion of the earth. Thus in the *Dialogue*, replying to the claim that if the earth really moves, we should see a departure from straight motion in the free fall of a body from a tower, Galileo said:

Rather, we never *see* anything but the simple downward motion, since this

[12] *Ibid.*, pp. 264–65.

other circular one, common to the earth, the tower, and ourselves, remains imperceptible and as if nonexistent. Only that motion of the stone which we do not share is perceptible, and of this, our senses show us that it is along a straight line parallel to the tower. . . . With respect to the earth, the tower, and ourselves, all of which keep moving with the diurnal motion along with the stone, the diurnal motion is as if it did not exist; it remains insensible, imperceptible and without any effect whatever. All that remains observable is the motion which we lack, and that is the grazing drop to the base of the tower. You are not the first to feel a great repugnance toward recognizing this nonoperative quality of motion among those which share it in common.[13]

A good example of Galileo's effectiveness at persuasion through the use of terrestrial and celestial analogies occurs in the First Day of the *Dialogue*. His Aristotelian opponent cannot believe that the rough, dark earth could possibly shine in the sky as brightly as the moon. Galileo points out to him that the comparison must be drawn for the moon as seen in daytime, since that is the only time we can see the earth illuminated, and then proceeds:

Now you yourself have already admitted having seen the moon by day among little whitish clouds, and similar in appearance to one of them. This amounts to your granting that clouds, though made of elemental matter, are just as fit to receive light as the moon is. More so, if you will recall having seen some very large clouds at times, white as snow. It cannot be doubted that if such a cloud could remain equally luminous at night, it would light up the surrounding regions more than a hundred moons.

If we were sure, then, that the earth is as much lighted by the sun as one of these clouds, no question would remain that it is no less brilliant than the moon. Now all doubt on this point ceases when we see those clouds, in the absence of the sun, remaining as dark as the earth all night long. And what is more, there is not one of us who has not seen such a cloud low and far off, and wondered whether it was a cloud or a mountain—a clear indication that mountains are no less luminous than clouds.[14]

In conclusion, I should like to comment on two arguments for a motion of the earth that were particularly dear to Galileo's heart, arguments which simultaneously link mathematics, astronomy, and physics. It happens that those two arguments are often discredited by modern historians and that they fell on deaf ears in Galileo's time. Correctly understood, however, they were unanswerable at the time they were propounded. It may

---

[13] Galileo, *Dialogue Concerning the Two Chief World Systems* (Berkeley, 1962), pp. 163 and 171.
[14] *Ibid.*, pp. 88–89.

seem paradoxical to include, in an attempted explanation of Galileo's effectiveness, two arguments of his that were (and still are) ineffective. But I think that the paradox is only apparent. The prime source of Galileo's effectiveness was his bringing together of mathematics, astronomy, and physics in an inseparable relationship. Hence even a questionable example of such a relationship given by him was still capable of revealing to others what sort of thing should be sought after in constructing a scientific explanation.

First I shall mention Galileo's theory of the tides, which occupies the final section of the *Dialogue*. He says that, in his opinion, the ocean tides cannot be explained physically if we assume a perfectly stationary earth. So far he is quite right. He then asserts that the double motion of the earth, around its axis and around the sun, affords a purely mechanical means of explaining the tides. His argument is that the two circular motions are additive on one side of the earth and subtractive on the other, so that the extremities of any large east-west basin of water, such as an ocean, will be traveling with non-uniform velocity. Since direct experience teaches us that water in a basin is disturbed when its velocity is altered, Galileo thought that he had found a basic cause of periodic disturbance in the ocean waters.

Before we dismiss this theory, as some have done simply because they know it happens to be wrong and as others have done on very abstract principles of mathematical physics unknown to Galileo or any of his contemporaries, let us consider what kind of a model Galileo can have had in mind that appealed to his physical intuition. For Galileo plainly asserts that although an artificial basin cannot be constructed that will have unequal speeds at its extremities, he nevertheless could construct a model to illustrate his point. Such a model, using materials available to Galileo, might be made as follows. Drill two large holes near the circumference of a grindstone, about 30° apart, connected by a narrow channel. Partly fill this cavity with water and cover it with a glass plate. Set the grindstone in horizontal rotation, and the water will seek to get as close to the edge of the grindstone as possible. Now add a second, larger rotation, by swinging the whole device on a long cable from a distant center, much more slowly. The grindstone now represents the diurnal rotation and the cable the annual revolution. Clearly, the position of the water will shift cyclically as the holes approach and recede from the cable. Hence Galileo's intuition of the effect of a double circular motion is by no means absurd.

The principal objection was that only one high tide a day would be expected from this model, whereas there are approximately two. This did not escape Galileo's attention, but neither did it bother him. So long as he had a primary mechanical cause for periodic disturbance, he was content.

He pointed out that a wide variety of other factors would affect its progress, such as the length of the basin, its orientation, its depth, the shape of its coasts, the action of winds, and so on. Since Galileo had never observed tides except along the Adriatic and the Tyrrhenian seas, where they are not very impressive, it is not surprising that he thought such factors sufficient to account for the observed discrepancy of period.

On the other hand, he felt obliged to mention long-period variations in tidal effects that were related to the positions of the moon and the sun. These he accounted for in his theory by postulating changes in the earth's orbital speed, which would appear to us as related to the sun's position, and by drawing an analogy between the earth-moon system and a moving weight on a rotating rod driven by a constant force to account for tidal changes apparently related to positions of the moon.

This whole theory illustrates vividly Galileo's program of seeking mechanical explanations for all physical effects, including celestial appearances. It is usually said that Kepler had already given a tidal theory based on lunar attraction, which Galileo was wrong in rejecting. But that is not a fair statement. Kepler's theory was that the moon attracted the water nearest to it away from the earth; but that likewise would account for only one high tide a day, or slightly longer. Furthermore, a special attraction for water on the part of the moon appeared to Galileo to be the invoking of an occult cause, which was precisely what his whole program was against. No correct tidal theory was possible until Newton's general gravitational law, and if any choice is to be made between Galileo's theory and Kepler's, it seems to me better to have demanded a mechanical explanation which happened not to include attraction than to settle for a hypothetical attraction that did not even attempt to account for the double period of tides in a lunar day. Galileo's devotion to a tidal theory having a mechanical basis, with great discrepancies from observed phenomena for which he invoked further mechanical phenomena, was in keeping with his expressed principles. Thus his admiration for Copernicus was heightened by the latter's loyalty to heliocentrism in the face of apparently serious discrepancies in the apparent and theoretical size of Venus at opposition and conjunction, discrepancies which could only be reconciled by telescopic observations made after the death of Copernicus.

The other argument in which Galileo combined mathematics, astronomy, and physics in support of the earth's motion concerns the annual variations in the paths of sunspots. Twice in the year, the paths are straight and tilted to the ecliptic; twice a year, they are at maximum curvature with the ends lying in the ecliptic; at all times, they change cyclically in tilt and degree of curvature. Galileo pointed out that those variations are easily explained if the earth has the two motions attributed to it by Copernicus. But for an absolutely stationary earth, they can be explained only by ascribing to the sun a highly complicated set of motions. First, the sun must have a rotation about an axis tilted to the ecliptic, having a period of about one month. Second, in its annual revolution about the earth, the sun's axis must rotate conically about another axis perpendicular to the ecliptic, in circles having a radius measured by the tilt of the first axis. The period of this conical rotation must be annual. A still further daily motion of the sun's axis would be necessary to keep the paths of the spots undisturbed on the face of the sun during each day.

Although, as Galileo duly pointed out, it would be theoretically possible from a strictly geometric point of view to endow the sun with the necessary motions, that would by no means be as simple as Galileo's modern critics commonly suppose. For they habitually attribute by implication a diurnal motion to the earth, whereas Galileo was arguing against a perfectly stationary earth. For example, a modern critic gives this illustration, supposed to be self-explanatory:[15]

Fig. 6.

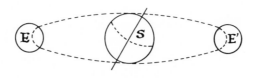

a. Copernican

b. Ptolemaic

But this ignores the fact that if the second figure is taken literally, for a perfectly stationary earth, it means that from a given point on the earth the sun remains invisible for half the year. Or, if the illustration is supposed to represent a single day (in which case the Copernican diagram is not truly analogous to the Ptolemaic), then the sunspot paths in the second figure

[15] See F. S. Taylor, *Galileo and the Freedom of Thought* (London, 1938), p. 134.

would not vary during the year, though they would twist about during every day. Any attempt to correct this leads to grave difficulties from the standpoint of dynamics. Now, Galileo was perfectly aware of the principle of conservation of angular momentum and of the inertial path of the axis of a rotating body, as mentioned earlier; thus, to him, these physical difficulties were very real. In a word, he knew that for the sunspot paths, as for the seasons and certain other observed phenomena, the assertion of geometrical equivalence among various world-systems was misleading. But for his contemporaries, who felt obliged to link mathematics with astronomy but not astronomy with physics, his argument remained inconclusive. Since Kepler, who had demonstrated the geometrical equivalence of the systems of Ptolemy, Tycho, and Copernicus but who nevertheless adhered to the last-named, died two years before Galileo published the sunspot argument, we do not have his comments on it.

Galileo's theory of the tides and his argument from the paths of sunspots are important in showing the manner in which he brought together simultaneously the fields of mathematics, astronomy, and physics. Whatever the defects of those two arguments for the earth's motion, they were founded on a fundamentally sound conception of the manner in which all physical phenomena, terrestrial and celestial, ought to be consistently explained. It was that conception which made Galileo's work more effective scientifically than the work of any one of this great contemporaries or recent predecessors. It was a conception which he himself thoroughly grasped and consciously applied to a definite program. It was a conception, moreover, which profoundly appealed to and exerted a recognizable influence upon many of Galileo's contemporaries, to whom it suggested a wide variety of further investigations once it had been broached. Since, finally, it is a unifying conception that was lacking in ancient science, the unsophisticated view that Galileo was the revolutionary founder of modern science is not to be rejected outright, but is to be reinterpreted in the light of our new knowledge that many of his particular discoveries and opinions had been anticipated by great thinkers of the Middle Ages and the Renaissance.

EMPIRICISM AND THE SCIENTIFIC REVOLUTION ❧ ERNAN McMULLIN

t is customary to speak of the development in science spanning the lives of Galileo and Newton as a "revolution." To what extent is this very strong term justified? Is it not the case that what seem like discontinuous episodes in the history of ideas turn out on closer scrutiny to have roots that go back indeterminably far? This is a crucial issue in historiography generally, but it is especially pressing here. Science is so central to all our living and thinking today that everything about its genesis is of immediate concern to us. In recent years, the traditional case for a discontinuity in the history of science around 1600 has been challenged, as more and more light is cast on the science of the medieval and Renaissance periods. I have tried elsewhere[1] to assess the present state of the evidence on the continuity-discontinuity question. What I want to do in this paper is to take up one strand in this complicated story and subject it to closer scrutiny than the limits of the earlier, more general, study permitted.

Those who argue for a "revolution" in science around 1600 nearly always base their case on an alleged change in the methods of science at that time. The "revolution" is assumed to have gone much deeper than the introduction of new theories, even theories as far-reaching in their significance as Copernican-Keplerian astronomy or Galilean-Newtonian mechanics. It is true that the term *revolution* has frequently been used by modern historians of science to signify major theoretical innovations, such as quantum theory, that have changed an entire intellectual landscape. But when it is used of the seventeenth-century watershed, it appears to carry more weight. The assumption that a fundamental methodological innovation took place at that time is not new; the pioneers of science in the seventeenth century themselves not only spoke of their work as a "new science" but also very frequently attributed its novelty to a methodology that seemed to them quite revolutionary. They each had different views as to what the new methodology ought to be, but they *did* agree that something important was about to happen in the domain of natural science and that it was connected with the procedures of inquiry and proof.

[1] See "Medieval and Modern Science: Continuity or Discontinuity?" *International Philosophical Quarterly*, V (1965), 103–29.

It is possible to trace, over the centuries preceding 1600, the growth in the use of experimental techniques involving specially designed instruments, the development in mathematical methods of analysis, the introduction of idealized models that simplified the unmanageable complexity of nature. These are some of the strands that went to the weaving of the methodology of Galilean-Newtonian science, and each of them has a somewhat different history. But in this essay, I want to probe somewhat deeper and ask not so much about methods as about the expectations that underlie methods. What kind of knowledge did the "new scientists" seek? More precisely, what would mark it off as *science* in their eyes? Was there any significant change in the ideal of scientific knowing between Oresme and Galileo? Between Galileo and Newton? Admittedly, Oresme's astronomy differed greatly from Galileo's. But were they both looking for the same sort of thing? *Science* (*epistēmē, scientia*) has always denoted the best that can be obtained by way of knowledge of a particular domain. And for Oresme, as for Galileo, the "science" (or the "philosophy," for the terms were not yet distinguished) of nature was the stablest and best-warranted knowledge man could obtain about nature. But had there been any change in the ideal of what could actually be achieved? This question can conveniently be subdivided into two closely interrelated ones: had there been a significant change in the kind of *certitude* science was supposed to give? was there any major change in what constituted *evidence* for the scientist?[2]

I. TWO THEORIES OF SCIENCE

Our discussion will turn, then, on the different views of what science itself is and especially on what constitutes scientific evidence. The history of science, from its Greek beginnings right down to the present, has been marked by a tension between two views concerning the nature of science,

---

[2] A further question would be: had there been any change in the sort of reasoning *back* of the choice of ideal? It can plausibly be argued that there was a movement away from an *epistemological* theory of science (that is, one derived from a prior theory of knowledge or metaphysics) to a much more *pragmatic* one (one governed by successful practice, where "success" could mean anything from adequacy in prediction to utility in technological design) between 1400 and 1600. The influences at work here have been discussed by Olschki, Strong, and others, but there is still much disagreement about their relevance to the growth of the "new science." This would be a topic for a separate article; I shall mention it only in passing in this essay. Our concern here is primarily to analyze successive ideals of scientific knowledge rather than to inquire into the general type of reasoning that would have been alleged in their support at any given time.

which we can call the *conceptualist* and the *empiricist* views.[3] The conceptualist supposes that a direct access to the essence or structure of natural objects is available, by a careful analysis of everyday concepts such as *motion* or *continuum*, for example, or through an insight on the part of a skilled investigator into the singular objects of sense, an insight sufficiently penetrating to allow the universal to be immediately grasped in them. The warrant for a scientific claim, in this view, is the self-evidence of the statement itself. If it be properly understood—and getting to understand it may be no simple affair—the connection between its concepts will be seen to be of a necessary sort. The warrant is intrinsic to the statement; once the latter is properly understood, there will be no need to look outside it for further evidence. The science reached in this way will be certain and definitive. And the natural world thus known must possess some kind of permanent and accessible structure of intelligibility (whether of essences or of relations or of some other kind).

The empiricist, on the contrary, believes that evidence for a scientific statement cannot be found in the concepts utilized in the statement, but only in the singular observations on which the statement rests. In this view, no mere inspection of the statement can tell one that it is true, no matter how well the statement be understood. One must look *outside*, therefore; one must gather extrinsic evidence. Two possibilities offer themselves: one may seek wider and wider generalizations by collecting more and more observations, according to some plan of organization (inductive validation). Or one may seek to verify predictions that serve to justify one's hypotheses indirectly (hypothetico-deductive validation). The science one reaches by either of these ways will be tentative, approximate, progressive. And the evidence on which it rests will be individual observations. The empiricist claims that the world of sense is too opaque to be directly penetrated by human insight in the way the conceptualist assumes. It may or may not possess an intelligible structure (empiricists will divide on this), but whether it does or not, the attainment of any sort of generalized knowledge about it is only possible on the basis of careful and persistent observation. For the conceptualist, the principle (or the theory) is prior, even in the order of evidence; for the empiricist, the observation (the "fact") is basic.

It is a question, then, of the way in which the arrow of evidence points.

[3] In a well-known essay, "Three Views Concerning Human Knowledge" (in *Contemporary British Philosophy*, ed. by H. D. Lewis [London, 1956], pp. 355–88), Karl Popper makes a distinction between *essentialism* and *instrumentalism* that parallels ours in some respects. But the basis of his distinction is not the type of evidence that counts in favor of a scientific claim, but rather the quality of the knowledge given by it. Popper uses his distinction to evaluate the methodological changes that took place in science during Galileo's period; his evaluation differs in some important respects from ours.

The conceptualist view is, of course, not independent of experience; the conceptualist can be a thoroughgoing "empiricist," in a broader sense of that term; i.e., he may, like Aristotle, insist on the primacy of experience, but see in the experience only the source for the clear concepts (or insights into the universal) on which science will ultimately depend. And the empiricist likewise will be dependent upon concepts, but he will treat them as purely instrumental, as having in themselves no capacity for generating scientific connections. In their most extreme forms, conceptualism becomes rationalism (when the role of sensible experience in science is altogether denied or at least seriously downgraded), and empiricism becomes positivism (when science is regarded as nothing more than a series of convenient correlations of individual observations, themselves having no sort of internal intelligibility).

These labels will now make it much easier for us to formulate our historical theses: (1) the dominant medieval theory of science was conceptualist, although it was severely challenged by the quasi-positivist views of the later nominalists; (2) early seventeenth-century science was still strongly conceptualist, though the beginnings of an emphasis on experiment can be seen, one that would later lead to an empiricist re-definition of what science itself is; (3) later seventeenth-century science became more and more empiricist in tone; in Newton's comments on method especially, we find a highly explicit empiricism, though his actual scientific practice was somewhat less empiricist; (4) eighteenth-century science moved back once again toward a more conceptualist notion of science, at least in the domain of mechanics; and (5) it is too simple, therefore, to define the "Scientific Revolution" in terms of a movement from conceptualism to empiricism.

II. ARISTOTELIAN CONCEPTUALISM

In his *Posterior Analytics*, Aristotle laid down the conceptualist ideal of science that would dominate most accounts of scientific method for two millennia. According to him, science consists of two kinds of propositions, first principles (which carry within them their own warrant) and conclusions that are deducible from such principles. The perfect example for him of such a science was geometry, with its organization into axioms and theorems. Natural science, when fully worked out in "demonstrative" form (i.e. in the form that would best manifest the articulations of conceptual interrelation on which its immediate or "self-evident" character depended) would be syllogistic. A typical major premise would connect essence and property in a way that the mind would *see* to be true, just as the mind could see the axioms of geometry to be true, once basic concepts like *line* and *point* were properly understood.

Aristotle was a very great biologist and physiologist. He carried through an almost unbelievable amount of empirical research, the records of which make vivid reading even still. What ideal of science did he hold up for himself in this domain? At first sight, one would be inclined to answer that he hoped that the organization of *genera* and *differentiae* he had inaugurated would, when complete, exhibit the same sort of demonstrative character that he believed geometry to have. The work of identifying the appropriate *differentia* that marked off the species within a genus was all the more laborious in that the connection between *differentia* and essence would have to be a "self-evident" one, that is, one that imposed itself on the mind once the essence and the *differentia* were described in the appropriate way. And such logically transparent connections were hard to find in the organic world. It has recently been argued by a number of writers that Aristotle, the biologist, despaired of the Platonic ideal of science set down for him by Aristotle, the methodologist.[4] The propositions in which he had so patiently summarized and arranged his vast empirical work of classification and dissection and observation of behavior were simply *not* transparent in the way that they should have been in order to be "scientific" in the rigorous sense he had prescribed in the *Posterior Analytics*. As time went on, did he begin to wonder whether a more modest ideal of natural science would have to suffice? We have no way of knowing, though one cannot help surmising that a mind as brilliant as his must have noted the almost irreconcilable tensions between his explicit and his implicit methodologies in the area of biology.

The conceptualism defined by the *Posterior Analytics* has three features worth noting in particular. It assumes a world of essences, arranged in hierarchical structures that can be exhaustively laid out in syllogistic form. The natural world thus has a completely intelligible structure of a basically teleological sort. Chance events may happen, but these are simply outside the order of intelligibility entirely; they are not to be taken as indications that our scientific concepts must remain partially inadequate. Secondly, the human mind has a quality of "insight" (Aristotle uses the term *"epagōgē,"* sometimes misleadingly rendered in English as "induction") which allows it to grasp essence via single observations. It is this power of seeing the universal in the singular that makes the premises of natural science into necessary truths that are *seen* to be such, without the need for further test.

---

[4] See, for example, the papers by D. M. Balme and I. During in the collection *Aristote et méthode*, (Louvain, 1961), pp. 195–221.

Thirdly, it is assumed that the premises can be verbally stated. The concepts of ordinary language, refined and clarified, will suffice for scientific purposes. These concepts have been derived by abstraction directly from the sensible instances, so that science is basically a *descriptive* enterprise. There is no element of hypothesis, of a distance between model and *explicandum* that has somehow to be bridged in terms of additional evidence.[5]

This ideal of science was, of course, inspired by Plato, and yet it was in Plato's work that its direct opposite was first defined also. Plato knew that *epistēmē*, true knowledge, involves grasping the articulations of the realm of Form. But he realized how difficult this was, not only because our powers of insight are limited (no *epagōgē* for him), but also because the natural world in his view does not have an intelligible structure, only the wavering image of one. So although he held up a rigorous ideal of a conceptualist science (whose warrant would derive directly from the world of Forms), he was willing to settle in practice for various compromise sorts of knowledge. Geometricians, he thought, did not really grasp the self-evidence of their premises as they should, yet he was content to allow their work some sort of status lower down the "divided line" that symbolized the various grades of knowledge. Positional astronomy, for example, was hardly science at all; it "saved the appearances," served useful functions of description and prediction without having in itself any sort of demonstrative transparency. This notion of a pragmatic "saving of the appearances" was historically the first version of a correlationist view of science. It was conceded by Plato as in principle only a second-best, though (in the domain of physics, at least) the best that could in practice be managed. During the Greek period, it was invoked mainly for observational astronomy. Because this was already a highly successful predictive discipline, it could not plausibly be dismissed as a form of knowledge, and yet it clearly could never be fitted in a conceptualist mold.

III. THE REVIVAL OF CONCEPTUALISM DURING THE MIDDLE AGES

When the conceptualist ideal of natural science was once more affirmed in the West in the thirteenth century, its impact was just as great as it had been in Greece more than a millennium before. To Arab and Latin alike, it seemed like a profession of faith in the "natural light" possessed by the

[5] For a fuller discussion of Aristotelian conceptualist methodology, see my essay "The Nature of Scientific Method: What Makes It Science?" in *Philosophy in a Technological Culture*, ed. by G. McLean (Washington, D.C., 1964), pp. 28–54. See especially Section 2: "The Failure of the Conceptualist Model of a 'Science' of Nature."

human mind. The sort of science it aimed for appeared to be within human reach, as the scope and power of Aristotle's own cosmology seemed to prove. True, it gave no predictive power; indeed, there was a rather shocking disharmony between the "scientific" astronomy of Aristotle and Ptolemy's highly effective "saving of the appearances." The former might be nearer to "science," but it was of no service to navigators or to others interested in concrete applications. The traditional "sciences" of physics and psychology seemed likewise to offer little aid to engineers or medical men, men with practical problems to solve. The gap between theoretical understanding and concrete application was discouragingly wide, though medical schools made constant and not altogether unsuccessful efforts to bring the two closer together.

During the twelfth century, the whole of Aristotle's *Organon* was translated into Latin, and the older translation made by Boethius also came back into use. Logic became the major constituent in the arts curriculum of the newly founded universities. Commentaries on the various treatises of the *Organon*, including the *Posterior Analytics*, became the major mode of publication of the numerous professors of logic. By 1200, the conceptualist ideal of science was taken for granted in most discussions of methodology. Robert Grosseteste, writing on optics around this time, tried to cast his work into a conceptualist mold; even though a Platonist in his metaphysics, he felt that the methodology defined by the *Posterior Analytics* was the one he ought to adhere to in his scientific researches. As the thirteenth century progressed, Aristotle's "natural works" also began to be studied in the universities, although they met with much opposition because of the obvious tension between them and certain of the tenets of Christian faith. They were first banned, but the ban proved ineffective; by mid-century, Aristotle's *Physics* was securely lodged in the university curriculum. Not much actual scientific work was going on, however. Albertus Magnus worked in geology and biology, Jordanus in statics, Theodoric in optics . . . But their work was little known or appreciated, and had no impact upon the teaching of science at the universities.

In the second half of the century, the situation of natural science could be summarized as follows: Aristotle's *Physics* and *De anima* have become standard texts; his biological treatises are less well known. There is much emphasis on Aristotelian science as providing an adequate general account of nature. Only a few people are trying to carry on scientific research of their own, although along Aristotelian lines, of course. At the universities, science is learned from the texts of Aristotle and his many commentators, not from empirical research of any sort. Courses are taught in scientific method, most of them centered around the *Posterior Analytics*. The approach is epistemological: from a general theory of human knowing, one infers how a scientist ought to proceed and what the status of his results will be. So little empirical research is going on that there is no serious challenge to the adequacy of the conceptualist model of science from this quarter; though medical men are beginning to make additions to the work of Aristotle and Galen, these are either absorbed into the general picture provided by the *De anima* or thought of as practical knowledge of an infra-scientific sort.

Nevertheless, by 1300, scepticism about the applicability of the Aristotelian conceptualist model of science to our knowledge of nature was beginning to make itself felt. It had many sources, three of which are worth noting in some detail. The situation in astronomy, where the Ptolemaic model conformed better with observation and was the basis of all planetary computation whereas the Aristotelian model was in harmony with the accepted physics, led people to think that "mathematical" astronomy could provide at best only a "hypothesis" about the motion of the planets. This "hypothesis" was to be taken as no more than a convenient fiction, whose only function was to "save the appearances." It had no truth-value, strictly speaking, since it was not *intended* to make any claim about the structure of the physical real. Ptolemy himself had sanctioned such an interpretation of his intent in the *Almagest*; later writers on natural science, like Proclus, Simplicius, Philoponus, continued to stress the special character of astronomical theory, so that when Aquinas and other Aristotelian philosophers of later centuries remarked that mathematical astronomy simply "saved the phenomena" without revealing anything of the real motions of planets, they were simply repeating what had come to be taken for granted by that time.

The grounds for this view were various. The epicyclic motions invoked by Ptolemy seemed physically implausible; there was a real incompatibility between Ptolemy's view and the more intuitive homocentric model put forward by Aristotle; the mathematical form of the model moved it further away from true "science," to Aristotelian eyes, at least. More fundamentally, however, the Ptolemaic model was not "science" because its warrant was *entirely* empirical. There was no intrinsic conceptual necessity in it that would give the hope that it could someday form part of a "demonstrative" system. Its purpose was a purely practical one, divorced from questions of truth. "Saving the phenomena," as this tradition developed, was not a way of arriving at "science"; in particular, there was no suggestion that successive approximations of this sort would lead one nearer to the *truth* of the matter.

It is necessary to stress this point in order to remind oneself that there is here no question of the "hypothetical" view of natural science characteristic of the modern period.[6] Ptolemy's model was *not* regarded as a hy-

[6] It is to Pierre Duhem that we owe the most thorough account of the development of these views about astronomy (*Système du monde* [10 vols.; Paris, 1913–59]). On oc-

pothesis in the modern sense. It was not taken to be an approximation to the truth; it was a useful fiction. It was not part of "science," nor even propaedeutic to it. There *was* a "science" of Nature, but it was arrived at in a totally different way, one that owed nothing to any deliberate or contrived saving of the appearances. It is clear, then, that there was no suggestion in all this that the method of "science" might be hypothetical: these writers spoke of "hypothesis" in a significantly different sense from ours, and they supposed that they *had* a "science" of planetary motion which was in nowise dependent upon hypothesis, either in their sense of the term or in ours.

But now two other trends must be noted, trends that led people to question whether indeed they *did* have such a "science" of physical motions. As the Aristotelian revival gained impetus in the thirteenth century, a theological reaction to its claim to provide "necessary demonstrations" about the physical real began to set in. It seemed to many theologians that such a claim unduly limited the freedom and omnipotence of God in His creation of the universe. As Duhem was the first to stress, the condemnation of Averroist Aristotelianism by the French bishops in 1277 was largely motivated by this fear. The condemnation and the controversies that accompanied it led everyone, except the Averroists, to wonder whether or not a strict "science" of nature in the Aristotelian sense could, in principle, be formulated, especially if it was supposed to extend to the lowest species. It was easy enough to take the most general analyses of motion via notions of matter and form to be "scientific" in the old sense, but it was an altogether different matter when one came to the detailed, apparently contingent, structures of the physical world.

The growth of nominalistic philosophy tended to reinforce this doubt from a different quarter. Ockham argued that we can have natural certitude only in a formal discipline such as logic. We can. be certain too of our apprehension of singulars. But when it comes to generalizing about the physical world, the best that can be managed is some sort of probability. Regular successions of cause and effect can be perceived, allowing generalizations to be formed, but necessary causal relations can never be demonstrated. The world is made up of singulars, the relations between which are contingent. Knowledge of these relations must, therefore, always be rooted in observation and observation alone. A truly "scientific" proposition is one which is true under all circumstances, and there are no such propositions about our

---

casion, Duhem tended to "modernize" the views he recorded, and recent scholarship has been particularly critical of his presentation of the medieval view of the nature of mathematical astronomy as though it were to all intents that of the eighteenth century.

physical world, which might well have been other than it is. It is possible to formulate a conditional "science" involving physical concepts; its theses will be true of nature to the extent that its principles are, but one can never be quite sure to what extent the latter *are* verified. The nominalist physicists of Paris, men like Buridan, went about their "science" in this spirit. Its principles were suggested by generalization from experience; it was then formalized in a deductive fashion. Such a discipline had the intrinsic "necessity" of a formal science; it was applicable to nature because of the empirical origin of its concepts and principles, but it was not applicable with certainty because its principles were no more than plausible. Thus insofar as it was a "science," it was not of nature.[7]

The nominalists were not, therefore, abandoning the Greek notion of "science"; they were setting more realistic limits to its applicability. But one can see here the beginning of a new realization: if "science" is to be the best attainable knowledge of some domain, then the "science" of nature cannot be described in the traditional Euclidean terms. This will ultimately involve a modification in the notion of science itself. But the nominalists were not yet ready to make this modification. "Science" in the sense of unchanging necessary truths still seemed to them the goal of man's desire to know; the fact that he could not know nature in this way was simply an unhappy circumstance. So Buridan would say, characteristically: there is no true "science" of nature; not: there is a "science" of nature, but it has to be differently defined *qua* "science."

The effect of this joint movement in philosophy and theology was to push the systematic investigation of nature further and further in an empirical direction by calling into question the possibility of establishing intrinsic, ultimately non-empirical links between the concepts of physics. If the world was—as the Augustinian theologian and the nominalist philosopher alike believed—a contingent could-have-been-otherwise affair, then the only way to find out what was there was to go and observe it. In other words, all genuine knowledge of nature had to be empirical. If the kind of "science" sought were still to be Euclidean, then it would have to be hypothetical in its application to the real world of contingents, and the scientist would proceed by trying different hypotheses and seeing which of them worked. Whereas if the stress were rather on the generalization of singulars, physics would be no "science," but only a set of probable approximations.

---

[7] For references, see Ernest Moody, "Empiricism and Metaphysics in Medieval Philosophy," *Philosophical Review*, LXVII (1958), 145–63; Edward Grant, "Late Medieval Thought, Copernicus, and the Scientific Revolution," *Journal of the History of Ideas*, XXIII (1962), 197–220.

IV. "COPERNICAN REVOLUTION"?

How did this empiricist drift affect the views on the nature of "science" held by the great figures of the "scientific revolution" of the seventeenth century? Nineteenth-century theorists of science, like Mach, were wont to claim all of the pioneers (except Descartes) for the cause of empiricism and to treat the "revolution" itself as an empiricist revolt against the alleged "apriorism" of the medievals. Historians have long ago rejected this as a fiction. Indeed, Grant has recently argued that Copernicus'

> conception of the function and role of an hypothesis is so radically different from that of his predecessors that it serves to symbolize his drastic departure from the scholastic tradition almost as much as his new cosmological system does. . . . Copernicus is really the initiator of a very basic attitude which came to be held in some form or other by most of the great figures of the Scientific Revolution— namely, that fundamental principles in the form of hypotheses or assumptions about the universe must be physically true and incapable of being otherwise. . . . Scholastics were most sophisticated and mature in their understanding of the role which an hypothesis must play in the fabric of science. They were not . . . deluded into believing that they could acquire indubitable truth about physical reality. But it is an historical fact that the Scientific Revolution occurred in the seventeenth century, not in the Middle Ages under nominalist auspices. Despite the significant achievements of medieval science . . . it is doubtful that a scientific revolution could have occurred within a tradition which came to emphasize uncertainty, probability and possibility rather than certainty, exactness and faith that fundamental physical truths—which could not be otherwise—were attainable. It was Copernicus who, by an illogical move, first mapped the new path and inspired the Scientific Revolution by bequeathing to it his own ardent desire for knowledge of physical realities.[8]

This long quotation may serve to show how far the pendulum has swung since Mach's time. But there is need for caution here, and consequently we shall subdivide Grant's thesis into different points, some of which need qualification.

First, though he is correct in saying that nominalist scepticism was prevalent in the fourteenth century, nominalism was only one among a number of competing scholastic systems. Aristotelians, both Averroist and Thomist, continued to believe in the possibility of a conceptualist "science" of nature. So it would be incorrect to characterize scholastic philosophers generally as not believing that one could acquire "indubitable truths about

[8] *Journal of the History of Ideas*, XXIII, 197, 220.

physical reality." Indeed, the majority of scholastic natural philosophers in the period immediately preceding Copernicus continued to assume that a strict "science" of nature was possible. This is not to underestimate the importance of the sceptical theological and philosophical currents discussed in the last section, but one has to be careful not to overstress the point.

Copernicus was, therefore, no "initiator" of a new attitude in this regard. His views as to what constituted scientific knowledge were, so far as one can tell, very like those of Aristotle, or even more, of Plato, with whom he had so many other affinities. There is no question, then, of arguing that he somehow introduced a new conceptualist ideal of science which by successfully counteracting the nominalistic loss of nerve made the "Scientific Revolution" possible. If this were correct, both Duhem (who extolled the nominalism of later medieval methodology) and Mach (who extolled the empiricism of the "Scientific Revolution") would be wrong. But the thesis goes too far. The ideal of natural science for which Copernicus stood was not different from that of the majority of natural philosophers of his day. What carried conviction about the "new" science was not any return to a conceptualist ideal, but rather the fact that it *worked* so much better, in so many different ways.[9] It was not because Copernicus and Galileo were so *sure* of their principles that they carried the day—they were no surer than their Aristotelian opponents were—but rather because their astronomy and mechanics gave a more exact and more fruitful account of the way the world was.

It would be misleading, then, to suggest that medieval science failed because of the fog of uncertainty cast over it by nominalism. The reasons why it failed are complex,[10] but nominalism can hardly be ranked high among them. As we have seen, the increased stress it laid on observation, on the contrary, led to important advances. One cannot hold that a change in the notion of what constituted "science" was of itself decisive in the "Scientific Revolution," if only because Copernicus' medieval predecessors were themselves completely at odds with one another on the matter, and Copernicus defended one of the traditional views. Neither the Aristotelians (with whom Copernicus agreed about the nature of "science") nor the nominalists (with whom he disagreed about the possibility of a natural "science") succeeded themselves in constructing a successful natural science, so whatever it was that made the Copernican "revolution," it was not primarily a thesis (either empiricist, after Mach, or conceptualist, after Grant) about the nature of "science."

[9] These ways are discussed in McMullin, *International Philosophical Quarterly*, V.
[10] See *ibid.* and a critical notice of J. A. Weisheipl: *The Development of Physical Theory in the Middle Ages*, in *International Philosophical Quarterly*, II (1962), 483–89.

Yet, it *is* true to say that Copernicus was making a methodological point. The point itself is easily misunderstood because of the ambiguity in the word *hypothesis,* around which it hinges. Copernicus maintained that his system was not "hypothetical," but that the earth really *did* move. Against whom was this directed? Against the nominalist thesis of the entirely hypothetical character of natural science? No, it was quite clearly directed against the Aristotelian division of astronomy into "physical" and "mathematical" and the attendant claim that "mathematical" astronomy provided only convenient fictions. Copernicus is making three separate methodological points, two of which were correct. First, the Aristotelian dichotomy between physics and astronomy lacks any real justification; consequently, there are no more grounds in astronomy than there would be in physics for supposing that one part of it enunciates truths about reality whereas the other simply "saves the phenomena." This point had been made earlier by several writers (such as Nicholas of Cusa), and it amounted to a declaration of the unity of science and of the unreasonableness of dismissing any part of natural knowledge as though it had no truth-content whatever. Indeed, Copernicus will try to turn the tables and suggest that instead of the astronomer's looking to the physicist for the "truth" about things, it might be more appropriate for the physicist to look to the astronomer.

Secondly, Copernicus vehemently protested against the Aristotelian treatment of astronomical hypotheses as *necessarily* fictions. As we have seen, it was assumed that astronomers' models had nothing to do with the reality of things and so were not so much "hypothetical" (in the modern sense of the term) as fictional. The term had so long been used in a derogatory fashion by Aristotelian philosophers that it was little wonder that an astronomer with a brilliant new idea should bristle at it. Indeed, it would seem plausible to say—and this brings us to the third point—that it was this reaction that led him to affirm the "reality" of his model with such vehemence. It was all very well for Ptolemy to concede to his Aristotelian contemporaries the fictional character of his circles. But Copernicus was in nowise willing to make the same concession, both because he rejected the dichotomy between the terrestrial and celestial realms on which the Aristotelian claim to a superior truth for the physical-terrestrial proof rested and because he felt that his model (unlike that of Ptolemy) had a perfectly plausible physical character.

He went too far in insisting on the certainty of his theory, but it is all-important to notice that he was making his point against an Aristotelian (who would hold that all astronomical models were inherently fictional, of no truth-value whatever), not against a nominalist (who would maintain that all theses about the physical world must be regarded as hypothetical in character). Consequently, what he was affirming, in the context of the astronomical debate in which he moved, was by no means an anti-nominalist declaration of the possibility of attaining to a true "science" in matters physical, but rather a much more modest (and more accurate) disavowal of the "fictionalist" view of astronomical theory. If the "realism" of his claim appears striking, it should be remembered that his Aristotelian opponents maintained an equally "realist" view. Copernicus is merely arguing that the grounds for their denial of physical relevance to mathematical astronomy are unsound. He assumes without question (just as his opponents did) that a realistic natural science is possible; the only question at issue is whether his astronomy should form a part of it or not.

In the light of all this, several conclusions can be drawn. One, the fictionalist account of astronomical hypothesis characteristic of medieval Aristotelianism was (rightly) rejected by Copernicus; this account ought not, however, be treated as somehow an approximation to, or direct predecessor of, later views on the hypothetical character of natural science. Two, Copernicus remained untouched by the nominalist critique of the conceptualist ideal of natural science; like his Aristotelian contemporaries, he continued to believe that such an ideal could be realized. And his way of realizing it was, methodologically, not notably different from that of Aristotle, who asserted in the *Metaphysics* that the number of the spheres is fifty-five, "neither more nor less."[11] Three, the effect of nominalism on the "new scientists" was, at best, a very indirect one, since hardly any of them were in sympathy with its principles. It would be better to say that the empiricist climate fostered by the earlier nominalistic physics helped to set the stage for the growing emphasis on observation and experiment in the "new science."[12] But it did this without, so far as one can determine, significantly influencing the "new science" in the direction of a more "hypothetical" approach to theory. Of course, it was the new empirical stress that would ultimately force the admission of a "hypothetical" methodology. But one came long after the other.

---

[11] "Let this then be taken as the number of the spheres, so that the immovable substances and principles also may probably be taken as just so many; the assertion of necessity must be left to more powerful thinkers. But if there can be no spatial movement which does not conduce to the moving of a star, and if further every being and every substance, which is immune from change and has attained to the best in virtue of itself, must be considered an end, there can be no other being apart from those we have named, but this must be the number of the substances." (*Metaphysics*, 1074a 14–21). See Ernan McMullin, "Realism in Modern Cosmology," *Proceedings of the American Catholic Philosophical Association*, XXIX (1955), 137–50.

[12] In this, of course, it was only one among many factors. See E. W. Strong, *Procedures and Metaphysics* (Berkeley, 1936), *passim*.

344

V. THE SEVENTEENTH CENTURY:
GALILEO AND KEPLER

In the period between Copernicus and Galileo, opinion among astronomers (whether Ptolemaic or Copernican) hardened in favor of the "realist" view of mathematical astronomy espoused by Copernicus. Two factors were responsible. One was the general decline of nominalist physics and the silencing of the doubts it had earlier raised, but far more important was the growing acceptance of Copernicus' argument that astronomy ought not be treated as methodologically different from the rest of natural science. As Clavius pointed out, there seemed to be only two consistent positions open: either astronomy was hypothetical (but then the rest of natural science would have to be regarded as similarly hypothetical, and this Clavius—like nearly all his contemporaries—would not concede) or else it was non-hypothetical, just like any other part of natural science.[13]

The dilemma is clear in Galileo's writings. On the one hand, his whole theory of science was a conceptualist one. The new mathematical physics would give an exact description of Nature because Nature was ultimately mathematical in its structure. Not only that, but the warrant for its principles would be primarily conceptualist: his frequent reliance on thought-experiments, his praise of Copernicus for disregarding facts that did not fit his model, indicate this clearly enough.[14] His theory of science was Aristotelian in this respect. Unlike Plato, he believed that a strict "science" of Nature could be achieved; unlike Aristotle, he gave mathematics a leading role in the achieving of it. Unlike Pythagoras (and Kepler), he never looked to mathematics for the *evidence* for the truths of physics; in his view, mathematics provided the *language* of physics, not its evidence.

But on the other hand, Galileo knew perfectly well that

to prove that the appearances may be saved with the motion of the earth . . . is not the same as to prove this theory true in nature. . . . [The Ptolemaic] system cannot give reasons for those appearances and is undoubtedly false, just as . . . [the Copernican one] may be true. And no greater truth may or should be sought in a theory than that it corresponds with all the particular appearances.[15]

[13] See Ralph M. Blake, "The Theory of Hypothesis among Renaissance Astronomers," in *Theories of Scientific Method: The Renaissance through the Nineteenth Century*, ed. R. Blake, C. Ducasse, E. Madden (Seattle, 1960), pp. 22–49.

[14] See Thomas McTighe, "Galileo's Platonism: A Reconsideration," in *Galileo, Man of Science*, ed. by Ernan McMullin (Basic Books, for 1967 publication). His thesis is that Galileo was more conceptualist in his theory of science than even Descartes was. See below.

[15] Notes made by Galileo for a reply to Bellarmine's letter to Foscarini. See *Discoveries and Opinions of Galileo*, ed. by Stillman Drake (New York, 1957), p. 169.

All that he can show is that the Aristotelian-Ptolemaic model does not fit the appearances and must therefore be rejected. He cannot prove that his own alternative, the Copernican one, is the *only* other possibility, and so he realizes that he cannot prove it to be "true in nature." Nevertheless, he constantly writes as though he had done just this. In fact, the passage cited above is almost the only one in which he concedes the methodological difficulties involved in his unqualified defence of the truth of the Copernican hypothesis.

Why did he write in this way? He knew that other alternatives (such as that of Brahe) were still open. He knew that different choices of a frame of reference would give mathematically different (though predictively equivalent) models of the planetary system. But he believed that natural science could, in favorable circumstances, come to make unqualified claims about the physical world, and he believed Copernican astronomy made such a claim. Yet he could provide no demonstration of this, not at least in the only sense of *demonstration* that his opponents would be willing to accept. The two arguments he most relied on (based on analyses of tidal motion and of the apparent motions of sunspots) were manifestly weak, and he had deliberately avoided trying to formulate a dynamic theory of motion, which might conceivably have helped to determine which of the two proposed reference-frames "really" was at rest. He was defending a non-hypothetical approach in the very area where it had always been recognized to be the hardest to defend.

To understand this famous controversy, it is vital to recall that it was *not* concerned with whether or not a truly certain and "scientific" claim about the motion of the earth could be made. Both sides agreed that it could. Galileo's opposition came from those, like Bellarmine, who maintained that mathematical astronomy could not decide such questions, though "physics" could. To counter this, Galileo found himself arguing that his astronomy *could* answer such questions—and did. It was an easy exaggeration in the context of a controversy in which both sides believed the question "is the earth *really* moving around the sun?" to be a perfectly meaningful one. The fashion of some historians of science to present Galileo's opponents as "relativists," almost in an Einsteinian sense, is profoundly misleading. These opponents thought that *they* could prove that the sun was really moving around the earth. This followed (they thought) from their physics; whereas Galileo thought he could show that his view followed from *his* physics, which, to his mind, was a better physics.

How could men who were so empirically inclined overlook what seems to us the obvious possibility of alternative theories? How could the "new science" be both empiricist and conceptualist? To get an answer, let us look at Kepler's writings, since he seems to have grasped the difficulty better than anyone else of his century before Newton.[16] In his response to the "fictionalist" view of astronomical hypothesis argued by Nicholas Baer, he analyzes the notion of hypothesis in natural science in considerable detail. He points out that Baer's error "is due to his perverse understanding of the term 'hypothesis,' which to him means the same as 'fiction.'"[17] It is possible, of course, for a true conclusion to follow from false premises, so that even if deductions from the hypothesis are verified, there is always the possibility that the hypothesis might still be false. But a false hypothesis will eventually betray itself by giving rise to a non-verified prediction and will thus be falsified—unless, Kepler adds, with considerable acumen, one were to permit its defender to keep adding extra hypotheses. But an end must come to this, and so if one keeps testing two alternative hypotheses, one of them must ultimately fail.

He rejects the possibility that they might give predictions that always agree; every hypothesis, he feels,

will exhibit some conclusion peculiar to itself and distinct and different from all others. For if in their geometrical conclusions two hypotheses coincide, nevertheless in physics each will have its own peculiar additional consequence. . . . Authors do not always consider this variety in the physical consequences, but more often confine themselves to the boundaries of geometry or astronomy, and raise the question of the possible equivalence of hypotheses within a single science, without realizing that diverse consequences in neighboring sciences might lessen or destroy the alleged equivalence.[18]

In practice, Kepler says, it almost never happens that the scientist has to choose between equivalent hypotheses; a "crucial test" can easily be devised to decide between the Ptolemaic and the Copernican hypotheses, for instance. In short, he does not believe it possible "that from the assumption of an unsound hypothesis there should follow a conclusion in all respects sound and conformable to the celestial motions." In particular, the astronomer in choosing a hypothesis does not take a deliberately fictional one; he tries, as best he can, to find a true one. Only a true one will have consequences that are true in all respects; thus, in principle at least, the truth is always discoverable in this way:

That astronomer well performs his office who predicts with the greatest degree

---

[16] See Blake, *Theories of Scientific Method*, pp. 38–43.

[17] *Apologia Tychonis contra Nicolaum Ursum*, in *Opera omnia*, ed. C. Frisch (Frankfurt, 1858), I, 242, translated in Blake, *Theories of Scientific Method*, p. 43.

[18] *Ibid.*, p. 240.

of accuracy the motions and situations of the stars; but he does better and is more praiseworthy who in addition furnishes us with true opinions concerning the form of the world.[19]

The Keplerian theory of "science" is clearly a bridge between old and new. Like Aristotle, he supposes that certainty is attainable in natural science and furthermore that the "scientific" character of a thesis is in direct proportion to its *certitude*. But he differs from him in one vital respect. He no longer maintains the conceptualist thesis that the evidence for a truly "scientific" statement ought to be intrinsic to the statement itself or to the other statements from which the first one could be demonstrated. He does not believe that an astronomical model can ever become "self-evident," in this technical conceptualist sense. He has taken the irrevocable step toward an empirical theory of "science": the justification from now on is not in any intuitive linking of fully grasped concepts but in the patient empirical observation and elimination of alternative hypotheses that precede the final triumphant affirmation. He still sees in such affirmations the *raison d'être* of scientific inquiry, and he assumes that in practice they can frequently be made, when (as in the case of the Copernican model) a great number of predictions have been verified and no further plausible alternative suggests itself. According to him, the role of the empirical is not just heuristic. The empirical is not discarded once the scientific thesis is formulated; it remains the permanent and logically adequate warrant for the thesis.

Bacon is usually cited as the empiricist, *par excellence*. Yet he had some interesting remarks on the necessity of employing both the "experimental and rational faculties" for any worthwhile result to be achieved:

The men of experiment are like the ant; they only collect and use; the reasoners resemble spiders, who make cobwebs out of their own substance. But the bee takes a middle course: it gathers its material from the flowers of garden and field, but transforms and digests by a power of its own. Not unlike this is the true business of philosophy: for it neither relies solely or chiefly on the powers of the mind, nor does it take the matter which it gathers from natural history and mechanical experiments and lay it up in the memory whole, as it finds it; but it lays it up in the understanding altered and digested. Therefore from a closer and purer league between these two faculties, the experimental and the rational (such as has never yet been made), much may be hoped.[20]

This is a rather striking definition of a "science" which would combine the conceptualist and the empiricist tendencies. He saw it as a knowledge

[19] *Ibid.*, p. 242.
[20] *Novum organum* i. aph. 95.

348

of the forms of simple natures (qualities) and supposed that it could be attained with certitude through following the elaborate tables of inductive procedures he provided. The role of hypothesis, and of the "rational faculty," in its attainment was something he never, however, succeeded in clarifying.

VI. THE CONCEPTUALISM OF DESCARTES

Descartes, on the other hand, has always been cited as the conceptualist, *par excellence*, the man who attempted to carry the Aristotelian ideal of "science" even further than Aristotle himself had tried. And the ultimate failure of Cartesian physics is usually attributed by historians of science to its *a priori* character. Yet Descartes himself performed many experiments and remarked that "they become so much the more necessary the more one is advanced in knowledge."[21] There is a paradox here, one with which historians of methodology are still wrestling. It will not do simply to say (as some writers recently have) that the conceptualist program of science was true only of the young Descartes of the *Regulae* and that he became a defender of a full-blown empirical and hypothetical method in his later works. A careful reading of the following passage from the last section of his mature work on method, the *Discours*, may give us a clue:

I have first tried to discover generally the principles or first causes of everything that is or can be in the world, without considering anything that might accomplish this end but God Himself who has created the world, or deriving them from any source except from certain germs of truths which are naturally existent in our souls. After that I considered which were the primary and most ordinary effects which might be deduced from these causes, and it seems to me that in this way I discovered the heavens, the stars, and earth and even on the earth, water, air, fire, the minerals and some other such things. . . . Then when I wished to descend to those which were more particular, so many objects of various kinds presented themselves to me that I did not think it was possible for the human mind to distinguish the forms of bodies which are on the earth, from an infinitude of others which might have been so if it had been the will of God to place them there, if it were not that we arrive at causes from effects and avail ourselves of many particular experiments. . . . The power of nature is so ample and vast, and these principles are so simple and general, that I observed hardly any effect . . . which could not have been deduced from the principles in many different ways; and my greatest difficulty is usually to discover in which of these ways the

[21] *Discourse on Method,* in *The Philosophical Works of Descartes,* trans. E. S. Haldane and G. R. T. Ross (Cambridge, 1931), I, 120.

effect does depend on them. As to that, I do not know any plan but to try to find experiments of such a nature that their result is not the same if it has to be explained by one of the methods [of deduction] as it would be if explained by the other.[22]

In this well-known passage, he is claiming (1) that many specific statements about the nature of the universe (e.g. about minerals) can be deduced with certainty from the first principles of his metaphysics and that they derive from "germs of truth" innate in the mind; and (2) that regarding more particular entities than these, it may be impossible to derive them in this way because of the multitude of possible causal lines leading to them, and thus one will have to have recourse to experiment as a means of deciding between the competing possibilities. What are these "more particular" truths?

All the bodies which make up the universe are constituted of one and the same matter . . . divided into parts which are diversely moved, and whose movements are in some fashion circular, conserving always the same quantity of these motions in the universe. But we cannot determine by reason alone how great are the parts into which this matter is divided, how rapidly they move, nor what circles they describe. For since these could have been ordained by God in innumerable ways, only experience can tell us which God has chosen above the others.[23]

In other words, what Descartes finds himself unable to work out demonstratively are the distributions and motions of matter in the real universe. These must be regarded as contingent: he is assured of this by the familiar theological dictum about the omnipotence of God. Note that he is *not* saying here that any of the laws of his physics (as we would call them today) ought, because of this, be regarded as contingent; those he mentions can be discovered by the "force of reasoning" alone, it would appear. The role of the experiment, then, is in the first instance not in confirming one physical theory or law against another one, equally compatible with the first principles. It is in the discovery of the factual contingent distributions of matter at a given time, knowledge of which (when joined to the knowledge of the *demonstratively* given "principles," i.e. laws and theories, of physics) allows predictions to be made. In other words, what Descartes is saying is that his physics ought not to be taken as providing by itself concrete predictions of what will happen and when it will happen. This is a perfectly unexcep-

---

[22] *Ibid.*, p. 121.
[23] *Principia philosophiae*, in *Oeuvres de Descartes*, ed. C. Adam and P. Tannery (Paris, 1905), VIII, 100–1.

tionable point. No one has ever really defended a "physics" from which the entire history of the universe could be derived, without adding at least one concrete specification of the state of the system. When Descartes is described as "rationalist" in his science, no one has ever (so far as I know) had this sense of "rationalism" in mind. Consequently, it seems mistaken to argue on the basis of passages such as these (as some recent authors[24] have done) that Descartes was not really "rationalist" but was advocating a hypothetico-empirical method. What he was advocating in these texts was not so much a hypothetical physics as a conditional application of a basically non-hypothetical physics to singular situations.

When Descartes speaks of the "demonstrative" character of his science, he can mean one or other of two quite different things, and this ambiguity was to haunt philosophy up to the time of Kant. He can mean that scientific theses are "demonstrative" in the Aristotelian sense of carrying their own intrinsic warrant or being directly derivable from theses that do. That he *intended* his science to be "demonstrative" in this sense seems abundantly clear.[25] He may also simply mean that his science links premises and conclusion in deductive fashion. The "necessity" here would be a conditional or logical one, of the same sort that one finds, for example, in any physical theory of today. That he frequently speaks in this latter way cannot be denied, but it remains true that the primary sense of *necessity* in the *Discours* is a far stronger one. In Section V, he begins his summary of his earlier work, *Le monde*, by remarking that in order to avoid all disputes, he resolved to ask only "what would happen in a new world if God now created, somewhere in an imaginary space, matter sufficient wherewith to form it," and then set it in random motion. Even a "bracketing" as total as this of the contingent existential order does not prevent him (he claims) from arriving at a complete mechanics; on the basis of it:

I pointed out what are the laws of Nature and, without resting my reasons on any other principle than the infinite perfections of God, I tried to demonstrate all those of which one could have any doubt, and to show that they are of such a nature that even if God had created other worlds, *He could not have created any* in which these laws would fail to be observed. After that, I showed how the greatest part of the matter of which this chaos in constituted *must*, in accordance with these laws, dispense and arrange itself in such a fashion as to render it similar to our heavens; and how meantime some of its parts *must* form an earth, some planets and comets, and some others a sun and fixed stars. . . . [A few lines

[24] See, for example, Gerd Buchdahl, "Descartes' Anticipation of a 'Logic of Scientific Discovery,'" in *Scientific Change*, ed. by A. C. Crombie (New York, 1963), pp. 399–417.
[25] For references, see for example Blake, "Experience in Descartes' Theory of Method," pp. 75–103, in *Theories of Scientific Method*.

later, concerning the bodies of men:] Since I had not yet sufficient knowledge to speak of them in the same style as the rest, that is to say, demonstrating the effects from the causes and showing from what beginnings and in what fashion Nature must produce them, I contented myself with . . . [he goes on to give a detailed empirical discussion of the circulation of the blood, following Harvey].[26]

It would hardly be possible to give a plainer definition of the conceptualist ideal of natural science than this.

It might be said, in summary, that the empirical plays three different roles for Descartes: (1) The clear and simple ideas whose necessary interconnections constitute "science" are, in many instances at least, derived in some not very clearly specified fashion directly from experience:

He who reflects that there can be nothing to know in the magnet which does not consist of certain simple natures will have no doubt how to proceed. He will first collect all the observations with which experience can supply him about this stone, and from these he will next try to deduce the character of that intermixture of simple natures which is necessary to produce all those effects. . . . He can then boldly assert that he has discovered the real nature of the magnet insofar as human intelligence and the given experimental observations can supply him with this knowledge.[27]

(2) As we have already seen, the application of physics to the behavior or structure of concrete singulars will ordinarily require some sort of prior empirical specification of the singular. (3) In those cases where the knowledge sought is too specific to permit of its direct deduction from first principles, it may still be possible to set up a hypothesis and then demonstrate that this hypothesis is, in fact, verified.[28] This may be done in two ways: by exclusion of other possible alternatives:

the experiments that I have deduced from it necessarily, and which cannot be deduced in the same fashion from any other principles, seem to me to demonstrate it sufficiently *a posteriori*[29]

or else by showing the fertility and range of its predictions and explanations (here we remember Kepler):

---

[26] Trans. Haldane and Ross, I, 108. Italics mine.
[27] *Regulae ad Directionem Ingenii*, reg. xii, trans. Haldane and Ross, I. 47.
[28] This is fully discussed in Blake, *Theories of Scientific Method*, pp. 92–95.
[29] Letter to Vatier, *Oeuvres*, I, 563–64. See also *Principia*, I, 300–1. The last sentence of the long quotation from the *Discours* some pages back suggests something close to an *experimentum crucis*.

Although it is true that there are many effects to which it is easy to fit different causes, one for each, it is not always so easy to fit one and the same cause to many different effects, unless it is the true cause from which they proceed; indeed, there are often effects which are such that it is sufficient proof that a cause is the true one if the effects can be clearly deduced from it. And I claim that all these of which I have spoken are of this number.[30]

It will be noted that in all of these ways of utilizing experiment and observation, it is assumed that the "science" thus arrived at (or applied) is ultimately certain, even though part-empirical in its origins.

VII. THE GROWTH OF EMPIRICISM: BOYLE

As the century progressed, empirical procedures brought in a great wealth of information, some of it, at least, in a more exact form than had ever before been available. Not only were there new instruments and a new emphasis on accurate reporting of results, preferably in mathematical form, but there was an enthusiasm for discovering the riches of Nature by simply going out and looking. The early history of the Royal Society illustrates this very well. Week after week, the *virtuosi* displayed the results of their indefatigable, but usually unplanned, inquiries into the regularities and the oddities of the natural world. They scrutinized all sorts of chemical changes, used the new optical instruments to bring in new wonders of the sky or of the minute structures of living creatures, tested the properties of lodestones and medical remedies, and so forth. From one Wednesday meeting to the next, one never knew what new discovery would be announced. There was little experiment in the strict sense of controlled and directed observation, but there was much amassing of data and much discussion of how they might be explained. In addition, scientists were beginning for the first time to form an international community, with frequent communications of results by letter or pamphlet, and so the data to be explained grew ever more diverse, and the inadequacies of a purely conceptualist approach ever more obvious. Yet this approach was still influential as we can see in the work of Robert Boyle, the most ingenious experimenter of his age. Rather than speculating about forms and natures, as his predecessors had done, he assumed that chemical change is due to the recombining of corpuscles of different shapes under the action of various mechanical forces. He well knew

---

[30] Letter to Morin, *Oeuvres*, II, 199.

353

that their existence could not be directly supported and that even their combinatory properties were hypothetical. But the mechanical conception that underlay his work, a very similar conception to that of Descartes, he thought to be certainly and immediately established:

I look upon (these) metaphysical and mathematical principles . . . to be truths of a transcendent kind, that do not properly belong either to philosophy or theology, but are universal foundations and instruments of all the knowledge we mortals can acquire.[31]

However, such principles were regulative only and did no more than point in the direction of the empirical work to be done. Unlike Descartes and Galileo, Boyle could

see no necessity that intelligibility to a human understanding should be necessary to the truth or existence of a thing, any more than that visibility to a human eye should be necessary to the existence of an atom or of a corpuscle of air.[32]

Thus the conceptualist ideal held out very little promise to him; the complexities of the natural world were too great. The world just does not have the transparency that the conceptualists had taken for granted. He remarks tartly that if scientists would tend to their work instead of to their reputations, they would see for themselves that the best service they could do mankind would be

to set themselves diligently and industriously to make experiments and collect observations, without being overforward to establish principles and axioms, believing it uneasy to erect such theories as are capable to explicate all the phenomena of nature, before they have taken notice of a tenth part of those phenomena that are to be explicated. Not that I at all disallow the use of reasoning upon experiments, or the endeavouring to discern as early as we can the confederations and tendencies and differences of things. For such an absolute suspension of the exercise of reason were exceeding troublesome, if not impossible.[33]

He goes on to note the utility of hypotheses in physiology in order "that by examining how far the phenomena are, or are not, capable of being solved

---

[31] *The Works of the Honourable Robert Boyle*, ed. T. Birch (London, 1672), III, 429. The Boyle quotations are given by E. A. Burtt, *The Metaphysical Foundations of Modern Science* (London, 1924), pp. 166, 179, 182.
[32] *Works*, ed. Birch, IV, 450.
[33] *Ibid.*, I, 302.

by that hypothesis, the understanding may, even by its own errors, be instructed." He is using the term *hypothesis* here in precisely the modern sense of something put forward tentatively to account for something else, with the realization that further evidence may force its modification or abandonment. A hypothesis may have some intrinsic coherence or aesthetic appeal, but the warrant for it lies in the extrinsic evidence that can be marshaled in its favor. Boyle concludes with an exceptionally clear-sighted admonition to his scientific colleagues that they should

forbear to establish any theory, till they have consulted with, though not a fully competent number of experiments such as may afford them all the phenomena to be explicated by that theory, yet a considerable number of experiments, in proportion to the comprehensiveness of the theory to be erected on them. And in the next place, I would have such kind of superstructures looked upon only as temporary ones, which though they may be preferred before any others as being the least imperfect, or if you please, the best in their kind that we yet have, yet are they not entirely to be acquiesced in as absolutely perfect, or incapable of improving alterations.[34]

While implicitly conceding that mechanics, at its lofty level of abstraction, might be derivable from a small set of directly grasped principles, Boyle saw that such an ideal would never work for the profusion of the more specific sciences, like chemistry, pneumatics, and physiology, sciences that his own researches were helping to open up. The waning of this ideal in the seventeenth century was in no small part due, then, to the growing successes of the parts of science other than mechanics. These new sciences for the most part utilized the general conceptual structure of the Cartesian mechanical philosophy, but in establishing their *own* theories, relied exclusively upon induction and hypothesis and testing. Aristotelian science had never got beyond the purely descriptive level in these areas; the verbal-phenomenal emphasis on qualities and forms excluded the formulation, as well as the testing, of explanatory hypotheses of any degree of generality. The relative success of the experimental-mathematical approach and the corpuscularian model of matter in the mechanics of Galileo and Descartes provided a new conceptual framework for the special sciences, one that immediately suggested all kinds of limited testable hypotheses (e.g. of chemical combination). The success of these hypotheses pushed scientific methodology far more effectively in the empiricist direction than any development in mechanics could ever have done on its own.

[34] *Ibid.*, I, 303.

VIII. THE ENIGMA OF NEWTON

The century between Galileo's *De motu* and Newton's *Principia* saw the gradual erosion of the Aristotelian world-view and the rise of a new mechanical-mathematical approach to the knowledge of Nature. Descartes attempted to provide a metaphysics from which this approach and even the very physical principles themselves would follow. But the convoluted series of hypothetical vortices in terms of which he tried to explain natural motion led to incorrect predictions, or more often were incapable of any sort of empirical test, so that even though his system had an attractive deductive completeness and explanatory transparency, it lacked empirical warrant, something that was coming to seem an important defect, especially to the hardheaded English Baconians of the Royal Society. Descartes' forced geometrization of nature rather obviously ignored the non-geometrical factor of weight (or mass), thus making dynamics impossible. Students of Galileo's work could see this plainly enough, though in France so seductive did the elegant rationalism of the Cartesian system appear that the voice of dissent was hardly heard for a century after Descartes' death.

What was needed was a system in which the enormous mass of empirical data available in astronomy and in mechanics could be interpreted in the light of a small number of plausible general principles, mathematical in form and physical in experimental reference. Newton was the one who provided this system, one in which all the best in the physical insights of a hundred years was brought to a sudden focus, one so sharp that men of science would not again be bothered by fuzziness in the basic domain of mechanics for over two centuries. The "Scientific Revolution" came to a triumphant conclusion in the *Principia*. With its writing, the "revolution" would seem to have been over. It is crucial to inquire, then, what part empiricism played in Newton's achievement. If the empiricist ideal of science is claimed as either cause or effect of the "Scientific Revolution," it ought to be clear in Newton's work, which was the most successful statement of what that "revolution" had set out to achieve.

There are several difficulties about discovering Newton's views on matters of methodology. He redrafted his work over and over, leaving a vast mass of manuscript material, much of it still unpublished. His major works were conceived, and notes for them made, years before they were actually set down in final form. On occasion, substantial modifications had occurred in his thought in the meantime, occasioning important revisions. He entrusted the editing of his later works to others and frequently made them his spokesmen, even when their views did not altogether coincide with his. Many of his most illuminating asides on method occur in his letters; one wonders what weight to attach to them, especially since they do not (as we shall see) form an altogether coherent pattern.

What shaped Newton's method? We have seen that scientific methodology before Galileo's time was defined largely in terms of a prior general theory of knowledge or a metaphysics. The procedures of test and proof one adopted in natural science were adopted not because of any proved successes on their part in this domain, but rather because it seemed to follow from the nature of man (or of knowing or of being) what procedures one should follow in such an inquiry. There is a definite movement away from "epistemological" methodologies of this sort during the seventeenth century and a growing tendency to seek methods that will be adequate to the solution of limited definite problems in terms of some relatively pragmatic notion of adequacy. Newton may have, to some extent, inherited a mathematical-mechanical metaphysics from Descartes and Barrow, one which served as a background for his own thinking. But it would be wrong, I think, to claim this metaphysics as a "foundation" for the science of Newton, as Burtt did in his classic *Metaphysical foundations of modern science*. One could as well speak of his science as the "foundation" of (that is, his warrant for accepting) the metaphysics. The two were concomitant in his thinking, as they had not yet been in the thinking of Galileo and Descartes.

Whatever of his metaphysics, it can hardly be questioned that Newton tailored his methodology to the actual procedures he found to be successful in reducing the domain of quantitative mechanics to an easily comprehended order. His use of mathematics or of deduction or of experiment was dictated not so much by a theory of knowledge as by a clear realization of what would constitute a solution to the problem of the unification of mechanics and astronomy, bequeathed to him by Galileo. So that when we ask what shaped Newton's method, the first answer is that it was shaped by a new kind of exact data and by the provisional successes that had already been achieved in mathematically correlating these data.

There are two other circumstances that must be kept in mind if we are to understand why Newton proceeded in his science in the way he did. His major opponent was Descartes, whose influence was dominant in physics and astronomy at the time Newton was a student. Thus Newton frequently tended to define his approach by an implicit contrast with that of Descartes, just as his predecessors (including Descartes) had done fifty years before when they contrasted *their* views on method with that of the Aristotelian orthodoxy of *their* day. And just as they had tended to exaggerate their differences, so Newton exaggerated those methodological features of his system which most set it off from that of Descartes; his frequent remarks on method were evoked, on occasion, at least, by the fear that someone would confuse his approach with that of his Cartesian opponents. In those crucial years of enforced retirement when the plague had closed his university (1665–66), the young student had been able to derive the inverse-square law of gravitation from Kepler's Third Law of planetary motion. It seemed to him a straightforward deduction,[35] although in actual fact it was not, since Kepler's Law was a purely descriptive correlation between the periods and the mean distances of the planets from the sun, whereas Newton's "Law" was, in fact, an explanatory hypothesis, introducing not only a new explanatory construct (*force*) but also a non-intuitive notion of gravitational action at a distance. Newton saw that his "Law" immediately brought planetary as well as terrestrial gravitational motions under a single explanatory principle, one that could also serve to predict and describe these motions to any desired degree of accuracy. This was precisely what Descartes had never succeeded in doing; his ether vortices had "explained" but were perfectly useless for prediction or even exact description.

Newton thus realized at an early stage that the fatal flaw in the Cartesian methodology was the encouragement it gave to "explanatory" hypotheses that were in no way amenable to empirical test, whether direct or indirect. And so, like Aristotle, but for a quite different reason, he tended to think of hypothesis as illegitimate, or at best as a temporary expedient in scientific inquiry. This led to serious distortions. He would, for example, separate off the "hypotheses" in his published work, tacking them on as speculative and inessential appendices and consistently underestimating the hypothetical aspect of his own central insights. The separation between explicit and implicit methodology, between what is said about method and the method actually practiced, is more troublesome in Newton's work than in that of almost any other scientific writer; it is oddly reminiscent of the separation in Aristotle, who was (as we have seen) far more tentative and empirical in his actual scientific researches than he was in his declarations about what scientific demonstration *should* be.

There was a second reason for this striking lack of fit between theory and practice. Newton was both introverted and proud, abnormally sensitive to criticism of his work, whether real or imagined. So great was the power of his extraordinary insight, unparalleled perhaps in the entire history of science, so far beyond the speculations of his contemporaries did his suggestions reach, that it is not really surprising to find such a man impatient of opposition and constantly exaggerating—even to himself—the certainty of his claims. This is perhaps the main reason why he could not bear to have

---

[35] Cambridge University Library, Add. Ms. 3968, No. 2, cited by A. R. Hall, *From Galileo to Newton 1630–1720* (London, 1963), p. 278.

others describe any part of his work as "hypothetical" unless he had himself expressly marked it off as such. The word suggested to him a lack of faith in, or an implicit criticism of, his scientific insight, something he found hard to tolerate, especially during his later years.

We may more easily understand his reluctance to allow his theories to be described as "hypotheses" if we look at the way in which he responded to criticism of his very first published paper. The sensitive young genius was deeply affected by the exchanges that followed the publication of his "Theory about light and colors" in the *Transactions* of the Royal Society in 1672. He was persuaded that what he had to report was "the oddest, if not the most considerable, detection which hath hitherto been made in the operations of Nature," which is no mean claim.[36] The paper describes his experimental work with prisms and concludes that light is a "heterogeneous mixture of differently refrangible rays," that colors are not produced by refraction or reflection but are "original and connate properties" of the light. He is tempted to discuss how light produces in us the sensation of color but decides not to "mingle conjectures with certainties." His new theory of light was controlled so well by the ingenious series of experiments he describes that he obviously thought of it as a "certainty."

But many others thought otherwise. The Royal Society's leading expert on optics, Robert Hooke, wrote a careful report on the young physicist's paper, in which he conceded himself to be pleased with the "niceness and the curiosity" of the experiments Newton had described, but quite rightly noted that Newton's claim about the composition of white light could not be said to be "demonstrated," since other hypotheses—Hooke's own wave-theory among them—might very well account for the same phenomena:

As to his hypothesis of solving the phenomena of colors thereby, I confess I cannot see yet any undeniable argument to convince me of the certainty thereof. . . . Nor would I be understood to have said all this against his theory, as it is an hypothesis. For I do most readily agree with them in every part thereof, and esteem it very subtle and ingenious. But I cannot think it to be the only hypothesis, nor so certain as mathematical demonstrations. . . . If Mr. Newton has any argument that he supposes an absolute demonstration of his theory, I should be very glad to be convinced by it.[37]

[36] He made this remark in a covering letter to Oldenburg, secretary of the Royal Society, *Correspondence of Isaac Newton* (London, 1959), I, 83.

[37] *Isaac Newton's Papers and Letters on Natural Philosophy*, ed. I. B. Cohen (Cambridge, Mass., 1958), pp. 110–15. All of the documents relating to this famous controversy are printed in this volume and are prefaced by an essay by Thomas Kuhn on "Newton's Optical Papers," which is very helpful in assessing the reasons underlying Newton's incessant controversies. See also the chapter "Concept and Experience in Newton's Scientific Thought," *Newtonian Studies*, ed. A. Koyré (London, 1965).

At this point, Newton could very easily have granted the older man that his original claim *was* a hypothesis, and then could have gone on to show why he thought it the most plausible one available. In his reply, he does in fact give some strong reasons for supposing that Hooke's wave-theory cannot account for straight-line propagation of light. But the main point of his response is that his own theory is *not* hypothetical but directly derived from the experiments, "without having regard to any hypothesis."[38] He rather disingenuously takes Hooke's argument to have been mainly concerned with a minor point in the first paper, a suggestion that light is "perhaps" corporeal. He is quite content to have *that* characterized as hypothetical, because he knows that the experiments can be "explicated not not only by that, but by many other mechanical hypotheses. And therefore I choose to decline them all, and to speak of light in general terms."[39] The implication throughout his reply to Hooke is that "hypothetical explications" must not be "acquiesced in" and that only theories with conclusive experimental evidence in their support can be allowed. He feels that the *"experimentum crucis"* which he had brought forward in his original paper was enough to establish his case with certainty.

A month after Hooke's report came another critique, this time from Ignatius Pardies, S.J., of Clermont, who spoke of Newton's claim as an "ingenious hypothesis" and raised some objections to it. Newton dealt with the objections easily enough, but ended his reply:

I do not take it amiss that the Rev. Father calls my theory an hypothesis, inasmuch as he was not acquainted with it. But my design was quite different, for it seems to contain only certain properties of light, which, now discovered, I think easy to be proved, and which if I had not considered them as true, I would rather have them rejected as vain and empty speculation than acknowledged even as an hypothesis.[40]

The implication is clear: Newton will not advance a scientific claim unless he knows it to be true; no half-way house of hypothesis will be allowed between truth and "vain speculation." However, in a second letter to Pardies, he suggests a somewhat less extreme methodology:

The doctrine which I explained concerning refraction and colours, consists only in certain properties of light, without regarding any hypotheses by which those properties might be explained. For the best and safest method of philosophizing

---

[38] *Newton's Papers and Letters*, ed. Cohen, p. 123.
[39] *Ibid.*, p. 119.
[40] *Ibid.*, p. 92.

seems to be, first to inquire diligently into the properties of things, and establishing those properties by experiments and then to proceed more slowly to hypotheses for the explanation of them. For hypotheses should be subservient only in explaining the properties of things, but not assumed in determining them, unless so far as they may furnish experiments. For if the possibility of hypotheses is to be the test of truth and reality of things, I see not how certainty can be obtained in any science, since numerous hypotheses may be devised which shall seem to overcome new difficulties. Hence it has been here thought necessary to lay aside all hypotheses as foreign to the purpose.[41]

The young scientist, at the beginning of his brilliant career, shows himself surprisingly Baconian. He seeks to know the properties of things; *explaining* the properties is a separate and apparently secondary affair. Hypotheses are not to be used in establishing properties unless they suggest useful experiments. Hypotheses cannot be admitted as anything other than provisional in science, since science (he says) is a search for certainty. And he betrays his impatience with all critics who cry "hypothesis" just because an alternative is logically open. The abstract possibility of such an alternative ought not be taken seriously, once the original position be adequately established by experiment.

A couple of weeks later, Newton began a letter to the publisher of the *Transactions,* Oldenburg, with these words:

I cannot think it effectual for determining truth to examine the several ways by which phenomena may be explained, unless where there can be a perfect enumeration of all those ways. You know, the proper method for inquiring after the properties of things is to deduce them from experiments. And I told you that the theory which I propounded was evinced to me not by inferring 'tis thus because not otherwise, that is, not by deducing it from a refutation of contrary suppositions, but by deriving it from experiments concluding positively and directly.[42]

The purpose of science, it would seem, is to "determine truth." A hypothetical approach will not suffice because ordinarily it will be impossible to enumerate all the possible alternative hypotheses and so the truth of one or other could not be determined with certainty.

It was not altogether implausible for Newton to represent his theory of dispersion as something "derived from experiments concluding positively," although it *did* have a hypothetical character, as later developments in optics would show. But when a few years later he attempted to handle diffraction phenomena, he could no longer stay at the descriptive level and

41 *Ibid.,* p. 106.
42 *Ibid.,* p. 93.

had to invoke an explicit hypothesis to explain how the colors of thin plates were formed. He postulated an aether in which a vibratory motion was set up by light-corpuscles, bringing about alternate "fits" of easy transmission and easy reflection. The mathematical model derived from this postulate was adequate for prediction; he supposed that he could then leave aside the postulate, retaining only its mathematical skeleton, on the grounds that it (unlike the explanatory hypothesis) had been completely verified. The hypothesis was thus to him no more than a heuristic device, possessing no genuine explanatory or evidential force of its own. He did not notice that the mathematical model was, in fact, still hypothetical; if it *had* been purely descriptive, the use of the wave-hypothesis would not have been necessary. The gap between model and data was small, however, and the tests of the model were ingenious and varied, which misled Newton into supposing that he had "deduced" his theory from the phenomena.

So great a violence does Newton seem to work on his own scientific procedures when he comes to describe the proper methods of science that we cannot help concluding that the explanation is not (as it was with Aristotle) a commitment to a prior theory of science, but rather

a paralyzing fear of exposing his thoughts, his beliefs, his discoveries, in all nakedness to the inspection and criticism of the world. "Of the most fearful, cautious and suspicious temper I ever knew," said Whiston, his successor in the Lucasian chair.[43]

It was probably this, more than anything else, that prompted him to describe science in terms of certainties and to flee from the vulnerability characteristic of hypotheses.

But one other note must be added. Newton was determined to avoid the unempirical jungle of hypotheses of the Cartesians and was encouraged by his early successes in discovering mathematically expressed generalizations of great power and scope. He wanted to insist on the empirical character of what he was doing and had no other way to describe this than the traditional language of deduction and induction. In his famous letter to Cotes, who was preparing the second edition of the *Principia* for publication, Newton (by then in his seventies) instructed him to add to the text:

For whatever is not deduced from the phenomena is to be called an hypothesis, and hypotheses, whether metaphysical or physical, whether of occult qualities or mechanical, have no place in experimental philosophy. In this philosophy, par-

---

[43] J. M. Keynes, "Newton the Man," *Newton Tercentenary Celebrations* (Cambridge, 1947), p. 28, quoted in Kuhn, *Newton's Papers and Letters*, p. 39.

ticular propositions are inferred from the phenomena, and afterwards rendered general by induction. Thus it was that the impenetrability, the mobility and the impulsive forces of bodies, and the laws of motion and gravitation were discovered.[44]

So that his last word on hypothesis was just as uncompromising as his first words had been. And he represents his own method as "deduction" followed by "induction." What other method of validation was open? Newton knew of no other and in particular could not conceive that the use of hypothesis might itself provide a new kind of empirical warrant. The word *hypothesis* had been used in a pejorative sense for too long, and his own sensitivities on the matter were too keen, for him to see that a new method of validation was in the making, in his own work and in that of many of his contemporaries. For what he and they were actually doing was to formulate hypotheses which were neither simple inductive generalizations nor deductions from the data, but explanatory models of a higher order; the next step was to "verify" the model by means of successive experimental tests of predictions made with its aid. This was not the familiar notion of verification, and though it was the basic type employed by Newton in his researches, he could never concede it in his methodological discussions, partly no doubt because he had no way of describing it that did not make it sound like the hated "method of hypothesis" of the Cartesians. And he was intuitively certain that his method was *not* the same as theirs.

Let us now try to gather the threads of this lengthy discussion. To what extent can Newton's approach to science be regarded as empiricist? The answer would seem to be that (1) it was more empiricist than almost anything that had gone before, but that (2) it was not as empiricist as Newton claimed it to be, and (3) its empiricism was of a new kind, importantly different from that of Bacon or even Boyle.

(1) For Newton, the primary warrant of a scientific claim was the set of experimental data that one could bring in its support. It was not its self-evidence; he realized that his theory of colors, for example, had no sort of intrinsic conceptual necessity about it and had to be derived from a finite and specially designed set of experiments. It is true that he frequently described this logical derivation as a "deduction," or an "induction" that concluded with certainty. In this respect, his approach differs from that of both earlier and later empiricism. But it was still much more of an empiricism than a rationalism or conceptualism; the grounds for his theories of gravitation or of dispersion did not lie in an insight into their conceptual necessity nor in their coherence as an explanatory causal structure. Rather it

[44] *Principia*, trans. A. Motte and ed. F. Cajori (Berkeley, 1934), p. 547.

was the way in which they accounted exactly for the mathematically expressed data of observations. That Newton was led to exaggerate the necessary character of the logical relation between data and theory in these instances does not show that he was imputing powers of penetration to the mind that it does not possess (after the fashion of rationalism). The logical relations he mentions are those of deduction and induction; he does not appeal to a direct justificatory insight into first principles. So what we have here is not (despite the appearances) rationalism so much as badly described empiricism with rationalist overtones, to which we shall return in a moment.

It is also true that Newton occasionally *did* admit a tentative character to scientific results, especially in his later writings. For instance, the fourth "rule of reasoning in philosophy" in the *Principia* reads:

In experimental philosophy we are to look upon propositions inferred by general induction from phenomena as accurately or very nearly true, notwithstanding any contrary hypothesis that can be imagined, till such time as other phenomena occur by which they may either be made more accurate or liable to exceptions.[45]

And again in the final *Query* he added to the second edition of the *Opticks* not long before his death:

Although the arguing from experiments and observations by induction be no demonstration of general conclusions, yet it is the best way of arguing which the nature of things admits of, and may be looked upon as so much the stronger by how much the induction is more general. And if no exception occur from phenomena, the conclusion may be pronounced generally. But if at any time afterwards any exception shall occur from experiments, it may then begin to be pronounced with such exceptions as occur.[46]

One could scarcely wish a clearer statement of the tentative and progressive character of science due to the limited character (and thus the inevitable corrigibility) of its empirical warrant. Yet it would, I think, be incorrect to say in consequence that this represents Newton's "real opinion throughout" his career.[47] There are far too many instances (some of them cited earlier in this section) where he claims for his theses a status much stronger than this. Part of the trouble lay in the absence of any distinction between law and theory in the writings of the day, so that there was no way of dis-

---

[45] *Ibid.*, p. 385.
[46] *Opticks* (Dover edition; New York, 1952), p. 404.
[47] Blake, in *Theories of Scientific Method*, p. 123. Blake cites in support of his contention a passage in a letter to Oldenburg of 1672 in which Newton distinguishes the "certainty of mathematical demonstrations" from the "physical certainty" of such sciences as optics and says the former is greater than the latter. But it is not clear that he means by this that physical conclusions are provisional or not quite certain.

tinguishing between the corrigibility appropriate to statements of empirical regularity (which is what Newton refers to in the passage from the *Opticks* above), and the corrigibility of theories (such as the theory embodied in the three so-called laws of motion), a corrigibility which Newton implicitly denied. He wanted to say that experimental laws (such as the law of dispersion) were *not* hypothetical, did *not* involve explanatory suppositions. There is a sense in which he was correct in this. But, because he did not distinguish such laws from the theories or "principles" which were the heart of his enterprise, he was forced into an implicit falsification of the status of these theories by arguing for the non-hypothetical lawlike character of "good" science, thus implying that science, when done properly, would involve *only* mathematical correlations between data, with no supporting skeleton of hypothetical explanation at all. It is fortunate that this thesis, so characteristic of the positivism of our own century, affected Newton's actual practice so little. Had he taken it literally, the *Principia* could not have been written.

(2) Newton supposed that the conceptual structure of his mechanics could be derived directly from the data at his disposal. If someone had asked him whether his three "Laws" were tentative or open to later revision, it seems most unlikely that he would have conceded this as a possibility. More specific laws, yes. But the general conceptual structure underlying the *Principia*, no. His three "Laws" are not statements about regularities observed in nature, though he may have thought they were. The first two are best taken as interlocking definitions of inertial motion, of mass and of force; and even the third is not empirical in the sense that one could specify how it could be separately tested. The second "Law" defines force in terms of acceleration produced. But acceleration has to be measured relatively to some frame of reference; if one chooses a different frame moving non-uniformly relatively to the first, the measurement of acceleration relative to it will give a different answer. Thus, the measure of force is also dependent on the choice of reference-frame. But force was quite obviously for Newton a real influence acting between bodies, an absolute. He was thus forced by his "Laws" to postulate an absolute time flowing in an absolute space, in which absolute motions (i.e. the ones in terms of which real forces could be defined) could in principle be determined.

This highly sophisticated conceptual structure clearly had no simple logical relationship with Newton's observational data. Not only was it not deduced from them; it was in no way a generalization from them either. What Newton had done was to take a set of familiar concepts (*force, matter, space, time*) and stipulate sharp definitions for them in terms of one another in such a way that empirical predictions could be inferred from them when they were applied to concrete situations. But in doing this, he was not

365

merely summarizing a set of data in some convenient mathematical way. He was making use of categories whose resonances were far wider, experientially, than any set of observations. His notion of mass, for instance, depended in a complex way upon a concept of matter which he had inherited from Greek philosophy and whose warrant was of a much broader sort than any he could derive from a set of predictions.[48]

The point here is not just that the conceptual structure underlying the *Principia* derived from very complicated sources in Newton's own thinking, sources much more basic than Kepler's Laws or anything of that sort. Rather it is that the warrant of this structure, i.e. what made Newton himself defend it with such firmness, was not simply that it allowed him to make correct astronomical predictions. There was an evident coherence and a satisfying completeness about the way in which it gave clear definitions to all of those concepts that originate in our primitive perceptions of motion and the causes of motion. Whether or not one wishes to put this as Burtt does by speaking of the "metaphysics" which Newton brought, already partially formed, to his mechanics, it seems clear at least that his assurance of the certainty of his "Law" was not of the sort that empiricists like Bacon or Hume could have accepted. But it *was* of the sort that would afterwards lead Kant to propose the basic structure of Newton's mechanics for the exalted status of a synthetic *a priori*. Though Newton's influence on science was primarily that of an empiricist, in some important respects he not only departed from the narrow correlationism he liked to preach but even exhibited an assurance about the ultimacy of the basic concepts of his mechanics that was more than a little reminiscent of his great rationalist adversary, Descartes.

(3) Newton's empiricism (i.e. his manner of appealing to experience) was of a more fruitful kind than any that had preceded it in natural science. It combined a Galilean stress on exact mathematically expressed data with a Baconian stress on observation and constant progressive testing. Newton emphasized the observational origin of his specific mechanical laws (like those governing gravitational or rotational motions) in a way Galileo had not; he emphasized the importance of a quantitative structuring of experience in a way Bacon had not. This combination provided what amounted to a new method of scientific inquiry: the formulation of hypotheses on the basis of empirical data sufficiently exactly stated to allow the hypotheses to be stringently tested. Thus to the old types of "proof" enumerated by Aristotle, deduction and induction, a third was added, that of hypothetico-deductive validation. It had been foreshadowed by many writers, from

---

[48] See the essays in Part 3 of *The Concept of Matter*, ed. E. McMullin (Notre Dame, 1963). See also Chaps. 5 and 6 of *Concepts of Mass*, by Max Jammer (Cambridge, Mass., 1961).

Plato onward, but not until Newton were the data with which it could operate sufficiently exact to allow it to be used with any great success. It was this "method" that ultimately detached natural science from its parent philosophy, by providing it with a new sort of empirical ground on which to stand.

An extraordinarily clear statement of this method can be found in the preface written by Newton's friendly adversary, Huyghens, for his *Treatise on light* in 1690. In optics, he notes, one finds

a kind of demonstration which does not carry with it so high a degree of certainty as that employed in geometry, and which differs distinctly from the method employed by geometers, who prove their propositions by well-established and incontrovertible principles, while here principles are tested by the inferences which are derivable from them. The nature of the subject permits of no other treatment. It is possible, however, in this way to establish a probability which is little short of certainty. This is the case when the consequences of the assumed principles are in perfect accord with the observed phenomena; more especially, when these verifications are numerous; but above all, when one employs the hypothesis to predict new phenomena and finds his expectations realized.

Not only is Huyghens, in this much-quoted passage, giving the three criteria for "scientific" knowledge that would most often be cited today, but he is also asserting that "science" must be seen to consist of hypotheses, verified to a greater or lesser degree according to these criteria. There is no longer to be any question of a thesis becoming "scientific" only when it goes "over the top" in certitude, so to speak. Knowledge of nature is not to be divided into a "science" which is certain and demonstrative, and probable knowledge which, on the basis of future research, may sometimes become "science." The "science" of nature is a unified entity, supported by a complex web of rational and empirical supports. It is the same web which supports the established physical law and the provisional hypothesis. The only difference is in the degree of support. There are no theses which float free of the web, which somehow carry their own intrinsic and necessary warrant. The only "necessity" that occurs within the structure of physical science is the purely formal one that links the premises and conclusion of any valid argument-form.

IX. CONCLUSION

It might have been expected that after such a successful turn to empiricism on the part of the greatest scientists of the 1650–1700 period, empiricism would have become a basic and permanently accepted theory of

367

natural science, and that all that was left to do was to get rid of the remaining conceptualist idiosyncrasies in Newton's description of what science should be. But in actual fact, the story took a very different turn: a movement back toward conceptualism in physics began, and continued for a century or more. The reasons for this are complex and would take us beyond the scope of the present essay. They lie in the oddly conceptualist character of mechanics itself, in the specifically conceptualist mold in which Newton had cast the *Principia*, in the fact that Newtonian mechanics had soon come to be regarded as the model for scientific knowing generally. It is easy to mistake mechanics for a branch of mathematics, stated in terms whose experiential origin and reference can be disregarded for practical purposes of deduction and mathematical manipulation. To Galileo it had seemed self-evident that the distance through which an accelerating body falls *must* be proportional to the square of the time elapsed; once one knows the mathematical definition of acceleration, the rest follows—or seems to. To the Newtonians it seemed equally self-evident that the inverse-square law could be derived directly from the geometry of a Euclidean space and was thus in *no* sense a provisional empirical law. Huyghens had drawn a perceptive distinction, on the basis of the kind of evidence one would call upon, between geometrical axioms and theorems on the one hand and empirical "laws" stated in geometrical terms on the other. But his distinction was forgotten in the face of the simple certainties of eighteenth-century Newtonian physics, certainties that even Newton himself, wary though he was of the "hypothetical," would have had some hesitation in accepting.

We can now conclude, by returning to the question posed at the beginning of this essay. To what extent is the "Scientific Revolution" of the seventeenth century definable in terms of a shift from medieval conceptualism to a new empiricism? If our discussion has shown anything, it has surely shown that the term "Scientific Revolution," as applied to the Galileo-Newton period in the development of science, is too vague to allow this question to be answered in any unequivocal way. If, for example, one supposes Galileo to typify the "new scientist," then empiricism must not be emphasized as a component of the "new science." If the "Revolution" is supposed to have set a definitive new style for science and not just to have brought about some basic new theories in mechanics and optics, then once again this new style cannot be defined in terms of empiricism, since eighteenth- and early nineteenth-century science is very restrained in its enthusiasm for empiricism. Going back to the two criteria in terms of which we defined the conceptualism-empiricism dichotomy at the beginning of this essay, we can say that (1) the empiricist view of science as provisional and progressive made no real impact until the nineteenth century. Newton supported it only with regard to low-level empirical laws, and the weight of his influence could be counted here much more strongly on the conceptualist side; and (2) the empiricist view of evidence as a sequence of controlled observations gradually made headway between Galileo's time and Newton's; in Newton's *practice* (though not in his own understanding of his practice) it crystallized as a specific method of validation for the hypotheses which are the growing points of science. Even though Newton's successors went back to an older conceptualist ideal of science when speaking of mechanics as though it were a completed enterprise, in their own scientific inquiries, in chemistry, optics, biology, and the rest, they gave testimony to a new and immensely powerful way of harnessing the soaring powers of insight and hypothesis through a planned use of exact observation.

PART III: HISTORY

THE RENAISSANCE INTEREST IN HISTORY[1]

ᘒ⳺ FELIX GILBERT

I

n the Renaissance, knowledge of the past was not believed to be of primary significance—neither for explaining the structure of the social world nor for understanding or judging human action. History was not highly esteemed.

This is well illustrated in Erasmus' beautiful adage entitled "Spartam nactus es, hanc orna."[2] In this piece Erasmus tried to demonstrate the senselessness of wars and conquest, and as an example of the wastefulness of war he pointed to the death on the battlefield of the young Prince Alexander of Scotland, Archbishop of St. Andrews, who had been one of Erasmus' most promising pupils. Using this adage to write an obituary for the young prince, Erasmus described the studious way in which Prince Alexander lived while Erasmus was his tutor. From Erasmus Alexander learned Rhetoric and Greek; he was also studying law, and in the afternoons music and singing. Even at mealtime, he would listen to the reading of "some serious book, for instance, the Decretals, or St. Jerome or Ambrose. . . . After dinner there might be stories told, but short ones, and with a literary flavour. In this way there was no part of life without its contribution to study, except what was given to divine service, and to sleep. For if there could have been any time left over—and really there was hardly enough time for all this varied programme of study—if by chance there were any extra time he spent it on reading history." The time allotted to history in the study program of the young prince suggests that Erasmus hardly considered history to be an essential part of man's education.

Erasmus expressed a view shared by many of his contemporaries. The historical writings characteristic of the Renaissance are the work of the

---

[1] For the problems discussed in this article, see, in general, Myron P. Gilmore, "The Renaissance Conception of the Lessons of History," *Humanists and Jurists* (Cambridge, Mass., 1963), pp. 1–37; Eugenio Garin, *L'Umanesimo Italiano* (Bari, 1952), Chap. 7; Herbert Weisinger, "Ideas of History during the Renaissance," *Journal of the History of Ideas,* VI (1945), 415–35; August Buck, *Das Geschichtsdenken der Renaissance* ("Vorträge des Petrarca Instituts"; Köln, 1957).

[2] The following translation is taken from Margaret M. Phillips, *The 'Adages' of Erasmus* (Cambridge, 1964), pp. 300–8.

humanists. Yet, in their *oeuvre* historical writings were neither very numerous nor did they occupy a prominent place.

There are several reasons why the humanists assigned to history a limited value among the fields of human knowledge. One of the most proudly held assumptions of the men of the Renaissance was that they lived in a time different from the past, in a new and better world; and the humanists regarded themselves as the protagonists of a new intellectual order. Emphasis on the value of novelty, however, casts a shadow over what is old. A modern psychologist has said, "New ideas, new movements, new countries, all act first as if they were superseding history."[3]

Moreover, an interest in history was not considered a necessity. Such a lack of commitment to a concern with the past is alien to our point of view, for the historical dimension forms an integral part of our thinking. Our actions are circumscribed by surroundings which were formed in the course of history, and our mental attitude is a product of the cultural experiences of the past. We are always conscious that our freedom of action, our possibility of patterning the future, is limited by the historically given situation in which we are placed. It is characteristic of our attitude toward history that, if we say of someone that he acted in an anachronistic way, we imply that he acted ineffectively; anachronism is a pejorative term.

In the Renaissance the past was not conceived of as a unified process determining the possibilities of the present. It is true indeed that we owe to the Renaissance the chronological division of world history into three successive periods: ancient, medieval, modern. But despite this periodization of history the men of the Renaissance were not interested in establishing a historical scheme which organized the past from the beginning to the present. To the men of the Renaissance, the Middle Ages were the Dark Ages; the period of light and splendor which had existed in the ancient world and the period of a rebirth of this era of light which the men of the Renaissance believed their own time to be were separated by an era without interest or value. Petrarch declared that he was not willing to give guidance by his writings through such darkness ("tantas per tenebras").[4] History did not appear as a unified continuing process to the humanists; for them history offered an un-co-ordinated mass of material from which they could select relatively isolated actions and periods for study, dismissing those which seemed without relevance to their own interests or values.

Thus, the historical process as a whole had no more immanent logic to the men of the Renaissance than to the men of the Middle Ages. And fre-

[3] Erik H. Erikson, *Insight and Responsibility* (New York, 1964), p. 211.
[4] See Theodor E. Mommsen, "Petrarch's Conception of the Dark Ages," *Speculum,* XVII (1942), 234.

quently the emotional mood with which the men of the Renaissance looked upon the theater of history was not very different from that of the Christian writers of the Middle Ages who saw history as a demonstration of man's impotence and of the misery of the human condition. Although Poggio Bracciolini lived when the humanists were making exciting discoveries of classical texts, and although he himself took great pride in his part in these achievements, he was inclined to emphasize that history pointed out the insecurity of human existence, that the results of historical actions were ephemeral and that those who acted in history became involved in crimes and sins. "Omnia enim praeclara et memorata digna ab injuria atque injustitia contemptis legibus profecta."[5] Even Rome was not excluded from this condemnation: it "flagellum non solum Italiae sed totius orbis verissime dici potest."[6] In his treatise on *De varietate fortunae*, Poggio used the events of his own time to illustrate how princes are tied to the wheel of fortune, which raises them up and brings them down; and one of his prime examples was Richard II: "Huius regis fortunam nulla certa antiquorum tragedia superat vel varietate rerum vel regni magnitudine." Poggio was a man of moods, and everything he saw was colored by the mood that possessed him. Even a century after Poggio, however, the same pessimistic attitude permeates Guicciardini's *History of Italy*, full of the theme of the fall of princes and of Time's revenges. When Shakespeare composed the tragedy which Poggio had found in the life of Richard II, it was in harmony with this pessimistic mood: Shakespeare's Richard II saw in history chiefly

. . . sad stories of the death of kings:
How some have been depos'd; some slain in war;
Some haunted by the ghosts they have depos'd;
Some poison'd by their wives; some sleeping killed;
All murder'd. . . .

Nevertheless, in spite of a failure to discover evolution or order in the historical process, a concern with the events of the past was not considered a useless activity. The ancients had said that "history teaches by example," and this aspect of the usefulness of history was recognized in the Renaissance as it had been in the Middle Ages. But in the Renaissance the nature of the lessons which man was supposed to learn from history began to change: the lessons were no longer, as they had been in the Middle Ages, exemplifications of religious truth; they became instructions in moral philosophy. The humanists, rhetoricians rather than original thinkers, were more con-

[5] This quotation comes from Poggio's *Historia Tripartita*.
[6] From Poggio's *De Miseria humanae conditionis*.

cerned with presenting ancient moral philosophy in a comprehensible, relevant form than in advancing a new and original system of thought. Thus, the categories and concepts of classical ethics penetrated their discussions of the lessons which could be drawn from past events. Emphasis was placed on the role which man could play in the world: on his potentiality for action. The world became a struggle between man and fortune. Views might differ about the chances of man in this struggle. But even a person as skeptical as Poggio admitted that virtue must test its strength against fortune and that man had a duty to try to realize his goals: "Virtus in actione consistit."[7] History can encourage man by calling into his mind the achievement of the *viri illustres*. Nevertheless, history in the Renaissance was not important *per se;* it provided illustrative material for the teaching of moral philosophy.

Since history in the Renaissance was not an autonomous field of knowledge, there were no historians in the sense of scholars who devoted their lives exclusively to the writing of history.[8] The humanists who did write histories regarded such tasks as a secondary activity; it is scarcely surprising that few humanist histories evince great scholarly or literary merit. Leonardo Bruni Aretino, whose *History of Florence* is the outstanding achievement of humanist historiography, is famous as a translator and as the popularizer of moral philosophy. Valla's *History of Ferdinand of Aragon* is lively and amusing but hardly distinguished by those features which established his fame: philological method and audacious philosophizing. Pontano was a great literary figure, a poet and philosopher of elegance and originality, but his contribution to history, his *Bellum Neapolitanum*, is rarely mentioned. It is true that some governments or rulers—the Sforzas in Milan, the Florentines, the Venetians, and the Genoese—appointed public historiographers, but these public historiographers—Navagero, Bembo, Machiavelli, to name a few—were not chosen because they had previously distinguished themselves by writing histories; nor after they had received their commissions did they feel obligated to devote themselves exclusively to their assignments.

Because the humanists felt no strong commitment to historical work, they showed little concern about the theory of history. Works on the theory of history were few and, with very rare exceptions, brief and not very significant. They were usually limited to a rehashing of traditional issues: to a re-examination of Aristotle's views about the relation of poetry and history, and to a discussion of the merits of ancient historians in order to establish a pattern for imitation. The latter issue was usually resolved by the choice of

[7] From Poggio's *De nobilitate*.

[8] For a detailed discussion of the character of humanist historiography, see Felix Gilbert, *Machiavelli and Guicciardini* (Princeton, 1965), Chap. 5; this book contains also a bibliography.

Livy. La Popelinière, succinctly summarizing the results of many discussions, called Livy "captaine en chef et général des historiens."[9] More substantial work on the theory of history was done only in the second part of the sixteenth century, and culminated in Bodin's *Methodus ad facilem historiarum cognitionem*. But by then the situation for historical work had changed.

Imitation of the ancients, the basic principle of humanist scholarship, came to be applied to historical work with positive and negative effects on the development of historiography. Following the classical usage, the humanists distinguished between full histories written from a distance and commentaries of participants in which they described the actions they had directed or played a part in. Historical writings were to deal with a well-defined and circumscribed subject: with a particular war or with the history of a particular city-state. They were to narrate events in strictly chronological order and present a connected story. But if these were advantages, the imitation of classical patterns had also the inhibiting consequence that political events and military actions alone were considered worth being preserved for posterity.

Imitation was not only a question of literary form, of organization, and choice of subject matter. It was predicated on the assumption that the events of the past can be treated in the same way as those of the present because the factors which determined the actions and thoughts of men in the past are identical with those which determine them in the present. If the humanists were receptive to any general theory of history, it was the cyclical theory of the ancients, which postulated that history was repetitive and that everything that had happened before would happen again.[10]

This assumption of the fundamental identity between the present and the past was a further reason for the humanists' lack of interest in questions of historical method: a special historical method for the establishment of historical facts was not needed. Their procedure in writing history was the following: they would pick out an event, an action, or a historical figure, and they would present their subject so that it became immediately relevant to the present. They were not averse to stylizing and embellishing the event or the person about whom they were writing, for their intention was to make the lessons which history taught as clear as possible. Their historical accounts were kept on a consciously high, generalizing, and idealizing level. The humanists were not concerned with the individually unique features of an event or a personality. A man was described and judged according to

[9] La Popelinière, *L'Histoire des Histoires* (Paris, 1599), I, 324; see also Peter Burke, "A Survey of the Popularity of Ancient Historians, 1450–1700," *History and Theory*, V (1966), 134–52.
[10] See Frank Manuel, *Shapes of Philosophical History* (Stanford, 1965), particularly p. 51.

the traditional scheme of cardinal virtues and vices. When Lorenzo Valla inserted realistic details into his history, he was vehemently criticized for spurning convention.

A striking illustration of humanist methods and procedure in the writing of history is provided by the manner in which the public historiographers proceeded in composing their official histories. They were not expected to undertake research of their own. The collection of material, the setting down of a factual account, was done by others, frequently by men employed in the chancelleries; the assembling of these annals was regarded as a necessary but secondary activity. The man of literary distinction who was appointed public historiographer was expected to transform these materials in a literary work that would be a monument to the spirit which had guided the citizens in the past and which ought to guide them in the future.

The almost unavoidable consequence of this procedure was that emphasis shifted from content to form and that the humanist histories became vague and empty. This was recognized in the sixteenth century by Montaigne, who in his famous essay "On Books" subjected historical writings of the humanist genre to a biting criticism:

> They undertake to choose things worth being known and often conceal from us such a word, such a private action as would better instruct us; they omit as incredible the things that they do not understand, and also some things because they cannot express them in good Latin or French. Let them boldly display their eloquence and their reasonings, let them judge according to their fancy; but let them, however, leave us something to judge after them, let them not alter nor arrange by their abridgements and selections anything of the substance of the matter, but rather deliver it to us pure and entire in all its dimensions.

> For the most part, and especially in these latter ages, persons are culled out for this work from among the common people upon the sole consideration of skill in the use of words, as if we were seeking to learn grammar there! And having been hired only for that, and having put on sale only their babble, they are right to be chiefly solicitous only of that element.[11]

II

Montaigne was by no means entirely negative. Historical writings were his "favorite dish," and some histories he valued very highly. These were works "that were written by the persons themselves who commanded in the

---

[11] The translation is taken from Montaigne, *Selected Essays* (Modern Library Edition; New York, 1949).

affairs or who participated in the conduct of them, or at least who had the fortune to conduct others of the same nature." According to Montaigne, experience in practical politics gave two great advantages: histories written by statesmen would be more correct factually because their authors knew intuitively which one of several accounts "is the more likely to be true"; and historical writers with political experience were able to explain the motives which caused princes to act as they did. But for Montaigne, there were few "truly outstanding historians." He thought that "almost all the Greek and Roman histories" showed the required knowledge of practical politics. And he believed also that some of the historians of his own time deserved praise. First among them was Francesco Guicciardini.

In pointing out the affinity between the historians of the ancient world and Francesco Guicciardini, Montaigne indicated that he was aware of a significant Renaissance development in the field of history: politics and history moved closer together. It is worthwhile to consider in detail the factors which brought about this event.

There are many forms in which concern with the past can be expressed, and written narrative accounts are only one of them. It is natural—almost unavoidable—that man places his own life into the framework of the life of his family and sees himself as a link between his ancestors and his descendents. Most men will keep lists of their ancestors; a proud and venturesome prince such as the Emperor Maximilian might even assemble around his tomb statues of the great rulers and the great saints of all ages and claim them to be members of his family. For his heirs a man will record his own achievements or the strange things he saw or heard in the course of his life; he will explain how his property was acquired and he will collect the documents that prove his proprietary rights. But when, in such ways, man places himself in relation to the past, he places himself inevitably into a social world; "awareness of the past is a social awareness."[12] Man acts within a social organization or institution and his life merges with that of this larger social body. A monk's or priest's interest in the past will frequently be expressed in a history of his monastery or of some ecclesiastical institution. When humanists began their lectures interpreting the text of a classical poet or philosopher or historian, they often started with a brief survey on the development of poetry, philosophy, or historical writing; they provided themselves with a professional family and ancestry. Concern with the history of one's family became concern with the history of the social organization in which man lived; and this led into politics. Particularly in the Italian city-

---

[12] See J. G. A. Pocock, "The Origins of the Study of the Past: A Comparative Approach," *Comparative Studies in Society and History*, IV (1961), 211.

state the fate of an important family was intertwined with the political developments in the city. Guicciardini's *Florentine History* grew out of a book of family memoirs; the role of the members of the Guicciardini family remained a focal point of attention throughout the entire narrative. In Gasparo Contarini's book on Venice, references to actions of earlier Contarinis served to illustrate Venetian attitudes.

But although the interest in the development of one's own society necessarily formed an integral part of man's concern with the past, humanism tied history and politics still more closely together.

By using Livy as the prime pattern for historical writing, the humanists reinforced the notion that the history of one's own city was a topic of supreme importance. Moreover, the example of the ancients prescribed that such histories ought to be written from a purely political point of view. In order to become histories, chronicles and memoirs had to be purged of all extraneous non-political material. The exclusive emphasis on politics had the effect of narrowing the lessons of history from broad prescriptions of moral philosophy to rules of political conduct. History became the handmaiden of politics. The very popular writings about "the perfect ruler," "the perfect citizen," or "the perfect republic" contained an immense amount of illustrative historical material in order to establish rules for the art of government and political action. This can be most clearly seen in the writings of Machiavelli. He drew "from the experiences of contemporary history and from the writings of ancient history about the actions of great men" the rules which a prince ought to follow. While Machiavelli's imperatives were new and original, his method of using history for establishing rules for political practice was not.

Historical argumentation remained an accepted procedure in the writings on politics. The extent to which concern with history became subordinated to the purpose of teaching lessons of politics can be seen from Bodin's *Methodus ad facilem historiarum cognitionem*. Only a few years after the appearance of Bodin's book a contemporary remarked that Bodin's work "should be more appropriately regarded as a treatise on republics, or on affairs of state than as a work on historical method."[13] Not without justification, this criticism implied that Bodin was more interested in the political lessons which can be distilled from history than in the methodical problems which confront a writer of history.

Nevertheless, the marriage of history and politics, the integration of a

[13] La Popelinière, *L'Histoire des Histoires*, II, 28. Nevertheless, *methodus* in the sixteenth century is not identical with our term *method;* see Neal W. Gilbert, *Renaissance Concepts of Method* (New York, 1960).

pragmatic approach to the past with humanist notions of history, served not only the interest of politics: it also gave form and purpose to writings on contemporary events, thereby making possible the one outstanding achievement of Renaissance historical literature. In the Renaissance accounts of contemporary events became chronological, ordered around political themes and described as causally connected processes. If Giovio was not a great historian, he was at least an interesting political reporter, and as full of details as Comines' memoirs are, they tell a connected story with a political meaning. But, as Montaigne indicated, the one great historian of the Renaissance was Francesco Guicciardini. He is the prototype of the statesman-historian in whom knowledge of the intellectual attainments of his time was joined with pragmatic political interest.[14] For centuries his family had been connected with Florentine politics, and his own career had led him to extend his view over the entire Italian and European scene. When Guicciardini was writing his *History of Italy*, the foreign invasions which had begun with the campaign of Charles VIII in 1494 had ended with the establishment of the Spanish hegemony over Italy. In Guicciardini's eyes these Italian wars were a finite topic and could be treated, therefore, in accordance with the prescriptions which the humanists had distilled from ancient histories—as the political and military history of one isolated war. But with the fall of the Florentine republic and the disappearance of Italian independence, Guicciardini's political career came to an end. Thus he looked upon the political events of his life with a certain distance and objectivity. Moreover, because he believed his own final lack of success not due to personal failures or miscalculations, he saw fortune ruling the affairs of men and he regarded with skepticism the possibility of controlling political events. He had little confidence in the effectiveness of concrete political rules, and therefore his history was more than a handbook of politics. History did not teach special political techniques but taught man about himself, about his strength and his weakness.

Guicciardini's *History of Italy* was written in a unique political and a unique personal situation. As a historian Guicciardini has few equals and these only in earlier or later centuries. None of the Italian writers who followed him ever reached his level; their social world and therefore also their awareness of the past narrowed to smaller political circles. Moreover, the emergence of princely rulers all over Europe made politics an arcanum which those who directed affairs were reluctant to reveal. History came to be written by court historians who considered that their task was to docu-

[14] See Felix Gilbert, *Machiavelli and Guicciardini*, Chap. 7.

ment how things had come to their present settled state. Their connection with rulers raised doubts about their truthfulness, and scholars questioned whether it was possible for historians to be objective.[15]

The intellectual developments which influenced and transformed the writing of history in the Renaissance did not create a new or secure place to the historian. Within the system of university scholarship, historians had gained no space of their own; if history was taught, it was as an auxiliary science. The appearance of a statesman-historian was accidental and ephemeral; social conditions became increasingly unfavorable for the continuance of such an approach.

III

Despite the superficiality of much of its historical writings and the paucity of great historical works, the Renaissance has exerted a decisive influence on the change in man's historical consciousness and in stimulating a new attitude toward history. We have said that the men of the Renaissance thought of themselves as creating a new world; but this new world was a world of the past. In itself the fact that they looked toward the ancient world for guidance did not increase the importance of history. For the ancient world was not viewed as the first link in a historical process of which the present was the ultimate link; classical times were regarded as floating above and outside history, the one great realization of a perfect life. "What else then is all history if not praise of Rome?" Petrarch exclaimed,[16] and with the statement that everything that could happen in history had already happened, he denied the modern view that history was a steadily accumulating process.

Nevertheless, the admiration of the Renaissance for the classical world had important consequences for man's historical outlook. Because the ancient world was raised as a norm above all later human affairs and was separated from the contemporary world by the dark Middle Ages, "the

---

[15] For these doubts, see Julian H. Franklin, *Jean Bodin and the Sixteenth Century Revolution in the Methodology of Law and History* (New York, 1963), pp. 89–103. On the question of the decline of historiography after Guicciardini, see William J. Bouwsma, "Three Types of Historiography in Post-Renaissance Italy," *History and Theory*, IV (1965), 303–14. A characteristic document of historical thought in the late sixteenth century is the "Dialogo della Istoria" by Sperone Speroni, printed in his *Opere*, II (Venice, 1704), 210ff.

[16] See Mommsen, *Speculum*, XVII, 237.

classical past began to be looked upon from a fixed distance."[17] At the same time the demands of imitation necessitated a belief in the reality of this distant world. When, in 1485, a marble sarcophagus with the well-preserved body of a girl was discovered, it was immediately believed that this was Julia, the daughter of Claudius, and that she was more beautiful than any living human being. Awareness of both the remoteness and the reality of the past are basic features in the modern interest in history.

We might briefly point out some of the new departures which followed from this powerful concern with the ancient world. First, there is the beginning of a division of history into different cultural epochs. Because the men and events of the ancient world were seen in "perspective distance," they were firmly placed into their own world, separate from that of modern times. The terms *moderni* and *antichi* were regularly used to designate the difference between the classical world and the contemporary scene.[18] The ancient world appeared as a whole to which not only statesmen and generals but also poets and philosophers belonged. But with increasing knowledge, particularly on the basis of their careful analyses of language, scholars began to distinguish various epochs within the ancient world. Accordingly, the modern world or single periods of the modern world were viewed as entities; in the praise of the *vir illustris* a writer frequently included the intellectual and artistic luminaries of his subject's time. In lauding the greatness of the Laurentian age, Ficino, Politian, and Michelozzi had their place next to the Magnifico himself.

Then too there is the beginning of insight into the importance and complexity of methodological issues. It is certainly true that a historian like Guicciardini was remarkably careful in the assembly and the use of sources. But Montaigne is probably right in suggesting that Guicciardini's awareness of the biases and partiality of writers and observers was the result of his political experience. Conscious adoption of a critical method arose from the desire to bring the classical world to life again. The precondition for imitation was knowledge of what the classical world had been—not only a knowledge of its literature and wisdom but also of the character of the human beings who produced the literary works. Petrarch with his letters to the great personalities of the ancient world expressed well this need for communicating with the men of the ancient world as individual human beings. Poggio's life provides an example of the zeal with which the search for the manuscripts of classical authors was pursued and of the delight with which

[17] Erwin Panofsky, *Renaissance and Renascences in Western Art* (Copenhagen, 1960), p. 108.
[18] For *moderni* and *antichi* see Mommsen, *Speculum*, XVII, or, as an example, Benedetto Accolti, *De praestantia virorum sui aevi*.

every new find was greeted—not only because the codex of classical wisdom was enlarged but also because the personality of each classical author became fuller and more real. For a time classical perfection was accepted without question. Every statement of a classical author was authoritative and accurate. The intrepidity of a Valla was needed to raise questions about the veracity of passages written by Livy. Such questions were raised with increasing frequency, and there were issues which forced upon the humanists a critical approach. For instance, it became fashionable for every Italian city or city-state to trace its beginnings back to the times of the Romans or the Greeks; the humanists were anxious to please the rulers by confirming these myths. And so they assembled from the different classical sources all the relevant material; naturally, they discovered contradictions which forced them to compare and to evaluate the various reports. It is characteristic that in the humanist histories of a city-state the account of its foundation frequently demonstrated a critical acuteness that is lacking in the later parts of the works. Likewise, a principal activity of the humanists—the establishment of a classical text in its original purity—involved comparisons and confrontations which revealed contradictions. The fervently sustained admiration for the classical world soon extended beyond manuscripts to other material—to coins, inscriptions, monuments—and these materials had to be related to literary sources. When Sigonius published his work on the *Fasti consulares*, he became aware that these bronze tablets represented a source superior to literary sources, one that provided a means for judging the reliability of classical writers. Thus Sigonius could show that Livy had erred, and he could set his own opinion against those of classical authors. The classical world was no longer what classical writers reported, but what a scholar reconstructed after having evaluated all relevant material on the basis of exact methodical principles. Philologists and antiquarians developed a critical method for the evaluation of historical material long before the adoption of such a method became a basis for the study of history in the nineteenth century.[19]

The more the ancient world became known, the more it became evident how much more had to be learned: the distance seemed never to shorten. And imitation remained a possibility to be accomplished in the future rather than a task of the present. In two areas, however, it was believed a definite answer could be given to the question of whether imitation of the ancients could be achieved. In art imitation was thought to have been successful and the greatness of ancient times to have been resurrected. In the

[19] See Arnaldo Momigliano, "Ancient History and the Antiquarian," *Contributo alla Storia degli Studi Classici* (Roma, 1955), pp. 66ff; see also H. J. Erasmus, *The Origins of Rome in Historiography from Petrarch to Perizonius* (Assen, 1962).

field of law the opposite conclusion was reached: that the modern world was so different from that of the ancients that the applicability of ancient law was not feasible.

In Vasari's *Lives* we find a striking and extraordinary use of the term "modern."[20] In general, *moderni* and *antichi* were placed in contrast to each other. But when Vasari spoke of the *"maniera moderna,"* he indicated that it was identical with the *"maniera anticha."* In Vasari's view the artists of the Renaissance had actually resuscitated classical art and had become the equals of the ancients. There are indications of a cyclical theory behind Vasari's outline of art history. Art had slowly grown in the ancient world, and when it had reached maturity, it was destroyed. Through a development for which Vasari has no rational explanation, art began to revive in the fourteenth century until, in Vasari's own sixteenth century, the "progresso della Rinascità" brought about an art which was not only equal but even superior to that of the ancients. Vasari regarded the artistic greatness of his time with some concern because he feared this might indicate the imminence of decline; but he was not without hope. The art of the ancients had vanished because of destruction by barbarians and not because of any natural decline. Thus, there was no proof that art was subjected to the natural cycle of birth, maturity, and death. And Vasari regarded as one of the purposes of his book to provide later generations with information which would help them to avoid decline. It would be wrong to ascribe to Vasari a systematic theory of development or progress, or to suggest that he regarded history as a causally linked process. But he evidently believed that, in the course of history, an accumulation of knowledge took place which gave man power to fend off the corroding impact of time. Through work in the realm of intellect—through the accumulation of knowledge and "advancement of learning"— the past could become for man more than an agglomeration of isolated events; he could begin to conceive of larger causally connected developments extending over long periods of time.

Whereas art was believed to have died at the end of antiquity and to have been resurrected in the Renaissance, law was believed to have remained in force since Roman times.[21] But to remain applicable, it had been glossed

[20] For Vasari, see Erwin Panofsky, "The First Page of Giorgio Vasari's 'Libro,'" *Meaning in the Visual Arts* (New York, 1955), pp. 169–225; the most characteristic passages can be found in Vasari's "proemio delle Vite." See now also John Shearman, *Mannerism* (London, 1967), pp. 172–73. Buck, *Das Geschichtsdenken der Renaissance*, p. 26, emphasizes too exclusively the cyclical element in Vasari's concept of history.

[21] For the following, see Franklin, *Jean Bodin and the Sixteenth-Century Revolution in the Methodology of Law and History*, J. G. A. Pocock, *The Ancient Constitution and the Feudal Law* (Cambridge, 1957), particularly the first chapter, and Beatrice Reynolds, "Shifting Currents in Historical Criticism," *Journal of the History of Ideas*, XIV (1953).

and commented upon. The refinements of philological method in the Renaissance made apparent the mistakes which the glossators and commentators had committed. Budé and his followers began to show that the Roman law itself was a composite. Its various parts had been written at different times; it was neither entirely consistent nor could its terms and meaning be understood without using other contemporary sources for interpretation. Thus the advocates of the "Mos Gallicus" regarded the books of Roman law as historical documents with limited applicability. Cujas, when asked about the bearing of Roman law on a contemporary issue, was supposed to have replied, "Quid hoc ad edictum praetoris?"

The idea of one archetypal law from which every particular legislation should be derived was not abandoned, but it was recognized that this natural law was not identical with the Roman law. Thus, in deducing basic legal principles, the laws of other nations also deserved to be taken into consideration. Such a concept widened man's historical outlook. Man needed to know more than the history of Rome or of the ancient world.

This does not mean that the histories of various countries were seen as being linked together and dependent on each other. They were still treated as separate topics. Nevertheless a knowledge of all known peoples was thought essential because the possibility for comparisons from which general principles might be derived was predicated upon it. If in the earlier years of the Renaissance the Christian division of world history into eras of successive empires had been disregarded, at the end of the Renaissance historical interest turned again toward a universal knowledge of history.

*   *   *   *

This widening of the historical outlook was influenced by the discoveries which had revealed the existence of new worlds. The decline of the Renaissance was a complex process, and there are many reasons for it. But clearly an important factor was the realization that the geographical area in which the classical civilization and the Roman Empire had ruled was only a small part of the world and that outside this center in which the ancient world had flourished other worlds had existed.

The developments in the field of historical thought reflect something

---

471–92; also Neal W. Gilbert, *Renaissance Concepts*, pp. 79ff. See now also the two articles by Donald R. Kelley, "Legal Humanism and the Sense of History," *Studies in the Renaissance*, XII (1966), 184–99, and "Guillaume Budé and the First Historical School of Law," *American Historical Review*, LXII (1967), 807–34.

of the importance of these experiences. They reflect still more the emergence of a society split up into different social and professional groups and varied in its intellectual concerns. For the Renaissance view of history can be presented neither as emanating from a unified system of thought nor as exhibiting a strictly linear development. We must be content with realizing that in the Renaissance history meant much less than it means to us—but perhaps also more. People looked to history not for one thing but for many things: for the moral adhortations of a Petrarch, for the dispassionate political intelligence of a Guicciardini, and for the scholarly earnestness of a Budé.

CLAUDE DE SEYSSEL AND NORMAL POLITICS IN THE AGE OF MACHIAVELLI ❧ J. H. HEXTER

nly two writers of the Age of the Renaissance produced books directly concerned with politics that are nowadays widely read. Indeed they produced the only two books on the subject written in that era that have been persistently and widely read ever since. The books were written within three years of each other, and both were written, so to speak, by accident. That is to say, they came into being in the midst of and in relation to odd conjunctures of circumstances in the lives of their authors, and there was little in the previous careers of those authors to suggest that, had it not been for those conjunctures, they would have produced such works. They had both been too busy, and neither had foreseen the interruption to his busyness that turned him to a sustained literary effort. Both had had time for occasional bits of literary scribbling; both had literary and even poetic pretensions. They had been too busy, however, for that steady concentrated contemplation of their fragmentary insights that enables a man to sense the general pattern into which his insights are falling, to bring out sharply the details of that pattern, and to press to their conclusion the implications the pattern suggests.

Suddenly and accidentally, through no intentions of their own, both of these busy men found themselves with time, and nothing but time, on their hands. The circumstances of their involuntary idleness led one of them certainly, the other possibly, to turn their very sharp vision most intently on to the problem of the general structure of politics in their own day.

The certain case is that of Niccolò Machiavelli, age forty-six, for fifteen years an official of the Republic of Florence, unemployed as of November 7, 1512. Others than he lost positions of authority or teetered precariously in them when, with a Spanish army running interference, the Medici returned to Florence after a long exile. Many, however, were able quickly to arrange for themselves a peace, or at least a truce, with the new order. After all, with the great Florentine houses which had served the Republic during their exile it was well for the Medici to avoid trouble. But what trouble could the Machiavelli make, that shabby-genteel family whose scion Niccolò, by serving as general factotum to Piero Soderini, Gonfaloniere of Justice, leader of the late anti-Medici regime, had achieved a political position rather above what his condition in life warranted.[1] He never would be missed.

More overwhelming to Machiavelli perhaps than the troubled state of his outward fortune—arrest on suspicion of complicity in a plot against the Medici, imprisonment and torture, forced retirement to his modest house in the country—was the utter disarray in which the torrent of events had left his notions about politics.[2] For Machiavelli was not a political hack who cared for nothing but a safe city job. Throughout his career he had been an ideologist at heart. Although he had never put them together in an orderly way, he had a number of rooted notions about politics. His recent experiences had been such as to shake his faith in all his deepest convictions, and as with good ideologists his need to restore order among his notions was as urgent as, perhaps more urgent than, his need to get his political career back on its tracks. Since his political career never did get back on the tracks, he ended up with plenty of time to think things over. Machiavelli was not to know that, however, when in 1513 he made his first essay at restoring some order to his badly shaken political cosmology. That essay—*Il Principe* —was written at considerable speed wth the stench of failure, a falsified faith, and hopes burnt-out still reeking in his nostrils and a hunger for the political life still gnawing at his vitals. The auspices for a detached impersonal effort were poor; and indeed *Il Principe*, whatever the appearances, is not such an effort. It is a book both extraordinary and extravagant.[3]

So is the other book, finished three years later by another man, that was to take its place alongside Machiavelli's exercise in catharsis among the classics of political literature. The circumstances of its composition, however, were not so devastating. Thomas More had larger literary pretensions than Machiavelli and, at the age of thirty-eight, not much to show for them. He too, had lacked time. Some of it, as with most Londoners of his eminence, had gone to that unpaid or ill-paid public service that the Crown and the City extorted from men who were capable, conscientious, and well-to-do, service not in the great world of diplomacy where Machiavelli had lived, but in the intimate one of local government. Most of More's time, however, seems to have gone to making enough money to support a considerable

[1] For details on Machiavelli's activities and circumstances during the crisis that preceded, accompanied, and followed the return of the Medici to Florence, see Roberto Ridolfi, *Vita di Niccolò Machiavelli* (Rome, 1954), pp. 183–232.

[2] Letter to Soderini, in Niccolò Machiavelli, *Lettere,* ed. Franco Gaeta (Milan, 1961), pp. 228–31.

[3] For a somewhat fuller attempt to come to grips with Machiavelli's state of mind when he composed *Il Principe*, see J. H. Hexter, "The Loom of Language and the Fabric of Imperatives: The Case of *Il Principe* and *Utopia*," *American Historical Review*, LXIX (1964), 950–52.

family plus a certain number of peripheral and occasional spongers in a style which without being lavish or luxurious managed to be very expensive indeed.[4]

From the rush of legal business More was suddenly becalmed in the most undemanding of human pursuits, that of the emissary abroad without a mission. Negotiations in Bruges with the agents of Prince Charles, ruler of the Netherlands, having been suspended for talks at a higher level, More had abundant time for contemplation.

What he seems to have contemplated was the contrast of the biblical Christian way of life that the humanists in his circle of friends held forth as the standard for man's imitation with the actual ways of life he had seen about him—from the low life that as deputy sheriff he had observed weekly for several years in the London sheriff's court all the way to the very high life that he occasionally encountered because of his friendships with men in the inner circle of the court of Henry VIII. The confrontation of a going society with biblical Christianity always has latent explosive possibilities, and in More's case these were realized when it led him to ask himself what the temporal conditions for a true Christian commonwealth actually were. Before he had quite finished writing what he intended to write about what a Christian commonwealth ought to be, he was offered a high position at the English court on terms both honorable and financially attractive. The comparison thus abruptly forced on his attention between the conditions necessary for a true Christian commonwealth and the conditions which actually prevailed in the putatively Christian commonwealth ruled by his native prince, Henry VIII, magnified the explosive force of his confrontation of the world as it was with the world as scripture said it should be. The result was a book, *Utopia*, that was both extraordinary and extravagant.[5]

I have emphasized the extravagance, the abnormality, of both *Il Principe* and *Utopia* because, as a result of their incorporation into the body of great literature on politics, we are prone to forget how completely they prescind from the normal view of politics in the early sixteenth century. Their authors, however, were well enough aware of what they were doing; they were indeed quite explicit about it. Machiavelli made his intention clear very early, in Chapter II of *Il Principe*.

---

[4] For a convenient general account of More's life up to his thirty-eighth year, see Raymond W. Chambers, *Thomas More* (London, 1935), pp. 48–144.

[5] For successive attempts to relate *Utopia* to More's immediate circumstances when he wrote the book, see J. H. Hexter, *More's UTOPIA: the Biography of an Idea* (Torchbook edition; New York, 1965), pp. 11–30, 99–110, and a correction of these views, 161–65; Thomas More, *The Complete Works of St. Thomas More*, Vol. IV: *Utopia*, ed. Edward Surtz, S.J., and J. H. Hexter, xv–xlix.

In hereditary states . . . the difficulties of maintaining power are considerably less than in new principalities. . . . If their prince possesses an ordinary amount of energy he will always maintain himself in power.[6]

So much, then, for hereditary rulers and what concerns the governance of the lands they rule. So much, in effect, for the normal politics of all the great monarchical states of Europe and for the problems of ruling that faced every viable political society in the Western world of Machiavelli's day except Venice and the Swiss cantons. So much for any place, indeed, where politics was something more than the smash-and-grab game that it had become from the Po Valley to the Kingdom of Naples since Machiavelli entered on his political novitiate at the turn of the century.

In the climactic chapter of the *Il Principe*, Chapter XVIII, he points to his intention in a more telling way; but one must read closely to understand just what he has done. He begins this chapter on "How Princes Should Observe Good Faith,"

Everyone understands well enough how praiseworthy it is in a prince to keep his word, to live with integrity and not by guile. Nevertheless, the experience of our times teaches us that those princes have achieved great things who have looked upon the keeping of one's word as a matter of little moment and have understood how, by their guile, to twist men's minds; and in the end have surpassed those who have rested their power upon faithfulness.

You ought to understand therefore that there are two ways of fighting, the one by the laws, the other with force. The first is proper to men, the second to beasts; but since in many instances the first is not enough, it is necessary to have recourse to the second. A prince, consequently, must understand how to use the manner proper to the beast as well as that proper to man. . . .

Since, then, a prince must of necessity know how to use the bestial nature, he should take as his models from among beasts the fox and the lion; for the lion does not defend himself from traps, and the fox does not defend himself from wolves. One must therefore be a fox to scent out the traps and a lion to ward off the wolves. Those who act simply the lion do not understand the implications of their own actions.[7]

It is all done so smoothly that even a careful reader may associate "laws," which are "proper to men" and therefore implicitly preferable to the "force . . . proper to beasts," with the prince who lives with integrity,

[6] Machiavelli *Il Principe* 2. All subsequent references to this work are by chapter.
[7] *Principe* 18. The wording of the legend of Chiron inserted between the two passages quoted above confirms the interpretation just offered.

keeps his word, rests his power on faithfulness. Not at all: *Le leggi* like *la forza* are a *generazione de combattere,* a way of fighting, a mode of conflict. The line of Machiavelli's argument comes clear when we compare it with the passages which it echoes.

Those passages from Cicero are as follows:

> . . . since there are two ways of settling a dispute: first, by discussion; second, by physical force; and since the former is characteristic of man, the latter of the brute, we must resort to force only in case we may not avail ourselves of discussion. . . .[8]

> While wrong may be done, then, in either of two ways, that is, by force or by fraud, both are bestial: fraud seems to belong to the cunning fox, force to the lion; both are wholly unworthy of man, but fraud is the more contemptible.[9]

Machiavelli's paraphrase must be one of the most efficient perversions of a writer's intention in the long history of literary hocus-pocus. He has slightly altered the passages paraphrased, he has conflated two passages separated by several pages and dealing with two different topics, and he has removed both passages from contexts inconvenient to his line of argument. In the Florence of Machiavelli's day the striking language of his paraphrases must have wakened in many minds memory echoes of their Ciceronian origin. Thus by his three-way bit of verbal legerdemain he makes it appear that his view has the support of the highest classical authority on the right conduct of politics.

To make it seem that he stood shoulder to shoulder with the man who in the eyes of the Renaissance was the noblest Roman of them all was quite a feat. The first passage from Cicero has as its context the relations of a state with its potential enemies. It is preceded by this sentence: "Then, too, in the case of a state in its external relations, the rights of war must be strictly observed."[10] It is followed by the sentence, "The only excuse, therefore, for going to war is that we may live in peace unharmed."[11] Thus Cicero is concerned with the conditions of rightful and humane warfare with *enemies,* while the over-all context of Machiavelli's observations is "the way a prince ought behave to *his subjects and allies.*"[12] It is against them he must act both as a lion and a fox, preying on them lest he be preyed on by some of them. In the second passage, while Cicero's attention is focused on the relative position of two vices on the moral scale, Machiavelli, in an attitude of moral neutrality, is simply weighing up the expediencies for a prince of

[8] Cicero *De Officiis* i. 11. 34.   [9] *Ibid.,* i. 13. 41.   [10] *Ibid.,* i. 11. 34.
[11] *Ibid.,* i. 11. 35.   [12] *Principe* 15.

using force or fraud as the occasion demands. And finally, while both passages from Cicero are part of his general discussion of justice, "the crowning glory of the virtues on the basis of which men are called 'good men,' "[13] in *Il Principe* Machiavelli's paraphrase has nothing whatever to do with justice but only with what "it is necessary for a prince wanting to maintain himself" to do.[14]

According to Machiavelli then law and force are separate species, but for the prince they are of the same genus; they are the two types of weapon that he needs to acquire and hold power, to avoid losing it or having it taken from him.[15] The distinction that Machiavelli is making between the use of laws proper to the human and of force proper to the bestial in men has recently found a succinct statement in the Beggar's Song from *Three-Penny Opera*:

When the shark bites with his teeth, dear,
Crimson billows start to spread;
Fancy gloves, though, wears McHeath, dear;
They don't show a trace of red.

The prince must know the arts of using both natures of man, the bestial and the human; but the difference between the practice of the two arts is not that between Richard III and St. Louis, between Alcibiades and Pericles; it is the difference between the shark and Mac the Knife.

Thought is bound to have its impact on vocabulary and syntax. The extravagance of Machiavelli's thought in *Il Principe* leaves a sharp impress on certain key words he uses and their relation to each other. From the time of Aristotle and especially from that of Cicero for fifteen hundred years, several such words had provided the hard currency by means of which men exchanged political ideas at all levels of abstraction higher than that of positive substantive law. Among such words were *nature, experience, law, right, good, custom, justice, reason, virtue,* and *order.* The considerable stability of the value of those words in relation to one another, and the intricate symbiosis by which all sustained each, had for centuries effectively guaranteed the general intelligibility of serious political discourse, although they did nothing to assure its practicality or immediate relevance. In *Il Principe* Machiavelli radically altered the values of several of these words, well-nigh abandoned one or two, and thus thoroughly snarled up the structure of values (both exchange and moral) that they had hitherto supported.

[13] Cicero *De Officiis* i. 7. 20.
[14] *Principe* 15.
[15] For the remarkable role of the verbs *acquistare, mantenere, tenere, perdere,* and *togliere* in the rhetoric of *Il Principe,* see J. H. Hexter, "*Il Principe* and *lo Stato,*" *Studies in the Renaissance* IV (1957), 119–20.

The operation Machiavelli performed on the common sense of the language of politics of his day was quite drastic. It involved the dislocation of virtue and reason and nature from their customary relations with each other, and the literal abandonment of justice. Without alteration medieval writers had taken over from antiquity the assumption and conviction that justice was a virtue discernible to man by means of that right reason with which Nature and "Nature's God" had endowed him. This capsule statement of twenty centuries of "the conventional wisdom" becomes utterly meaningless if one assigns to its single terms only such meaning as Machiavelli assigned them in *Il Principe*. Yet he assigned *reason, nature,* and *virtue* meanings quite common in his own time. *Natura* is linked with *qualità, naturale* with *ordinaria*. The *natura* of things is the way they actually are, the *natura* of men what they actually think or do or want; it is just the spectrum of their observed traits, so that no observed trait of men can be described as unnatural. Thus it is natural, not unnatural, for men longer to resent the loss of their property than the slaughter of their fathers.

The *ragionavole*, the reasonable, is merely "what figures," what is explicable; it is also paired with *ordinaria*. To *ragionare* is "to figure," to discuss or discourse, to explain. And there is no *ragione*, no Reason, in *Il Principe*, only the *ragioni*, the reasons, or explanations, or the *ragione*, the reason or explanation, the yield as it were, of figuring. Thus Machiavelli's *ragione* is instrumental, concerned wholly with means; in the language of Weber it is not *wertrazional*, concerned with the orderly arraying of ultimate goals. And since his *natura* has nothing to do with ultimate ends either, but with what men happen to want, to *ragionare* can help men, be auxiliary to their *natura*, but no more than force and fraud can. By making *ragione* and *natura*, again in Weber's terms, *wertfrei*, practically divesting them of normative value, Machiavelli has effectively separated them from *virtù*, from virtue or the virtues, in a common classical sense and the most common Christian sense. But he further ensures his subversion of the common sense of his day by only rarely and most equivocally using *virtù* in any such sense. Except in those rare and equivocal instances he keeps *virtù* pretty strictly within the connotational ambit of *prowess* or *valor*, or, to offer a, perhaps closer, recent slang equivalent, *moxie*. Amid the *Götterdämmerung* of so many semantic values, *giustizia* for all practical use disappears until the last hortatory chapter of *Il Principe*. There it crops up when Machiavelli describes the aim of ridding Italy of the barbarous as an act of justice, somewhat beclouding the issue, however, by equating the just with the necessary. In *Il Principe* justice does not amount to much.

The heart of the matter (or perhaps one might say the trick) then is that Machiavelli frequently uses most of the key words of the normal thought about politics of his day and for fifteen centuries before, and he uses them in quite common senses readily intelligible to his readers. But he almost never uses them in the particular senses that they had been shaped to in order to fit with each other in normal political discourse. The one term that was resistant to the trick was *giustizia,* since from Plato's day justice was what political discourse had mainly been about. Its effective banishment from the consideration of politics in *Il Principe* is a dead giveaway.[16] Scant wonder then that generations of scholars seeking a central core of normal politics in *Il Principe* have got hopelessly lost without ever finding what they seek. They are men wandering through a labyrinth in quest of a center that is not there.

What they look for in vain in *Il Principe* appears clearly and perspicuously in *Utopia.* There most of the key relations sanctified by centuries of common use are packed into two pages: man's highest good, happiness, is attainable only through the practice of virtue; virtue is living according to nature; to this end men were created by God; to follow the guidance of nature is to obey the dictates of reason; right reason directs men into the course suitable to their God-given natures, the distinguishing virtue of which is humanity; to foster these relations, laws were instituted among men, deriving their just powers from their promulgation by a just ruler or from the "consent of the people, neither oppressed by tyranny nor deceived by fraud"; in a society where justice rules, the laws promote human happiness, and the public good becomes the concern of reasonable men.[17]

Thus More deploys the traditional vocabulary and syntax of politics rapidly and with practiced ease in *Utopia,* the book. But according to that book the relations explicated by that vocabulary and syntax are actually to be found only in Utopia, the place. That is, they are to be found no place since Utopia is literally no place. Utopia is not merely not-Europe; it is anti-Europe, the society More knew and lived in turned upside down and backward. Reason, nature, virtue, and justice stood in right relation to one another in the land of Utopia because the law made the necessary provisions for maintaining that relation. The necessary provisions were the equality of goods, the abolition of private property, and the banning of the use of money. In the so-called commonwealths of Europe the law, based on and

[16] The preceding remarks are based on a reasonably careful search for all occurrences in *Il Principe* of the terms analyzed and a consideration of the meaning of each *in context.* For two earlier experiments of this sort, see Hexter, *Studies in the Renaissance,* IV, 113–38, and *American Historical Review,* LXIX, 945–68. To the best of my knowledge the method has not found many takers among historians, perhaps because the output of knowledge it achieves has appeared trivial in proportion to the input of labor.

[17] This pastiche of assertions occurs mainly in *Utopia* 163–67, supplemented from *Utopia* 105 and 195.

concerned to maintain private property, was imposed on the people, oppressed by tyranny and deceived by fraud, through a conspiracy of rich and powerful men parading themselves as the commonwealth. These men managed to dress out the self-serving decrees by which they robbed and bled the poor with the false appearance of popular approval, the specious approval of the very people the decrees themselves were designed to exploit.[18] In such commonwealths virtue does not receive its due, and instead of naturally imbuing men with public concern, reason naturally drives them to a ruthless pressing of private advantage.[19] The result is not the best ordered and most just society, enjoying universal happiness, but a fierce and worse than bestial scramble for power, wealth, and worship culminating in universal wretchedness and misery. And the lives of most men are poor, nasty, brutish, and short.

It is not an accident that in the language pattern of *Utopia* two terms with a high value charge ordinarily conjoined with virtue, reason, nature, and so on are left out or oddly treated. They are *custom* and *experience*. In the common view custom and experience were the links between the ultimate values of men and their day-to-day practices. It was through custom that those values found their expression in actual law, and law was itself a reflection and embodiment of collective experience. In *Utopia*, however, appeal to the custom and experience of More's day and his land is either irrelevant or provides only negative instances, since the experience was of evil and the custom was corrupt. The only relevant experience and custom was that of Utopia, which is to say that it was wholly the product of More's imagination.

Utopia and *Il Principe* are books very different from each other in structure, in tone, in outlook, and in impact. But in one respect they are remarkably alike. Following widely divergent paths, their authors, one in 1513, the other three years later, arrived at the same pinnacle, a point from which they saw in the same lurid light the society in which they lived. For a society without justice is a mere den of robbers. One must recognize it for what it is, since only thus can one rationally make up one's mind about one's relation to it, whether, like the princes, one seeks to get hold of and keep hold of as large a share of the loot as possible or whether, like the Christian-humanist philosopher Hythloday in *Utopia*, one decides to opt out.

Because in their extraordinary books Machiavelli and More saw the world in which they lived in much the same way, they could not or would

[18] *Ibid.*, 241.   [19] *Ibid.*, 239.

not try to bring it into any useful connection with the traditional set of relations that was the indispensable frame of reference for normal thinking about day-to-day political actualities in their time. To do this, a man had to be willing to accept the greyness of ordinary political life, the mixture of human motives, the ambivalent tensions between group interest and public concern, the existing structures of power, wealth, and status. Having accepted them, one had to squeeze them somehow into the traditional frame of reference. The job could never have that neat and surgical precision which Machiavelli displayed in cutting day-to-day politics free from the frame; but anyone who attempted it might hope to gain something in relevance for what he lost in effective force; political prefrontal lobotomies are desperate measures even with very sick bodies politic.

Fortunately, one seeking at least an inkling of the normal look of politics in the age of Machiavelli can find some reflection of it in the eyes of a man far less remarkable than Machiavelli or More. In the space of three years which separated the former's loss of high office and the offer of an important official post to the latter, another thoughtful man came to a point of change in his political career; but that change had almost nothing of a crisis character. A young prince had succeeded to the throne of the old king Louis XII, whom Claude de Seyssel had served more than fifteen years, but even before Francis I became ruler of France, Seyssel, well stricken in years, had announced his intent to turn his attention from the service of his earthly to that of his heavenly king. It was about time. Priest and pluralist since 1503, in 1515 he had already been for five years Bishop of Marseille, a diocese into which it seems he had not yet set foot. Indeed, his communication with his flock had been so exiguous that once the cathedral chapter inadvertently elected another man bishop under the erroneous impression that Seyssel was dead.[20]

The most serious crisis of Seyssel's career occurred some nine months before his birth about 1450, since he was conceived on the wrong side of the blanket. For an offspring of the upper aristocracy, a Seyssel of Savoy, however, even the bar sinister was not a disastrous impediment to advancement, but it may have led Claude to seek that advancement with a rather wider and deeper education than a more conventional entry into the world would have required of him. With some of the humanist's embellishment that then accompanied such an education, he was well trained in law; for several years he lectured on both the Digest and the law of fiefs at Turin; and

[20] For the details of Seyssel's biography, see Alberto Caviglia, *Claudio di Seyssel (1450–1520): La vita nella storia di suoi tempi* ("Miscellanea di Storia Italiana," Vol. LIV; Turin, 1928). For a brief account, see the modern edition of Claude de Seyssel, *Monarchie de France*, ed. Jacques Poujol (Paris, 1961), pp. 11–19.

he finally left his academic post to shuttle for several years between the service of his native duke and that of Charles VIII of France. In 1498 he finally
settled into a career of office-holding in the service of Louis XII, in which
he remained throughout the reign. An early member of the newly constituted Grand Conseil, counsellor in the Parliament of Toulouse, then that
of Paris, man of all works and some wisdom during the French administra-
tration of Milan, he also undertook a number of diplomatic missions,
notably to England, the Emperor, and the Swiss cantons. He also found
leisure to translate from Latin into French a number of classical histories—
Xenophon, Appian, Justin, Diodorus, Siculus, Plutarch, Thucydides—not
merely as a pastime but with the conscious intention of enriching French, as
Latin had been enriched before it, by absorbing into its vocabulary some of
the riches of the Greek lexicon.

In Louis XII's last years Seyssel had had the hard job of restoring peace
and amity in Franco-Papal relations, badly upset by the King's ill-advised
and wholly unsuccessful effort to drive the Pope into line with French policy
in Italy by summoning a Church Council to threaten his ecclesiastical authority. Apparently he did his work with considerable skill, for he remained
in Rome as a member of the Lateran Council, which the Pope had gathered
to counter Louis XII's move. Seyssel returned to France for the coronation
of Louis's successor, Francis I, but as he explained, "I wish to withdraw to
the service of God and my church as my present situation (*état*) and age
require." He felt it his duty at the end of his career to set down for Francis
to consider at leisure what he had learned from the matters he had dealt
with during his long period of service.[21] What he set down he entitled *La
Monarchie de France*. His equipment for producing a great work on politics
was excellent. Seyssel had in abundance that "long experience of modern
affairs and continued reading about those of antiquity"[22] on which Machiavelli claimed he based *Il Principe*. Perhaps, however, he lacked the monocu-
lar intensity of vision that his sudden catastrophic uprooting gave to Machiavelli, that the peculiar circumstances of his rapid transfer from busyness to idleness and back to busyness again gave to More. Or, as seems more
likely, the most extensive education and experience is never an altogether adequate substitute for being terribly bright in the way Machiavelli and More were terribly bright. In any case *La Monarchie* has
neither the force nor the fascination of *Il Principe* and *Utopia*.

It is nevertheless an interesting little treatise. At the crucial point
where *Il Principe* and *Utopia* are extravagant and heterodox, *La Monarchie*

[21] *Monarchie* 95–98.
[22] *Principe* dedicatory epistle.

is orthodox. The old statesman constructs his book of advice to young Francis I on the rooted traditional values of the relations among reason and nature and experience and virtue and justice and custom and law. He does not so alter several of them that they no longer can function in the traditional exchange system, which is what Machiavelli did with *ragione* and *natura* and *virtù*, or omit one altogether, which is what Machiavelli did with *giustizia*. Nor does he preserve the values intact only to maintain that their relevance is confined to Never-Never Land or Utopia. Their relations are what they had long been, and they are relevant to the rule of the kingdoms of this world, and therefore to the rule of France.

To say only this of *La Monarchie*, however, is to say of it nothing that is not equally true of scores of dreary, cliché-ridden books of advice to princes and reworkings of the pseudo-Aristotelian *Secreta Secretorum*, written during the preceding three centuries for rulers who, if they read them, must have had an egregious tolerance for sustained ennui.[23] Seyssel, who was no fool, was inclined to suspect that they had not read them. In any case he resolved not to produce another highly dispensable increment to the already massive literature of banal moral exhortation to rulers. *La Monarchie* is not such a book.

It is not such a book because of traits which, transcending their differences, it shares with *Il Principe* and *Utopia*, traits which distinguish all three books from medieval writings that concern themselves with ruling in the city of this world. We sometimes speak of men who do not see the forest for the trees, and we might equally well speak of men who do not see the trees for the forest. Medieval writers on civil rule seem to fall into one or the other of these categories, and it is at least probable that in this matter medieval categories of writing fairly represent the topography of medieval thinking. Much of what we now classify as medieval political theory was given over to general discussion of the nature, the origin, the substance, and the rightful extent of political authority on the one hand and of political obligation on the other, and to the related consideration of the nature, the source, and the ultimate sanction of law. From such discussion were directly deduced the unconditional divinely sanctioned rights and duties of subjects and rulers. Such writings stand so far off from the day-to-day bread-and-butter activities of officials, ecclesiastical and temporal, that only rarely and then with some difficulty, as in the investiture controversy, can we see how on occasion they were brought to bear on those activities.

On the other side are the officials carrying out the rule-bound work

[23] For full information about this literature, see Allan H. Gilbert, *Machiavelli's PRINCE and its Forerunners* (Durham, N.C., 1948); Desiderius Erasmus, *The Education of a Christian Prince*, ed. and trans. Lester K. Born (New York, 1936), pp. 94–130.

of their offices—receivers of petitions, counselors, stewards of estates, administrators of the receipts and expenditures of princes, judges of many
varieties. Within their limited spheres they knew what their offices required
of them, what their duties were, and what dereliction of duty was. All this,
however, was a long way off from the abstract value patterns of the medieval
philosophers and moralists, and if the men who handled the affairs of society
were aware of and understood what the men who handled its ideology were
talking about, they might well have wondered what in reason they could
be expected to do about it. They could no more see the forest for the trees
than the theorists could see the trees for the forest. For centuries that middle
distance where one could see both trees and forest and contemplate them
together seems to have been virtually a no man's land. In effect the area of
thinking about politics that devotes itself systematically to considering the
problems created by the relations or lack of them between ultimate principles and actual practice was waste.

It is hard to say where, when, or why this situation began to change;
and given the wide circulation in the late Middle Ages of Aristotle's
Politics, standing solidly in the designated area, the more interesting question might be why it had not changed at least two centuries earlier.[24] What
medieval men seem to have been deficient in was a living awareness that
any set of normative political goals could only be achieved or approximated
through adequate institutional means. Therefore they did not feel impelled
to analyze the means or resources necessary to achieve those goals, which
they simply prescribed as categorically imperative. Even less did they feel
impelled to examine the means actually available for the attainment of such
goals. And so they had no coherent way of formulating a policy either for
providing the necessary means or for adjusting the goals to bring them
within shooting distance of the existing means.[25] "The means actually avail-

[24] What appears in the text as the silence of the Middle Ages may in fact be merely an
indication of the ignorance of the author. It is based only on such acquaintance with the
secondary literature on medieval political thought as he has acquired over the years, and
that affords but a fairly superficial and shaky foundation. If it is indeed the case that for
two or three centuries a knowledge of Aristotle's *Politics* with its elaborate comparisons
of the constitutions of ancient states did not generate imitation in which contemporary
constitutions were so considered and compared, that fact would be important to keep in
memory during any attempt to inventory the content and character of "the medieval
mind."

[25] This may help to explain why so much of medieval law was "secreted into the interstices of procedure," while so much medieval legislation was *vox et praeterea nihil.*
Law got inserted into the interstices of procedure because officials who had to get on with
the business of governing had to make the kinds of rules adapted to the situations that
faced them; so the rules stuck. Much medieval legislation (and a good bit of modern
legislation, if one comes down to it) embodies the higher aspirations of the rulers, their
putative commitment to be good rulers, without much reference to the availability of
institutional structures for translating that commitment into political actuality.

able," "the existing means," are the institutions of government, the laws, the material resources, the social clusters and interest groups, the religious convictions, the habits of thought and modes of living of the people to whom the policy is to be applied. On such matters medieval writers on politics have little to say.

Before the end of the fifteenth century Sir John Fortescue in his *Governance of England* and his *De laudibus legum anglie* and Philippe de Commines in his *Mémoires* seem occasionally to have glimpsed the importance of relating the general rules of conduct for princes to the brute local facts of life. What is somewhat tentative in Fortescue and Commines has become a steady disposition of the mind with Machiavelli and More and Seyssel. Supporting its seemingly sudden appearance in three separate places and three very different men within the brief span of three years was no doubt a great deal of previous political talk, still sporadically audible in surviving fragments—in the minutes of the *Pratiche* in Florence and in Masselin's journal of the French Estates General of 1484.[26] It seems as if the articulate citizen of those days, whose English epiphany Arthur Ferguson has recently investigated,[27] was articulating some interesting views, few of which found their way into the surviving historical record.

Be that as it may, Machiavelli and More and Seyssel are more precise and more perceptive in tracing the interrelations of the political and social facts of their day and in exposing the connections of those facts with the theoretical and actual imperatives of ruling, "the way things actually are and the way they ought to be," as Machiavelli puts it,[28] than any writers on politics during the preceding fifteen hundred years. I hardly need to dwell here on Machiavelli's exercise in describing the modulations a new prince must make in his policy to suit both the varying circumstances under which he acquires power and the condition of those over whom he acquires it. Scarcely less well known is More's brilliant feat of tracing the failure of criminal justice in England to a legally maintained set of social and economic arrangements which by locking men into their poverty drove the poor to steal out of despair, and then, by establishing the death penalty for theft, made it only prudent for the thief to murder his victim.[29]

It was the same sort of clear-eyed awareness of what went on in the world that enabled Seyssel to make sense of a social situation of which, for

[26] Felix Gilbert, *Machiavelli and Guicciardini: Politics and History in Sixteenth Century Florence* (Princeton, 1965), pp. 28–78; Jehan Masselin, *Journal des États Généraux . . . en 1484 . . .*, ed. and trans. A. Bernier (Paris, 1835).
[27] Arthur B. Ferguson, *The Articulate Citizen and the English Renaissance* (Durham, N.C., 1965).
[28] *Principe* 15.
[29] *Utopia* 61–75.

at least half a millennium before and sporadically for several hundred years after, almost all European social critics made nonsense. The movement of men out of the place in the social hierarchy into which they were born, whether that movement was up or down, was for centuries a standard subject for professional and amateur viewers-with-alarm, the pet bogeyman through the display of which they elicited the appropriate shudders from the ruling elite. Thus Piers Plowman:

Bondsmen and bastards . and beggars' children,
Thus belongeth to labor . and lord's kin to serve
Both God and good men . as their degree asketh; . . .
But since bondsmen's bairns . have been made bishops,
And bastard children . have been archdeacons,
And soap-sellers and their sons . for silver have
*   been knights,*
And lord's sons their laborers. . . .
Holyness of life and love . have been long hence,
And will, till it be worn out . or otherwise changed.[30]

Such was the time-hallowed response of centuries of European social moralists to the facts of what sociologists today call social mobility: it was a Bad Thing to be exorcised by assuming a posture of appropriate horror.

To this response Seyssel pays no attention. Society, in his day, he knows, is indeed hierarchical, but in his very analysis of its structure he rejects the pat and standard procedure. For him the effective divisions in society are not the traditional ones of medieval literature or even of medieval law: the clergy who pray, the warrior aristocrats who fight, and the rest who work for those who pray and fight. He does not regard the clergy as an estate at all. The actual estates are the nobility, the well-to-do, and the common people. Having thus revised the customary medieval social taxonomy, he makes a far more radical and realistic observation:

Everyone in this last estate can attain to the second by virtue and by diligence without any assistance of grace or privilege. This is not so in going from the second to the first, for to attain to the estate of nobility one must secure grace and privilege from the prince, who renders it readily enough when he who asks it has done or is about to do some great service to the commonwealth. Indeed, in order to maintain the estate of nobility the prince must do this whenever there is legitimate reason. This estate is always being depleted because in the wars in which it engages great numbers are often killed and because some nobles become so impoverished that they cannot maintain their station in life. He must

[30] Quoted in Ferguson, *Articulate Citizen*, p. 60.

403

also do it to give to those of the middling estate the hope of arriving at the estate of nobility and the will to do so by doing virtuous deeds, and to those of the innumerable popular estate the hope of attaining to the middling, and through the middling of then mounting to the first. This hope makes every man satisfied with his estate and gives him no occasion to conspire against the other estates, knowing that by good and rightful means he himself can attain to them and that it would be dangerous for him to seek to make his way by any other route. If, on the other hand, there were no hope of mounting from one to the other or if it were too difficult, overbold men could induce others of the same estate to conspire against the other two. Here, however, it is so easy that daily we see men of the popular estate ascend by degrees, some to nobility and innumerable to the middling estate. The Romans always maintained this same order, for from the common people one rose to that of the knights and from that of the knights to that of senators and patricians.

In France the church offers another means, common to all the estates, for attaining to a high and worthy station. In this matter the practice in France is and has always been that by virtue and knowledge those of the two lesser estates may attain to great ecclesiastical dignities as often as or more often than those of the first, even to the rank of cardinal and sometimes to the Papacy. This is another great means to satisfy all the estates and to incite them to train themselves in virtue and learning.[31]

Surely this combination of the lessons of antiquity with the quick and accurate reading of the economic and demographic facts of life as they affected the aristocracy, this perception of the church as a useful by-route for a social mobility which was itself a means of releasing and diffusing social pressures, surely all this in face of the contemporary habit of thought which almost automatically triggered a spasm of horror at the mere notion of such mobility, entitles Seyssel to join More and Machiavelli among the select group of pioneers who by confronting contemporary political credos with contemporary political facts strengthened their own political vision and therefore ours. It was this habit of confrontation shared by all three that enabled them to see through and beyond the stale hortatory moralism which had hitherto spared those who wrote for rulers about ruling the inconvenience and discomfort of examining and exploring or even being aware of any serious political problems connected with what they were writing about. In the very first sentences of his preface to *La Monarchie*, Seyssel renounces the serene delights of this sort of intellectual lotus-eating.

Several philosophers, theologians and other wise men . . . have disputed,

---

[31] *Monarchie* 125–26. Seyssel's taxonomy of the social orders, heterodox in his day north of the Alps, probably reflects his education in Italy, long accustomed to distinguish between the *popolo grosso* and the *popolo minuto* (Seyssel: *peuple gras, peuple menu*).

written, and dogmatized on what the government of the commonwealth in general ought to be, and among the several forms of government which is the best . . . and on these matters have been made many treatises and great volumes hard to read and to understand. It would be harder still, however, to put them into practice, for, in writing, men set down what is desirable and what reason and natural sense quite readily teach; but human weakness is so great that no men are so wise, virtuous, and prudent as those the learned described, nor is there any city or republic, great or small, ruled entirely by moral and political reason, and few are without more imperfections than perfections. Therefore, to recite the arguments, reasons and opinions of the authors treating these matters would be repetitious, prolix, and . . . very difficult, and would make a book the mere size of which would frighten off anyone who wanted to read it, unless he had a great deal of leisure. Even if he were willing to take the pains, after he had read it, he would remain confused. . . .[32]

The renunciation is somewhat less caustic but no less decisive than the more notorious one of Machiavelli:

It remains now to consider the manner in which a prince should conduct himself toward his subjects and his friends. I know that many writers have treated this topic, so that I am somewhat hesitant in taking it up in my turn lest I appear presumptuous, especially because in what I shall have to say, I shall depart from rules which other writers have laid down. Since it is my intention to write something which may be of real utility to anyone who can comprehend it, it has appeared to me more urgent to penetrate to the effective reality of these matters than to rest content with mere constructions of the imagination. For many writers have constructed imaginary republics and principalities which have never been seen nor known actually to exist. But so wide is the separation between the way men actually live and the way that they ought to live, that anyone who turns his attention from what is actually done to what ought to be done, studies his own ruin rather than his preservation.[33]

Such firm and unflinching confrontation with actuality and such an end of idyllic innocence are likely to have an explosive impact on the mind of the man who achieves them, especially if he is among the first in centuries to do so. Seyssel, More, and Machiavelli did so, and, as we have seen, their achievement blew the thinking of the latter two right out of the normal orbit of political speculation as it was practiced by their contemporaries. The relation between moral political imperatives and actual political conduct was so hard to discern and the human habitudes and institutional structures through which any linkage between the two had to be traced were so dense

[32] *Monarchie* 95.
[33] *Principe* 15.

and tangled that, having coped successfully with the first problem, they turned their backs on the second and thus freed themselves to write their brilliant and extravagant tracts *Il Principe* and *Utopia*.

Seyssel did not have the will to do what his two great contemporaries did, or he did not have the imagination, or, most likely, he did not have either. His only option then was to try to bring that new sense of political actualities which he shared with them into some fruitful relation to the traditional view that in the city of this world virtue and nature and experience and reason and justice and custom and law ought somehow to be conjoined to provide the framework of rules by which men lived together in civil societies. We have already witnessed the first decisive move Seyssel made; it was to jettison the abstract and vacuous moralism which permeated most books on governance by medieval philosophers and poets and Renaissance humanists, and to do this in the interest of what Machiavelli would have called *la verita effetuale de la cosa*:[34] "Human weakness is so great that no men are so wise, virtuous, and prudent as those the learned describe, nor similarly is there any . . . commonwealth . . . ruled entirely by moral and political reason, and few are without more imperfections than perfections."[35] The consequence of this sharp deflation of expectations was to render Seyssel's enterprise possible, for by means of it he could avoid the despair that a comparison of inflated expectations and dreary reality engendered in More and Machiavelli.

The next thing Seyssel did was to give an entirely new orientation to two common species of political treatise—the treatise *de optimo reipublicae statu* and the treatise *de regimine principum*. (Note that *Utopia* was a wild mutant of the former, *Il Principe* of the latter species.) As a number of his predecessors had done, Seyssel combined the two species in *La Monarchie*. But in dealing with the question of the best form of government, Seyssel does not limit himself to an abstract rehashing of abstract arguments derived from Aristotle's *Politics* about the relative merits of democracy, aristocracy, and monarchy. He runs fast through that argument in the first chapter of *La Monarchie*, brings it to an inconclusive end,[36] and promptly turns to an examination of the merits and defects of the democratic regime of the ancient Romans and the aristocratic regime of the contemporary Venetians.[37] Having shown to his own satisfaction that these two regimes, each the best of its kind,

---

[34] *Ibid.*, 15.   [35] *Monarchie* 95.   [36] *Ibid.*, 103–4.
[37] *Ibid.*, 104–10. The chapter on the Venetians contains a digression on the rise and decline of all bodies politic; see *ibid.*, 108–9.

have dangerous defects, Seyssel states that reason both divine and human, natural and political, shows that one head is better than many in dealing with difficulties and dangers, while experience demonstrates that both in ancient times and his own day monarchical regimes are more stable and peaceable than the others. Among monarchical regimes experience and reason also show that hereditary monarchies will prosper better than elective ones.[38] Thus having whipped in jig time through the customarily pavane-like motions of this ancient abstract argument for monarchy, he turns to tell why among hereditary monarchies that of France is best.[39] For the rest of the first part of his treatise Seyssel examines those elements of the French political and social order which in his view indicate and are responsible for the superiority of the French monarchy. The proportions of this first part are striking. Less than an eighth of it is given over to theoretical discussion of the best political form in general and the best general form of a monarchy.[40] The remaining seven-eighths goes to concrete institutional analysis of Roman democracy, Venetian aristocracy, and French monarchy,[41] and of that remainder Seyssel devotes almost three-quarters to a consideration of the structural strong points of France in his day.[42]

This trait of *La Monarchie*—its focus on the particular as against the general, the concrete as against the abstract—is even more distinctly visible in the sections that in form are a treatise *de regimine principum*. It starts conventionally enough with a chapter entitled "Of the Education and Instruction of Princes and Monarchs in General."[43] The title is the only conventional thing about the chapter. In the chapter itself Seyssel notes that if kings were always good and wise, there would be no abuses to correct, but adds that since they are not, more particular instruction seems necessary. Then instead of going on to offer the sort of instruction that fills the ordinary prince book, he produces rather surprisingly a short bibliographical essay on treatises *de regimine principum* and pays his small respects to the whole genre by pointing out that the advice so lavishly dispensed in such exercises "is hard to retain because of the fragility and imperfection of human nature." He ends his observations with the wry remark that "if they are wise and virtuous, . . . men of high degree are ordinarily occupied with great affairs so that they scarcely have idle time to spend on reading lengthy writing, and if young and willful, they are given over to lewdness and other vain and voluptuous doings."[44]

[38] *Ibid.*, 111.   [39] *Ibid.*, 112.
[40] *Ibid.*, 103–4, 110–11.   [41] *Ibid.*, 104–10, 112–28.
[42] *Ibid.*, 112–28.   [43] *Ibid.*, 130.   [44] *Ibid.*, 132.

With an almost palpable sigh of relief Seyssel then turns from monarchs in general to "Particular Instruction . . . for the King . . . of France," in which he says he will concern himself with "what seems most necessary for the present maintenance and increase of this monarchy" dealing with "several points which call for special attention because it has been observed during the lives of those still living that on them hinges the good or the ill, the prosperity or the misery of the realm."[45] And he is as good as his word. It is precisely with such matters—policy and administration, social and economic and military, internal and external—that he thereafter deals for the rest of *La Monarchie*.[46] Again it is the relative allocation of space that is startling—three pages to a rather cool evasion of the standard banalities of treatises *de regimine principum* followed by eighty pages of reasoned and reasonably detached suggestions about royal policy based on a consideration of recent French history, things "observed during the lives of those still living." Overall then in *La Monarchie* one-twentieth of the work concerns itself rather casually with the standard topics of the standard works in its putative genres, nineteen-twentieths with the *modus operandi* of three actual political societies, one of the past and two of Seyssel's own day. And nine-tenths deals with France, the political society which he knew best from that long experience in a variety of posts by which Machiavelli set such great store.

The consequence of Seyssel's intense focus on France was that *La Monarchie* dealt although less vividly yet a bit more sensibly than *Il Principe* and *Utopia* with a number of matters which preoccupied men concerned with politics in that day—the use of mercenary soldiers,[47] for example, and the problems of counsel for rulers.[48] But most significant is Seyssel's handling of the crucial dilemma of politics in the age of Machiavelli. Lack of governance, weakness at the center, had prevailed in the era that preceded, and having tasted its bitter fruits, men in the politically advanced lands of Europe wanted an end to it. They wanted to be ruled, and if they were to have what they wanted, they must allow their rulers the power needed to govern. If the prince were truly to rule, his will must be effective, and to be effective, it had to be law or something very like law. But in the common opinion law ought never to be the mere expression of any mortal's mere will; it ought to be inextricably bound into the complex pattern of perdurable relations in which justice, reason, experience, custom, nature, and virtue were indispensable terms. The pressing problem was how to keep law

[45] *Ibid.*, 133. [46] *Ibid.*, 133–221.
[47] *Monarchie* 169–73; *Utopia* 149–51; *Principe* 12–14.
[48] *Monarchie* 133–41; *Utopia* 55–59, 85–103; *Principe* 22–23.

bound into that pattern of relations and at the same time allow rulers the power to rule, which meant in effect the power to make rules with the force of law.

In France, Seyssel argued, this problem was as near to being solved as —given the weakness and wickedness of men—it could be, because there the hazards of abuse of power on the part of the king were limited by three bridles. The term itself is significant. A bridle is not to stop action but to regulate it, to subject it to rule. The verb Seyssel uses in describing the function of the bridles is never *dominer;* it is always *regler.* The king cannot be *dominé;* he is *absolu;* no man or body of men can exercise dominion over him. The bridles are not to command his action, but to conform it to order, to prevent it from becoming *desordiné,* inordinate; *volontaire,* merely willful; *dereglé,* unruly.[49]

The bridles that perform this work are religion, justice, and police. Medieval and more recent writers on the duties of rulers had long harangued monarchs on their obligation to maintain religion and execute justice; indeed in their coronation oaths most kings solemnly undertook to do both. The trouble was that while good rulers hardly needed written exhortations about their duties, bad ones were not likely to pay much heed to them, or, for that matter, to coronation oaths. In speaking of the bridles, Seyssel had something more in mind than mere exhortation; according to him religion, justice, and police actually functioned in day-to-day politics to prevent or limit the merely willful exercise of royal authority.

Because the French people are religious Christians, he points out, the king must make it clear to them *"by example and by present and overt demonstration"* that he too is a zealous Christian. But to do this, he will have to live like a Christian, which in effect means that he cannot act tyrannically. If he is so willful as to do so, he can be reprehended to his face by any mere preacher. Even should the king desire to take action against such a preacher, he would not dare to brave the popular disapproval that such a deed would generate. The effective sanction throughout is the wrath of a religious people against a manifestly sacrilegious ruler—the king's "fear of provoking the ill will and indignation of the people," his awareness that, if through his acts he appeared indifferent to his religion, "the people would hate him and perhaps obey him but ill, . . . imputing all the realm's troubles" to the king's religious defect. In the monarchy of France the bridle of religion is neither an abstraction nor a mere pious hope; it is the ruler's understanding and fear that, if in his management of public affairs he does

---

[49] For this set of rhetorical ploys, see especially *Monarchie* 113–15, 120.

not at least outwardly conform to the minimal demand of Christian conduct, he will suffer rejection at the hands of the French people.[50]

As with religion, so with justice. The justice that bridles the king is not a remote inoperative ideal; it is embodied in the ancient and rooted practices of the king's own courts of law, and especially of the parlements. The eminence and authority of the judges of these courts are such that in civil matters a petitioner can get a judgment against the king as well as against another person; that in cases at private law the king cannot intervene in favor of one party to the detriment of the other, since his writs are subject to judgment not merely with respect to their fraudulent procurement but with respect to their legality as well; and that in criminal matters, the parlement so thoroughly debates the king's acts of grace and pardon and puts those who have wrongfully procured them through such an interrogation, that few indeed are foolish enough to rely on such acts. Moreover since judges can be dismissed only on charges over which the sovereign courts themselves have jurisdiction, they are able to act independently on their view of the law of the realm without fear of removal by a willful king.[51]

The third bridle, police, works in a similar way. Police is "the many ordinances made by the Kings of France themselves and afterwards confirmed and approved from time to time, which tend to the conservation of the realm. . . ." Such, for example, is the rule against alienation of the royal domain except in case of necessity. The parlements, whose approval of such exceptional alienations alone legalizes them, stall so long while considering them that they simply ruin the market for them. At the same time the accounting court (*chambre des comptes*) reviews the king's outlays. Between the two they dampen the appetites of those who seek to fatten off special favor, diminish the need for the king to rely on extraordinary exactions in emergencies, and keep the monarchy strong by restraining the overgreat liberality of individual monarchs. In general, police serves as a bridle on the monarchy because, should French kings willfully derogate from the time-sanctioned ordinances of their predecessors, "their commands would not be obeyed at all."[52]

Such were the bridles on the kings of France to prevent their *puissance absolu* from becoming *volontaire* and *desordiné*. Let us make two further points about them. First, as a clear-eyed man who knew a good bit both about

---

[50] *Ibid.*, 115–17. On Religion, the first bridle of kings.

[51] *Ibid.*, 117–18. On Justice, the second bridle.

[52] *Ibid.*, 119. On Police, the third bridle. There is not space here to consider what Seyssel meant when he used the term—ubiquitous in *La Monarchie*—*la police*. I hope soon to deal with this problem in the introduction to an English translation of the work.

human orneriness, especially in high places, and about the history of the French monarchy, Seyssel did not delude himself that the bridles were perfectly and instantaneously efficacious. He did believe that they were tough, resilient, and durable, capable of surviving the heavy strains put on them by wicked rulers and by kings who were children or idiots, and of doing their work when these plagues of God had passed. They had their imperfections, and by the whim of an arbitrary ruler they "might be bent, but they have been maintained so long that they can scarcely be broken."[53]

Second, the so-called bridles were also the great reservoirs of royal power. The power the king needed to rule effectively in large part derived from his authority as protector of religion, as source of justice, and as initiator of ordinances. Thus Seyssel is not whistling in the dark or indulging in the usual make-believe when he states his judgment that without the bridles the absolute power of the King of France would be "worse and more imperfect," just as "the power of God is not less but rather more perfect because he cannot sin or do ill," and that the bridles rendered the king's power "more perfect and complete and also more firm and lasting."[54] Finally it is precisely in the area of religion, justice, and police, taken quite concretely as we have seen, and in the ruler's role as guarantor of a historically specific social order that the relation between law as the emanation of the king's will on the one hand and law and custom as a part of a value pattern that includes reason, nature, experience, virtue, and justice on the other is firmly fixed, fixed by links that connect both of them with the actual needs and operative institutions of the government and society of sixteenth-century France. It is precisely at this point that Seyssel does in *La Monarchie* what neither More nor Machiavelli tries to do or cares to do in their extraordinary and extravagant little books. He finds a way to bring the empirical observation of political actuality into effective relation with the conventional wisdom about the ends of political association and therefore with the perennial aspirations of man, the political animal. However involutely, however fortuitously, the policy of the absolute King of France is therefore bound to reason and nature and virtue and justice and bound to make at least a pretense of conformity to them.

Between 1513 and 1516, then, in three small books three modern modes of thinking about politics attained maturity so abruptly that two of them at least give the impression of being born like Athene, full grown from the head of Zeus. The first mode, the predatory, found its first clear expression in

[53] *Ibid.*, 118.   [54] *Ibid.*, 120.

Il Principe; the second the utopian in *Utopia*; the third, the constitutional, in *La Monarchie*.[55] None of the books could have been written a century earlier, because one characteristic common to thinking in all three modes had not yet developed adequately. All three demanded a close scrutiny of the general principles men professed as the norms of political conduct in intimate conjunction with the networks of actual practice manifested in established institutions and recurrent patterns of political behavior. Since in the time before the age of Machiavelli the second element was either hopelessly fragmented or wholly absent, one of the necessary conditions for the emergence of all these modes of thinking was not fulfilled.

But it is not an accident that of the three early exemplars of the predatory, the utopian, and the constitutional modes of thinking, two are classics of politics, still current today, while the third is remembered by few and read by fewer. For if there is a an unbridgeable chasm between the norms of political conduct and actual political practice, only two active options are open to a writer who would concern himself seriously with politics. One is to prescind entirely from the avowed normative principles and treat politics as an exercise in predation, in which those who seek domination treat the body politic as prey. Then the only relevant question is how to acquire the prey, how to keep one's hold on it, how to avoid losing it or having it taken away by a rival predator. In this game reason, virtue, justice, and law, along with force and fraud, are merely instrumental to the wills of the predators;

---

[55] The author has misgivings not so much about his taxonomic scheme here as about the labels he has applied to it. He has least concern about the label *predatory*, mainly because it is not at the moment widely current among the words of art for describing political posture, but also because it so neatly fits the posture prescribed for would-be rulers in *Il Principe*. The word *utopian* creates far more difficulty, so much so indeed that in 1965 a more than ordinarily intelligent group of twenty-odd scholars spent two days chasing it around in hopes of pinning it down. The author hopes that he made it clear that in using *utopian* he does not refer to the form of political fantasy which constitutes the bulk of the "literature of Utopias," but rather to the political posture of those whose contemplation of the fabric of imperatives, professed by their own society or deemed possible by them, has profoundly alienated them from the actual life lived in that society. (For a more elaborate discussion, see Hexter, *American Historical Review*, LXIX, 945–68.) *Constitutional* has even worse drawbacks. It has been used to describe some forms of political theory in the Middle Ages and the political theorizing of part of Machiavelli's *Discorsi*. As here used it does not fit either of these—not the first because medieval writers did not exhibit publicly (although they may have possessed) the necessary grasp of the relations between theory and practice; not the second because therein Machiavelli seems mainly concerned with the stability and expansive force of political structures and scarcely at all with their legitimacy or capacity to achieve justice. The term *political*, with its reference to Aristotle's conception of the *politeia*, would have done better justice to sixteenth-century usage than *constitutional*, a word coined more than two hundred years later. Unfortunately, however, the present-day connotations of *political* are too all-embracing, so *faute de mieux*. . . .

they are the weapons used in the struggle; and the name of the game is power. Or as it was once put with perhaps a trace of poetic license:

Force should be right, or rather right and wrong, . . .
Should lose their names, and so should justice too.
Then everything concludes itself in power,
Power into will, will into appetite,
And appetite a universal wolf . . .
Must make a universal prey,
And last eat up himself.[56]

Though by no means inevitable, it is still intelligible that the first man who was bright and bold enough to envisage this extravagant option at all should be also bright and bold enough to grasp it fully. And this was what Machiavelli did in *Il Principe*.

So, conversely, with the alternative option. If on analysis actual institutions appear to be means not of approximating the professed norms of a society but rather means of thwarting them in the interest of the rich and the powerful, then one may inquire what institutions and patterns of behavior must replace them in order to realize those norms. And again the first man bright and bold enough to grasp this option might be bright and bold enough to grasp it fully. And that was what More did in *Utopia*.

The position of the constitutional mode is different. In the first place the transition from earlier modes of thinking about politics is less abrupt and dramatic. Constitutional thought is not faced with a simple either/or option, but with the more prosaic task of shoving and hauling medieval normative theory on the one side and medieval legal thinking and administrative practice on the other into bridgeable distance of each other. Moreover this task requires the patient and tentative modulation of both of these earlier forms in the process. Perhaps this is why, although the quest for medieval and Renaissance precursors in any serious sense for *Il Principe* and *Utopia* has failed, we may discern precursors of *La Monarchie* in such diverse places as Fortescue's *Governance of England*, Commines' *Memoires*, Masselin's *Journal* and the debates of the Pratiche. Second, whatever the overt avowals of its practitioners, the constitutional mode of political thinking requires that they treat nature and justice and order and virtue and reason as not forever fixed with complete precision in the human understanding and that they treat the actual institutions of society and its particular current patterns of activity as subject to gradual modification and amelioration. And this means that the constitutional mode of thinking about politics will

[56] William Shakespeare, *Troilus and Cressida*, I, iii.

413

always lack the clarity, the incisiveness, and the finality of predatory and utopian thinking and that works written in a mode dominantly constitutional will suffer from a tonal greyness, a tendency to obsolescence, an element of the ephemeral.[57]

I say "dominantly constitutional" because these three modern modes of thinking about politics are not wholly discrete. In one direction constitutional thinking lies closer to utopian, in the other to predatory thinking. Yet it would be a mistake to conceive of the relation to each other of these modes of thought, whose origins we have just explored, as a straight line continuum with predatory thinking at one extreme, utopian at the other, and constitutional in the middle. A triangle provides a better model. Constitutional and predatory thought converge in their common concern with the range of maneuver possible within the bounds of an existing political structure and situation. But as we have seen in the case of *Utopia and Il Principe*, utopian and predatory thought also converge in their common despair of the constitutional enterprise of maintaining contact between virtue and reason and nature and justice on the one hand and the going political and social order on the other. Thus we may conceive that the constitutional mode is most secure when broadly based, when it retains for itself a comfortable range of maneuver amid the hard and resistant political actualities without surrendering its purpose of providing the means whereby men who aspire to a good life may attain as good a life as circumstance and the human condition permit. In this way it can encompass and possess within itself the means of adjustment to change. It also thus maintains a wide angle opposite to it between the utopian and the predatory modes of thinking. And it needs to do so, for the narrowing of that angle, the closing together of those modes of thinking, are symptomatic of a revolutionary situation.

This way of considering these three modern modes of political thinking whose beginnings we have just explored may require us to change our very conception of the nature of understanding and truth as it applies to the constitutional mode. We are habituated to the platonic dogma that political truth and understanding involve the discovery of a fixed set of rules or formulas establishing the right order among a changeless set of facts. Perhaps we should resign ourselves to the actuality that neither the current rules embodying the perdurable human aspiration to a world based on order and reason and justice and nature nor the facts embodied in time-bound, place-bound institutions and patterns of actions are or can or should be fixed or changeless. In the constitutional mode we can retain our concern for

---

[57] Something of the sort has happened to many sections of *The Federalist Papers*, for example, and to those parts of *Le Contrat Social* in which Rousseau tries to relate his utopian conception of politics to the actualities of Europe in the later eighteenth century.

nature and reason and justice and order in our social cosmos without binding ourselves to an impossible quest for final and definitive solutions. Truth about political life then ceases to be a goal that the human mind can attain once and for all. It becomes that measure of prudence and judgment and wisdom that men deploy in the ever-changing process of seeking to maintain living and effective contact between the realm of justice and reason and nature and order on the one hand and the sphere of men's daily doing in an actual political society on the other, a continual rethinking of our belief about the one and a continual reforming of our doings in the other. Whether such a notion of the meaning of truth and understanding in the constitutional mode of political thought would survive the rigor of philosophical analysis, I do not know. Nor do I much care, since I suspect that our very notions of truth and understanding may be on the verge of a period of ecumenical redefinition.

DONNE AND THE ELIZABETHANS

ৡ৵ WOLFGANG CLEMEN

ithin a series of lectures on aspects of the Renaissance, it may be fitting to devote one lecture to the change of style that occurred in English poetry toward the end of the sixteenth century. This change of style did not happen all at once, it was not brought about by one poet alone, and its various aspects did not even all point in the same direction. If we examine the various kinds of poetry written in any one single year,[1] we are struck by the great variety of styles which existed side by side for whole decades. Elizabethan poets like Willam Drummond, Richard Barnfield, William Browne (all of whom died well after 1620) continued the typical Elizabethan manner until late in the twenties, while various new forms of poetry had already emerged in the works of John Donne, Ben Jonson, the Cavaliers, and the purely devotional poets. Thus the change of style which marks the transition from Elizabethan poetry to the poetry of the seventeenth century turns out to be a gradual and complicated process stretching over twenty or even thirty years. Within this process, however, we are able to single out some particularly important and influential figures. John Donne naturally comes first to our minds, and I should like to concentrate on his role as innovator, considering the change of style as it appears in his poetry.

Close scrutiny of Donne's poems carried out in many books and articles during the last thirty years has enabled us to define the characteristics of his style and manner. But while we can perhaps give a satisfactory description of what this style of Donne's is like, we find it far more difficult to account for these changes, to say how and why this happened, why Donne chose to write in this particular new manner which strikes us as being so different from the typical Elizabethan style.

If we are to answer these questions, we should remember, at the outset of our endeavor, the warning pronounced by F. P. Wilson, speaking of Donne's break with Elizabethan poetry: "Trying to account for the break is no doubt as complicated an affair as trying to account for life."[2] Thus we

---

[1] For this purpose the use of Norman Ault's anthology *Elizabethan Lyrics* (London, 1960) may be recommended.
[2] F. P. Wilson, *Elizabethan and Jacobean* (Oxford, 1945), p. 56.

shall never be able to find *the* answer to our problem, for it appears to demand a method of approach rather different from the one used in describing and enumerating stylistic traits: instead of establishing reliable causal connections and stating exact findings, we have to proceed by "hints and guesses," by surmise and approximation. A change of style cannot be tied down to one or two or even three or four reasons. It is the result of a multiplicity of factors concurring on many levels and in many different fields. And since the extent to which each of these factors has become efficient in bringing about the change of style can never be accurately measured, we must content ourselves with pointing out the possible co-operation of these various impulses toward the formation of a new style.

Nor are the explanations thus put forward for the change of style of equal value and reliability. Let us review them briefly in order to clarify our own position. There is, to begin with, the theory that the disturbed balance of style mirrors the disturbed harmony of the Renaissance world-picture, that a general unrest, pessimism, and uncertainty about man's position in the world, a shattered confidence in the former scale of values, and the beginnings of a cleavage between belief and knowledge—these changes in man's attitude toward the world account for the new stylistic developments. This theory has been very popular, for it allowed a connection to be established between the history of culture and of ideas on the one hand and style on the other hand. But tempting though these parallels are, they remain vague and we can make use of them only in the most general manner; they do not lead to any specific discrimination. Moreover, they provoke criticism. For was the world-picture of the Elizabethans in 1590 really so much more balanced, consistent, and harmonious than it was in 1600, and can we believe in anything like this rapid change from an optimistic outlook in 1590 to a pessimistic one in 1600? This is not to deny that there exist connections between the intellectual and philosophic background of a period and the form and style of its poetry. But these connections are of an elusive kind, they defy our effort at precise demonstration, and they certainly do not allow of such simple and generalizing formulas as those I have hinted at.

However, the intellectual temper which emerges toward the end of the century yields us nevertheless one or two clues. For the mood of disillusionment and self-criticism, of analysis and introspection, which becomes evident when man looks into himself is at the root of Donne's poetry just as it informs the whole of Montaigne's writings and reappears in Shakespeare's dark comedies and his tragedies. This attitude fosters the tone of bitterness and cynicism, of mockery and irony, which, as Cruttwell has shown in his book *The Shakespearean Moment*,[3] forms a link between the dramatist and Donne. Here we can with a little more confidence suggest that the "Spirit of the Age" tends to express itself in the mood and tenor of its poetry. And we can note how this tone and mood transformed the style.

We discover more connections if we examine the milieu in which Donne was brought up, his mental training and his education, the reading public for which he wrote his poems. Here we come upon a whole mass of influences and conditions which help to determine some of the new elements which we observe in Donne's poetry. Robert Ellrodt, in his admirable book on the Metaphysicals,[4] has much to say on all these points. He has shown how through the combined influences of the Inns of Court and of the town, which replaced the former predominance of the Court, a new taste was formed to which Donne responded in his poetry. The society of versatile, witty, and worldly young men in which Donne moved and which he must have thought of as his readers fostered a masculine and flippant attitude, a more realistic view of the world and of the relationship between man and woman, a more sincere search for the undisguised truth, a rejection of affectation and false idealization and a predilection for improvisation, naturalness, and outspoken individualism. Ellrodt even goes so far as to suggest that the peculiar fusion of irreverence and seriousness, of witty entertainment and critical reflection, which we find in Donne's poetry was favored by this new milieu that was closely associated with the Inns of Court.

With the help of Ellrodt's findings, for which he gives much evidence, we can indeed detect several links between some of the basic attitudes emerging in Donne's poems and the spirit of this new society. But though we may be able to say that we have a better understanding of such and such a characteristic element in Donne's poems, this does not yet mean that we can account for the specific form, for the unique combination of theme and style in these poems. By examining the literary and social milieu of a new species of poetry, we can at best hope to determine the soil out of which a new plant is to grow, but we cannot determine its actual and individual shape. For this new soil could also have grown flowers of a very different kind.

We are on firmer ground if we look at the internal literary situation instead of at the "extra-literary factors." A major change of style never

[3] Patrick Cruttwell, *The Shakespearean Moment and its Place in the Poetry of the 17th Century* (New York, 1960).
[4] Robert Ellrodt, *L'inspiration personelle et l'esprit du temps chez les poètes métaphysiques anglais* (Paris, 1960), especially Part II.

happens unawares but occurs within a situation which has been well prepared for and already contains—though sometimes only "in the bud"—those tendencies of which the new poet will avail himself. Thus we could discover the germs of some of the new features Donne was to develop in poets like Chapman, Raleigh, or Drayton. Even Sidney, though representing Elizabethan poetry at its heights, could be quoted for a growing dissatisfaction with Petrarchan love conventions,[5] for introducing the colloquial style into the sonnet, and for attempts at dramatization of conventional situations. But an inquiry into what was going on, not in the field of lyric poetry but in neighboring literary genres, yields even more important and illuminating results. In order to account for a change of style occurring in one field we must always review the whole literary situation. For literature can be compared to a system of communicating pipes which, much more slowly than in the realm of physics but inevitably nonetheless, aim continuously at adjustment to one another.

Thus we find that English drama at a much earlier stage displayed these tendencies toward realism, satire, and colloquialism which were later to become characteristics of the English lyric. Elizabethan drama, which was a product of a great number of diverse influences, was not fettered by a firmly-rooted tradition to the same extent as was Elizabethan poetry. It was more open to the impact of the outside world, being itself to a higher degree a mirror of life than poetry. Thus Elizabethan drama was able to overcome its own artificialities and its remoteness from actual life at an earlier stage than was Elizabethan poetry.

On the other hand Elizabethan lyric poetry had preserved an almost insular position among the other literary genres. The tradition of Petrarchism was firmly rooted in the past, and in the hands of epigones it had grown to be a whole systematized code of clichés and formulas, although, along with it, a revolt against this attitude was developing even before Donne began writing. Pastoralism and Platonism had become similar conventions, based on idealization and fiction far remote from truth and reality, contributing to the establishment of a form of poetry which looked toward the past rather than toward the present and future. This poetry in fact avoided direct contact with man's actual experience, with his immediate needs and conflicts. Although Elizabethan lyric poetry, too, will reveal its darker undertones on closer reading and is not so golden and rosy as it is sometimes believed to be, yet its predominant temper is one of optimistic enjoyment and pleasant ease, of cheerful acceptance and harmony. The despair of unrequited love often expressed in the Elizabethan sonnet-sequences is

[5] See K. Muir, Introduction to *Elizabethan Lyrics* (London, 1952), p. 21, 40.

in itself a fiction and a convention and has little to do with an effort at getting closer to truth.

But if we look at the other kinds of Elizabethan poetry, at the epigram, the satire, the epistle, and the elegy, we discover a new tone and a new spirit. Sharp observation of man's nature, ruthless exposure of his weaknesses and contrarieties, defiance and even parody of outworn ideals and conventions, cynicism and mockery—these and similar attitudes alien to the spirit of Elizabethan lyricism emerge in the satires and epigrams written in the nineties even before Donne began his career. The increasing output and the growing popularity of this kind of poetry suggest that feelings and attitudes which could find no expression in the lyric poem were here given in other, more appropriate forms of poetry.

The growing influence of poets such as Martial, Propertius, Juvenal, and Persius, the tenor of whose writings reinforced this new vogue of epigram, satire, and verse-epistle, as well as the increasing impact of Ovidian love poetry on the Elizabethan ideals of love, are further indications that new developments were afoot and that even for the Elizabethan love-lyric a reorientation was imminent. The emergence of new models in antique literature replacing other Latin and Greek authors, whose influence is declining, has always been a forecast for a change of style. For behind these new foreign influences there is a dissatisfaction with the prevailing forms of poetry and a readiness to learn from other masters who stand for what the poets are themselves unconsciously striving at. Thus, because of the predilection for satire and epigram, reinforced by the Latin models of the silver period of Roman poetry, the satirical spirit was soon able to penetrate into the well-fenced garden of Elizabethan lyric poetry. And together with those new attitudes of disillusionment, sarcasm, realism, and the rejection of idealization came new stylistic qualities proper to epigram and satire: brevity, terseness and conciseness of expression, a personal conversational tone, dramatic presentation, the cultivation of paradox, surprise, improvisation and subtlety.

Thus what Donne did in transforming Elizabethan lyric poetry was to transfer the spirit, tone, and style not only of drama but also of genres like the epigram,[6] the satire, and the epistle into his love poems. He himself wrote satires and epistles and was looked upon by his contemporaries as a man who could have been, if he had chosen, the first epigrammatist.[7] It

---

[6] See Helen Gardner, Introduction to her new edition of Donne's *Elegies, Songs and Sonnets* (Oxford, 1965).

[7] "I think, if he would, he might easily be the best Epigrammatist we have found in English. . . ." Drummond of Hawthornden to Ben Jonson (*Critical Essays of the Seventeenth Century*, ed. J. E. Spingarn [Oxford, 1904], p. 216).

was only natural that he should have brought the lyric into closer contact with these other kinds of poetry, bridging the gap that had so far existed between different modes of expression and of style. At the same time, the attitude behind the satire and the epigram and their stylistic tendencies coincided with what we might call a revolt against the facileness, the superficiality and smoothness of Elizabethan poetry, which had become insipid and worn-out as a result of the too frequent imitation and repetition of well-known clichés. A change of style occurs as a reaction against existing modes of expression which have reached a certain perfection. Thus many traits in the poetry of Donne and his followers are exactly the opposite of what was cultivated by the Elizabethans.

Lastly we might refer to the anti-Ciceronian movement with regard to prose, and the general change in critical opinion which now demanded "more matter and less words," and defended an author for his "hardness," and even for his obscurity. These critical pronouncements, of which a great many have been collected during the last decades,[8] are an illuminating illustration of the growing awareness that new tendencies and ideals of style were developing. But it should be noted that most of these utterances appear *after* 1600. They reflect what was happening on the level of stylistic expression, but they did not *cause* these changes, although it cannot be denied that once these new principles of style had become articulate they themselves acted as a stimulus to the next generation of poets.

But while this recognition of certain trends in the literature of the last decade of the sixteenth century can to a certain extent explain what was happening in Donne's early poems, the real key to what we call the change of style, as I venture to suggest, lies in Donne's own personality. For he not only came from a different stratum of society and was brought up in a milieu different from any of his forerunners, but he was also a person of a marked individuality, of a more complex cast of mind and of wider interests, gifts, and contrasting faculties than they were. The unique combination of strong religious thought and unbiased intellectual analysis, of faith and skepticism, of worldliness and devotion, of probing search after truth and scoffing flippancy, of diplomatic shrewdness and downright honesty—all these peculiar blends, of which only some can be mentioned here, contributed to the formation of a character whose unmistakable voice speaks to us in each of his poems. It is clear that such a man could not go on writing in the same outworn, conventional, and pretty manner as the Elizabethan lyrists. Moreover, Donne, as a poet, was not a "professional" like Spenser, Daniel, and Drayton,

---

[8] See Wilbur Samuel Howell, *Logic and Rhetoric in England, 1500–1700* (New York, 1961); George Williamson, *The Senecan Amble* (London, 1951).

but an amateur writing for an exclusive circle of people, and he could thus claim special license to develop something new outside the poetic tradition. Thus although we may elicit from the literary situation as well as from altered social and cultural conditions a number of trends which point toward a probable turn, the ultimate and decisive reason why this happened in the specific way known to us lies in the personality of Donne himself.

For why does a great poet choose to take a new direction, to write in a manner rather different from his contemporaries? Certainly not because he falls under the influence of mannerism or of poets like Juvenal and Persius, or because he wants to adjust himself to a new fashion or a new public, or because he is influenced by the drama and the satire. All these are concomitant features of secondary importance. But the primary impulses are the poet's own choice of expression and his own need to express himself. A great poet writes differently and finds a new manner because he feels that the old manner will not do for what he has to say. New themes and aims demand new words, new attitudes require a new manner of proceeding and even of versification, and a changed conception of the purpose of poetry gives rise to a new method of composition. For behind every change of style there emerges a new conception of poetry and of the poet, a new poetics.

In assessing a change of style our task should therefore be to interpret the difference of style and method between Elizabethan and Metaphysical poetry in terms of contrasting attitudes and conceptions. We must inquire into the motive forces inherent in the means and forms of expression. This will also enable us to understand the interdependence of the different aspects of style. For the stylistic features of a poem form a complex of mutually interrelated elements, which cannot be treated separately. If one of these elements undergoes a major change, this will affect the other elements, too.

Such an inquiry, however, can only be carried out by means of comparison between individual poems. Only this method can help us to reduce the generalization to the level of concrete observations gathered from the text itself. However, to do justice to the achievement of John Donne as well as to the lyric art of the Elizabethans, at least a dozen poems should be compared with each other in order to bring out the great variety of characteristics on either side. This we cannot hope to do here. We have to confine ourselves to three poems only, and shall compare Donne's poem "The Sunne Rising" with one poem by Lodge and one by Sidney.[9] These three poems lend themselves to a comparison, for Sidney's and Donne's

---

[9] The poems are printed here in modernized spelling. The original texts are to be found in *The Complete Works of Thomas Lodge* (New York, 1963), a reprint of the 1593 edition; Sir Philip Sidney, *The Countess of Pembroke's Arcadia* (1598); Donne, *The Elegies, Songs and Sonnets*, ed. Gardner.

poems derive from the *aubade*, whereas Lodge's poem contains at least certain features of the *aubade*, the sunrise song in which two lovers take leave of each other.

Phillis (1593)

My Phillis hath the morning sun
 At first to look upon her;
And Phillis hath morn-waking birds
 Her risings for to honour.
My Phillis hath prime-feathered flowers
 That smile when she treads on them;
And Phillis hath a gallant flock
 That leaps since she doth own them.
But Phillis hath so hard a heart—
 Alas that she should have it!—
As yields no mercy to deserve,
 Nor grace to those that crave it.

 Sweet sun, when thou lookest on,
 Pray her regard my moan;
 Sweet birds, when you sing to her,
 To yield some pity, woo her;
 Sweet flowers, whenas she treads on,
 Tell her, her beauty deads one:
And if in life her love she nill agree me,
Pray her, before I die she will come see me.

              (Thomas Lodge)

Madrigal (1598?)

Why dost thou haste away,
O Titan fair, the giver of the day?
Is it to carry news
To western wights what stars in east appear?
Or dost thou think that here
Is left a sun, whose beams thy place may use?
Yet stay, and well peruse
What be her gifts, that make her equal thee;
Bend all thy light to see
In earthly clothes enclosed a heavenly spark;
Thy running course cannot such beauties mark.
No, no; thy motions be
Hastened from us, with bar of shadow dark,
Because that thou, the author of our sight,
Disdain'st we see thee stained with other's light.

             (Sir Philip Sidney)

The Sun Rising (after 1603)

 Busy old fool, unruly Sun,
 Why dost thou thus,

Through windows, and through curtains, call on us?
Must to thy motions lovers' seasons run?
      Saucy pedantic wretch, go chide
      Late school-boys and sour prentices,
   Go tell court-huntsmen that the king will ride,
   Call country ants to harvest offices;
Love, all alike, no seasons knows nor clime,
Nor hours, days, months, which are the rags of time.

      Thy beams so reverend and strong
      Why shouldst thou think?
I could eclipse and cloud them with a wink,
But that I would not lose her sight so long.
     If her eyes have not blinded thine,
     Look, and tomorrow late tell me,
   Whether both th'Indias of spice and mine
   Be where thou left'st them, or lie here with me.
Ask for those kings whom thou saw'st yesterday,
And thou shalt hear, "All here in one bed lay."

      She's all states, and all princes I;
      Nothing else is.
Princes do but play us; compared to this,
All honour's mimic, all wealth alchemy.
     Thou, Sun, art half as happy as we,
     In that the world's contracted thus;
   Thine age asks ease, and since thy duties be
   To warm the world, that's done in warming us.
Shine here to us, and thou art everywhere;
This bed thy centre is, these walls thy sphere.

                                         (John Donne)

When we read the first stanza of Lodge's and Donne's poems respectively, the first thing that strikes our ears (for we are always apt to listen before we analyze intellectually!) is the difference of rhythm and tone.

The regular easy-flowing meter of the poem by Lodge is not striking in any way; it does not arrest our attention by concentrating it on one line or even on one word more than on others. There is no modulation of tempo and tone. The regular alternation of iambic lines of four and three beats which each time form one sentence falling into two distinct parts gives to this poem a very obvious and simple pattern, reinforced by the devices of symmetry, repetition, and parallelism of subject matter as well as of sentence structure.

The structural principle of the poem is that of enumeration and addition rather than of evolution. After listening to two pairs of alternating lines, we are almost lulled into expecting something similar in the next lines; we almost know that there will be no surprise. Moreover, some of these pairs (e.g. the one about the birds and the one about the flowers) could even be exchanged and their order could be altered without much damage to the intelligibility of the poem. The poem's second stanza is a symmetrical and obvious counterpart to the first stanza. It takes up the first three motifs in the first stanza which described Phillis in her surroundings and turns them into objects of address. But the address to "Sweet sun" is limited to a very simple and conventional formula: the lover addresses him as his ally: "Pray her regard my moan"; there is no scenic situation developed as in Donne's poem; no relationship is established between the sun's movement and the situation of the lovers; no tension is suggested between the sun and the lovers. The morning sun looking upon Phillis is no more than a pretty descriptive framework, the first item in a series of other conventional images. The whole poem falls into three clear-cut parts which are not linked by consistent argument or logic and which correspond to a conventional pattern. The description of the beloved lady is followed by the lover's complaint about her "hard heart," and this is followed by the lover's request addressed to the sun, the birds, and the flowers to help him in his wooing. The method of the poem is to amplify the basic motif by familiar images and phrases which repeat the same idea. The images therefore serve the purpose of decoration rather than clarification.

Thus the smoothly flowing regularity of the verse, which excludes any audacious, unusual, or harsh expressions from its vocabulary, the symmetrical, repetitive structure of the stanzas, the ornamental elaborations of conventional motifs, the obvious tripartite division of the poem and its easy intelligibility—all combine to give this poem a lightness of tone and mood, a graceful charm. They are an adequate expression for simple, straightforward, unproblematic subject matter. Poetry which uses these methods aims at pleasing the ear and delighting the mind by the presentation of familiar things in a pleasant way. But poetry of this kind makes no particular demand on the intellectual comprehension of the reader; it does not involve him in any complicated process of argument. It is a style suited to simple statement, and the enjoyment it produces is the enjoyment of the glittering, superficial aspect of things. Probably these Elizabethan poets, people like Lodge, Greene, and Peele, knew as well as Donne that love, real love, was nothing like this love which they painted in their poems. But they still paid lip service to the fiction that lovers should behave in this way and express themselves in those formulas which had been handed down from one generation to the next. The change of style which we detect in Donne's early poems, composed in the nineties while this Elizabethan love-convention was still in vogue, was also due to the fact that Donne rejected this artificial attitude of the dutifully languishing lover, an attitude which must have seemed to him entirely false.

There are more parallels and hence points of comparison between Donne's poem and Sidney's. Sidney, in respect of his stylistic accomplishment, is halfway between Lodge and Donne. Consequently we can even detect similarities between Donne and Sidney. For both poets give a fresh and dramatic treatment to the old Petrarchan motif: the superiority of the beloved over the sun. In both poems the lover addresses the sun throughout, so that a vivid dialogue with a partner is suggested although the partner himself remains silent. Not only Donne but Sidney, too, introduces colloquial speech into this dialogue (l. 12, "No, no . . ."), and they both begin their poem by putting, as it were, a personal question to the sun: "Why dost thou. . . ?" In both poems these questions are followed by imperatives, and the conclusion is in both cases an apodictic statement. The invitation addressed to the sun to stay and to look at the beloved is placed into the exact middle of both poems. We have, moreover, in both poems a complicated heterosyllabic metrical scheme which terminates in rhymed couplets and uses enjambment much more frequently than Lodge.

These and other similarities which bring the two poems close together demand more detailed comparison, and this will in turn show us their characteristic differences. Let us begin by describing the different impressions made on us when listening to the poems and by analyzing these differences in terms of syntactical, metrical, and linguistic observations. For despite the immediacy of the address, the verse in Sidney's poem still moves forward in a smooth and leisurely, indeed almost solemn, manner; the stresses of the iambic meter are regular except in the twelfth and thirteenth lines. Regular, too, are the alternation between short and long lines of six and ten syllables (up to the eleventh line) and the use of enjambment, which links the short line to the following longer line, thus forming a balanced syntactical unit. These units are again bound together by rhyme, so that we have a continuous flow without any striking pauses. Moreover, the rhyme binds together the meaningful words which we would normally stress; the beginning and the end of the poem are particularly emphasized by the rhyme, which joins the first and the last pairs of lines, thus giving, as it were, a symmetrical framework to the poem. There is parallelism in the sentence structure (three interrogative sentences of two lines each; two imperatives of two lines each); the sentences are complete and contain no elliptical abridgments; they are the proper instrument for the amplification, the periphrasis, and the tautology which we also note. If we consider the

427

effect of alliteration and assonance in Sidney's poem ("western wights"; "disdain'st"; "stained"), we find that their function is predominantly decorative and that euphony is cultivated for its own sake.

If we examine the same elements in Donne's poem, our ears are at first struck by a difference in the flow and sound of the verse. Here there is a halting and nervous staccato rhythm that grips our attention, provokes us into surprise, and compels us to listen with concentration, for we feel that the ebb and flow of the verse and the many modifications of speed and tone demand our co-operation and reflect the inner movement of the poem. The metrical pattern constantly supports, modifies, or counteracts the rhythm of speech. Our ears will find it far more difficult to recognize the structure of each stanza, to distinguish the individual line, to grasp the construction of sentences, which are split into separate parts and interrupted by insertions; enjambment is used irregularly and serves to veil the pattern of the stanza as well as the underlying symmetry in the sequence of lines. The meter alternates irregularly between iambic and trochaic beginnings to the lines, though the trochaic is preferred because of its more abrupt effect. There is often an accumulation of stresses, as in: "Why dost thou thus . . . ?"; "Nothing else is . . ."; "She's all states, and all princes I. . . ." The syntactical forms are concentrated and even elliptical: Donne's aim is condensation, not amplification. All these devices serve to bring out and to offset more sharply the thought on which the reader's attention should be focused. In this connection it is illuminating to compare the way in which the two poets make use of rhetorical figures. For the figure of *chiasmus* occurs in both poems. But whereas in the line "In earthly clothes enclosed a heavenly spark" this figure serves to produce the effect of a play with sounds (through the echo effect of the two words in the middle: "clothes," "enclosed"), Donne in his line "She's all states, and all princes I" strips the phrase of all verbal music; the two words "She" and "I," the cornerstones of the line, as well as the elliptical expression "and all princes I," give this line the quality of a naked and extremely condensed statement. Nowhere in Sidney would we find such condensation of meaning.

Further differences are revealed if we consider the vocabulary and diction of the two poems. The select poetical language of Sidney moves in the pastoral sphere and contains no expression which would shock or surprise us. We find the conventional epithets placed after the noun ("Titan fair"; "shadow dark"), the mythological periphrasis (Titan), the amplifying apposition ("the giver of the day"; "the author of our sight").

Donne's vocabulary, on the other hand, has a far wider range. It aims at precise expression and consequently does not shun the unusual, the blunt word; it refers to trades and far countries, to astronomy and alchemy, to low and to high objects ("rags of time"; "kings"; "princes"); it serves to express abstract equations as well as concrete and realistic detail.

Both poets begin their poems with lines addressed to the sun. We can thus inquire into the different manners in which they use the instrument of their language in this context. Donne plunges us, as it were, into the middle of a dialogue; he places the scolding, impatient address "Busy old fool, unruly Sun" before the question "Why dost thou thus. . . ?" Sidney, however, begins in a soft and leisurely manner with the question; the address follows only in the second line, and his whole further manner of speaking to the sun keeps within the limits of restraint and adoration. But no poet before Donne would have used such contemptuous insolent language toward the sun, chiding him as a "saucy pedantic wretch" and disclaiming (in the first two lines of the second stanza) all traditional reverence. Just as the poet has stepped down from the heights of idealization, the sun, too, is brought down from his celestial throne.

Donne uses the resources of his language and his versification to establish a closer and more direct contact, not only between himself and his poetical partner (here the sun, in other poems the beloved), but also between himself and his reader. In Donne's poem we ourselves feel that we are being spoken to; we cannot but be drawn into the process of thought and of event which (as in our poem) started even before the poem begins with its first line. Many of Donne's stylistic devices succeed in bringing about this unique effect of a real speaking voice, addressing its partner not from a distance but as in a natural conversation. Donne's introduction of the conversational tone into poetry was an important step toward putting an end to the distance, the remoteness, which had been characteristic of a great deal of Elizabethan love poetry. The personal tone that strikes us as a distinctive feature of Donne's poems is also produced by the fact that in his poems there is a constant dialogue: someone is constantly talking to someone else, trying to convince him or to scoff at him, to implore him or to warn him. The provocative address to the sun, however, is only the first step in Donne's reversal of conventional themes and elements. Both poets make use of the Petrarchan motif: a comparison between the sun and the beloved, establishing the superiority of the beloved over the sun. But whereas Sidney handles this convention in a fresh and vivid manner by giving it a dramatic turn, observing, however, the usual scale of laudatory comparisons (culminating in the assertion of "heavenly" beauty [1.10]), Donne turns the convention upside down. The superiority of *both* lovers (with Sidney as with all other Elizabethan poets it was only the *beloved lady* who was allowed to be superior to the sun) is first established by abusing the sun, while in the last stanza it is expressed by a gesture of pity. The end of the poem reverses

429

the initial relationship between the lovers and the sun: whereas in the beginning the lovers still felt in some way dependent on the sun, thinking him a tiresome intruder, now the sun is wholly dependent on the lovers, being addressed in a patronizing condescending tone. Altogether something has fundamentally changed in the course of the poem. As often happens in Donne, the standpoint which we have reached at the end of the poem is very different from that at the beginning. In Sidney's poem, however, attitude and point of view remain the same.

The middle stanza of Donne's poem gives us an example of the poet's characteristic manner of taking a convention "literally," thus driving it *ad absurdum*. Of the three conceits Donne uses here (I could eclipse the sun by closing my eyes; her eyes blind the sun's sight) the third reveals Donne's method of presenting the fabulous fiction of a conventional comparison (the lady's beauty equals the treasures of both Indias) as actual, concretizing its implications in a realistic situation (". . . or lie here with me"). In the lines which follow, too, Donne bases logical conclusions on illogical premises, pursuing his argument up to its utmost limit, and arriving (as in the last lines of the poem) at an entirely unforeseen conclusion. Donne thus brings home to us the preposterous aspect of a convention. The logic which has been noted by many critics as a structural element of his poems is in fact not genuine logic but pseudologic, a cunning play of wit by which the poet betrays us.

To say that Donne turned away from or indeed rejected the Elizabethan conventions is not an adequate description of what he really does. For he uses these conventions as building materials, as it were as set pieces within a new structure, disguising them in unconventional language and giving them a surprising and unexpected turn. Donne in fact analyzes more precisely than his predecessors the potential implications of a convention, exploiting its equations and relationships more ingeniously than the Elizabethan poets before him ever did.

What Donne does here with the motif of the sun and the lovers is a reversal, in fact a "deformation" of a convention, and would have called forth Ben Jonson's censure: "That which is tortured is counted more exquisite, nothing is fashionable till it is deformed!"

Both poems assert the supremacy of love. But the tone of solemn tenderness and lyric inspiration in Sidney's poem is very different from Donne's dry and witty manner which is not only more robust but also more complex, more intellectual and meaningful than Sidney's method. Donne's terse and masculine language serves to pitch down the praise of love from the heights of sententious declamation or beautiful poetic utterance to the intimacy of a private discourse hiding profound seriousness and intense emotion under the veil of sophisticated argument and a scoffing attitude.

This can be seen by comparing the two last lines in both poems. In Sidney's poem the conclusion is what we would expect after the preceding lines, for the sun's hastening away is explained by a conventional answer. Donne, however, arrives at his conclusion by a surprising and illogical turn of thought ("Since . . ."), using a bold image which defines the relationship with almost mathematical precision. Donne's last line does several things at once. It resolves the conflicting relationship between the two partners in the poem (lover and sun) into a final formula which not only crowns the argument but also crystallizes the dramatic situation of the poem into a last concentrated image. But it also gives a final triumphant expression to the victory of fulfilled love, in a language that is without pathos, without sentimentalism, without emotional exuberance. Donne appears to stir our intellect first before he touches our emotion. The theme of fulfilled love, of the *two* lovers united, had also been a major innovation of Donne's, and it goes without saying that this shift of emphasis, implying a change of inner attitude, helps to account for the transformation of the poetical language of love which we witness in reading Donne.

But what can be said about Donne's dramatic sense in comparison to Sidney? Sidney's address to the sun is dramatic, too, for the partner is continually present. (See the many instances of "thou," "thy," "thine.") But the dialogue keeps the same pitch throughout, whereas Donne's "speaking voice" displays the inflections, the up and down of personal talk, the abruptness of transition and the violence of chiding, that a dramatic speech may contain. Even more important is the concreteness by which the scene of the poem is suggested (windows, curtains, bed, walls). Elizabethan poems (with the exception of the narrative poems of the pastoral type) as a rule do not pay much regard to time and place. Sidney's reference to the "western wights" remains vague, implying rather the "nowhere" of Arcadia than any concrete locality. But Donne ties his poems to the here-and-now of a definite situation. Thus the sustained flow of thought and argument is woven into the dramatic scene, which in our poem is presented to us at the beginning of the poem and reappears at its very end, and which is itself simultaneously expanding. In speaking of the intellectual character of Donne's poetry, we must not forget that we are taken through a process not only of thinking, but of visualizing, observing, imagining, as well. For in spite of his subtle thought and his sophistication, Donne's poems are firmly rooted in the ground, with their wealth of concrete detail and realistic imagery from all spheres of life, which range from the "sour apprentices" to the sun's visit to India, to his sphere and his center. The unfortunate term *metaphysical* sometimes makes us forget this very physical aspect of Donne's poetry.

The purpose of our comparative analysis of three poems has been to examine the connection between poetic intention and style. We are now in a better position to answer the initial question as to *why* Donne chose to write in this new manner. We found that nearly every feature of style and method in Donne's poetry had its opposite in some trait of Elizabethan poetry and that in most cases it was possible to account for these differences. Allowing for simplification, let us again enumerate some of the more important opposites: in the Elizabethan poems we find verse that tends to be regular and fluent, maintaining the same dynamic level throughout and advancing in a leisurely manner; language that gives preference to the "poetic" expression, tends to be florid, profuse and periphrastic, decorative and descriptive, cultivating the beautiful and aiming always for the pleasing effect.

In Donne we find staccato verse which tends to be irregular and of changing pitch and speed, often bordering on the effect of prose or colloquial speech, a closely packed, elliptical, and concentrated use of language, of surprising complexity and range, cultivating strangeness rather than beauty. Whereas in Elizabethan poetry we generally find regularly built stanzas of an obvious pattern which tend to be self-contained, we have in Donne an over-all structure based on an intricate, continuous trend of thought. Consequently the Elizabethans make frequent use of the principle of addition and repetition, they elaborate on a set theme through the use of rhetorical devices, and they often aim for effects of sound which are designed simply to please our ears. The structure of their poems is frequently short-breathed, falling into small units, balanced by symmetry or antithesis. Donne, on the other hand, avoids the orderly disposition, the appearance of a well-planned, lucid structure, hiding his art of composition beneath an impression of spontaneity and improvisation. He subordinates sentence structure, versification, and the pattern of the stanza to the intense and sometimes naked expression of thought. We can continue to add further sets of opposites and use catchwords to point to wider issues: easy intelligibility, banality, and simplicity are replaced by complication, complexity, sometimes even by obscurity; fulfilled expectancy is replaced by surprise and the unforeseen development of thought; amplification by condensation; objective statement by paradox and ambiguous presentation of the case; idealization by realistic disillusionment.

Thus we can see the way in which Donne, by transforming the style of the Elizabethan lyric, forged a new instrument for himself. His purpose must have been to create a new kind of poetry which involved the reader in its very process of thought, placing him in a dramatic situation with which he was familiar. To this end Donne could no longer use the elaborate word schemes or the formal, self-contained lines of his predecessors; he had to reject their copiousness of diction, the sweet fluency of their verse, and their pleasing use of ornament. He had to dispense with mythology, with gorgeous description and stately declamation. For him this was no longer a valid way of seeking out the truth, of expressing personal experience of some complexity, not only with greater honesty but also with greater precision and conciseness than before. Donne therefore had to put demands on the reader's understanding, making it difficult rather than easy for him, forcing him to follow a complicated line of argument and often to read a poem several times over. But his readers were prepared to follow, to listen to the personal tone of a speaking voice, and to be involved in a dramatic dialogue that shortened the distance between themselves and the poem. They must have felt, as they are bound to feel today, that here they were being made witnesses to something which deeply concerned their own human nature in all its complexity.

CHARLES S. SINGLETON is Professor of Humanistic Studies and Director of the Humanities Center at The Johns Hopkins University. &ᴥ E. H. GOMBRICH is Professor of the Classical Tradition and Director of the Warburg Institute, University of London. &ᴥ JOHN WHITE is Professor of the History of Art and chairman of the department at The Johns Hopkins University. &ᴥ EDWARD E. LOWINSKY is Ferdinand Schevill Distinguished Service Professor in the Department of Music at the University of Chicago. &ᴥ WALTER H. RUBSAMEN is Professor of Musicology at the University of California, Los Angeles. &ᴥ LT. COMMANDER D. W. WATERS, R.N., is Curator of Navigation and Astronomy at the National Maritime Museum. &ᴥ GEORGE BOAS is Professor Emeritus of the History of Philosophy at The Johns Hopkins University. &ᴥ FRANCES A. YATES is a Reader in the History of the Renaissance at the Warburg Institute, University of London. &ᴥ MARSHALL CLAGETT is Professor of the History of Science at the Institute for Advanced Study, Princeton. &ᴥ STILLMAN DRAKE is Professor of the History of Science at the University of Toronto. &ᴥ ERNAN McMULLIN is Professor of the Philosophy of Science at the University of Notre Dame. &ᴥ FELIX GILBERT is Professor of History at the Institute for Advanced Study, Princeton. &ᴥ J. H. HEXTER is Professor of History at Yale University. &ᴥ WOLF-GANG CLEMEN is Professor of English at the University of Munich.

INDEX

Achellini, Alessandro, 284

Adam, Hermetic and Mosaic, 257, 267, 270

Agricola, Alexander, 172

Agrippa, Henry Cornelius, 258, 259, 262, 267

Alamanni, Antonio, 168, 172

Alamanni, Piero, 183

Al-Battānī, 282

Alberti, Leon Battista, 252

Albertus Magnus, 337

Albrecht von Brandenburg, 126

Alchemy, 255, 263, 264, 273. *See also* Chemistry; Hermetic tradition

Alcibiades, 394

Alexander VI, Pope, 113

Alexander of Scotland, Prince, 373

Al-Farghānī, 282

Alkindi, 284

Allegri, Alessandro, 169

Alpers, Mrs. S. Leontief, 12, 34

Alphonso V of Portugal, King, 205, 207

Alvarus, Thomas, 300, 302

Ambrose, St., 373

Amiens, 44

Andreas, Johann Valentin, 265

Animism. *See* Hermetic tradition

Anthoniszoon, Cornelis, 232 and *n*

Antico, Andrea, 172

Apelles, 10

Apollo, 11

Appian, 393

Aquilano, Serafino, 174

Aquinas, Thomas, 247, 296, 338

Archimedes, 259, 302, 307, 316, 320, 322

Architecture. *See* Sculpture

Aretino, Leonardo Bruni, 376

Aristotelian philosophy: and conceptualism, 341; and Condemnations of 1277, 275; descriptive nature of, 355; dichotomy between physics and astronomy, 343; opposed by Copernicus, 343–44; physics of heaven and earth, 277; rejected by Galileo, 320, 322, 346; theological reaction to, 339; mentioned, 268

Aristotle: Bacon on, 267; and Bradwardine rule, 283; commentaries on, 289–90, 305; conceptualist ideal of science, 334–36, 338; and Copernicus, 342, 344; *De anima*, 337, 338; deductive and inductive proof, 366; "demonstrative" theses, 351; and Descartes, 349; and Galileo, 345; contrasted with Kepler, 348; *Metaphysics*, 243–44, 344; on motion, 311; and Newton, 358, 362; *Organon*, 337; *Physics*, 296, 337; *Politics*, 401, 406; *Posterior Analytics*, 334, 335, 337; *Questions of Mechanics* (attributed to), 309, 320; quoted by Oresme, 276; versus animism, 273; views of the Infinite, 296; mentioned, 120, 247, 285, 376, 394

Aristoxenus, 120, 123

Arnold of Villanova, 284

Arnolfo di Cambio, 54*n*, 56, 61

Art: baroque, 263; Baroque, Early Renaissance, and High Renaissance, 60; and development of science, 44, 101–2; and dramatic narrative, 54; perspective in, 34, 41; rational foundations of, 10, 11; rendering of the human body, 14, 15, 18, 34, 35, 41; and theology, 54–55; unity of, 92. *See also* individual arts

Art, criticism of: creative power of, 38; standards, 3, 20, 24, 25, 27, 32, 38, 41; mentioned, 3–5, 10–11, 18, 28, 34

Astrology, 273, 319. *See also* Astronomy

Astronomy: Aristotle's, 337; and astrology, 319; Cartesian influence in, 357; Copernican, 277, 344–46; force physics and kinematics joined in, 277; Galileo's application of physics to, 322–30; and navigation, 193, 196, 197, 199, 200, 203, 204; positional, 336; physics and mathematics united with, 306–8; situation by 1300, 338; tradition before Galileo, 282, 310, 311; mentioned, 306, 331, 332, 344, 356

Augustine, Emperor, 124

Augustine, St., 272

Bacon, Francis: *Advancement of Learning*, 261, 262; against animism, 267, 268; apologist for Galileo and Newton, 250; compared with Dee and Fludd, 270; compared with Galileo, 307, 308; compared with Mersenne, 271; contrasted with Dee, 262; contrasted with Newton as empiricist, 363, 366; and Hermetic tradition, 266, 269; *New Atlantis*, 265; Newton as follower of, 361; *Novum organum*, 348; program for science, 241; rejection of Gilbert's magnet, 268; rejection of heliocentricity, 268; and teleology, 251

Baer, Nicholas, 347

Banco, Nanni di, 68*n*, 72*n*, 78*n*, 80